GAME THEORY

FOURTH EDITION

GAME THEORY

FOURTH EDITION

BY

GUILLERMO OWEN

Department of Mathematics
Naval Postgraduate School
Monterey, California

United Kingdom – North America – Japan
India – Malaysia – China

Emerald Group Publishing Limited
Howard House, Wagon Lane, Bingley BD16 1WA, UK

First edition 1968
Second edition 1982
Third edition 1995
Fourth edition 2013

British Library Cataloguing in Publication Data
A catalogue record for this book is available from the British Library

ISBN: 978-1-78190-507-4

Printed and bound by CPI Group (UK) Ltd, Croydon, CR0 4YY

ISOQAR certified
Management System,
awarded to Emerald
for adherence to
Environmental
standard
ISO 14001:2004.

Certificate Number 1985
ISO 14001

INVESTOR IN PEOPLE

Contents

Preface

Game theory once again proves to be a dynamic field. In less than 100 years since the name was coined, it has reached the status of a major branch of both mathematics and economics. It has also proved quite useful in several of the social sciences. Even biology and finance have been touched (and, I hope, improved) by game-theoretic analysis. The field continues to expand, and so a book that I last published in 1995 requires some modification. I am happy that Emerald Publishers is giving me a chance to bring out a new edition.

It would be very difficult to include all the new developments in the field. I have therefore merely added a chapter explaining some of the most important advances in game theory. In particular, I have tried to describe the developments that have been central to several of the Nobel Memorial Prizes in Economics.

I also wish to thank people who have helped me, by vigorous and stimulating discussion, to improve this book. These include Francesc Carreras, Hélène Ferrer, Gianfranco Gambarelli, Maurice Koster, Ines Lindner, Conrado Manuel, Gordon McCormick, Martha Saboyá, and Juan Tejada.

Finally, I am sad to mention the deaths of some members of the game theory community. Michael Maschler, my mentor and one of my best friends, was instrumental in the development of the bargaining set and related concepts. John Harsanyi, also a good friend, was one of the first to study games with incomplete information. I will always miss them both.

Guillermo Owen

Chapter 1

Definition of a Game

1.1. General Notions

The general idea of a game is that with which we are familiar in the context of parlor games. Starting from a given point, there is a sequence of personal moves, at each of which one of the players chooses from among several possibilities; interspersed among these there may also be chance, or random, moves such as throwing a die or shuffling a deck of cards.

Examples of this type of game are chess, in which there are no chance moves (except for the determination of who shall play first), bridge, in which chance plays a much greater part, but in which skill is still important, and roulette, which is entirely a game of chance in which skill plays no part.

The examples of bridge and chess help to point out another important element of a game. In fact, in a chess game each player knows every move that has been made so far, while in bridge a player's knowledge is usually very imperfect. Thus, in some games, a player is unable to determine which of several possible moves has actually been made, either by an opposing player, or by chance. The practical result of this is that, when a player makes a move, he does not know the exact position of the game, and must make his move remembering that there are several possible actual positions.

Finally, at the end of a game, there is normally some payoff to the players (in the form of money, prestige, or satisfaction) which depends on the progress of the game. We may think of this as a function which assigns a payoff to each "terminal position" of the game.

1.2. Games in Extensive Form

In our general idea of a game, therefore, three elements enter: (1) alternation of moves, which can be either personal or random (chance) moves, (2) a possible lack of knowledge, and (3) a payoff function.

We define, first, a *topological tree* or *game tree* as a finite collection of nodes, called *vertices*, connected by lines, called *arcs*, so as to form a connected figure which includes no simple closed curves. Thus it follows that, given any two vertices A and B, there is a unique sequence of arcs and nodes joining A to B.

From this we obtain

1.2.1 Definition. Let Γ be a topological tree with a distinguished vertex A. We say that a vertex C *follows* the vertex B if the sequence of arcs joining A to C passes through B. We say C follows B *immediately* if C follows B and, moreover, there is an arc joining B to C. A vertex X is said to be *terminal* if no vertex follows X.

1.2.2 Definition. By an *n-person game in extensive form* is meant

(α) a topological tree Γ with a distinguished vertex A called the *starting point* of Γ;

(β) a function, called the *payoff function,* which assigns an n-vector to each terminal vertex of Γ;

(γ) a partition of the nonterminal vertices of Γ into $n+1$ sets $S_0, S_1, \ldots, S_n$, called the *player sets*;

(δ) a probability distribution, defined at each vertex of S_0, among the immediate followers of this vertex;

(ε) for each $i = 1, \ldots, n$, a subpartition of S_i into subsets S_i^j, called *information sets*, such that two vertices in the same information set have the same number of immediate followers and no vertex can follow another vertex in the same information set;

(ζ) for each information set S_i^j, an index set I_i^j, together with a $1 - 1$ mapping of the set I_i^j onto the set of immediate followers of each vertex of S_i^j.

The elements of a game are seen here: condition α states that there is a starting point; β gives a payoff function; γ divides the moves into chance moves (S_0) and personal moves which correspond to the n players ($S_i, \ldots, S_n$); δ defines a randomization scheme at each chance move; ε divides a player's moves into "information sets": he knows in which information set he is, but not which vertex of the information set.

1.2.3 Example. In the game of *matching pennies* (Figure 1.1), player I chooses "heads" (H) or "tails" (T). Player II, not knowing player I's choice, also chooses "heads" or "tails." If the two choose alike, then player II wins a cent from player I; otherwise, player I wins a cent from II. In the game tree shown, the vectors at the

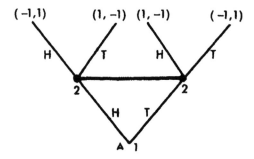

Figure 1.1.

terminal vertices represent the payoff function; the numbers near the other vertices denote the player to whom the move corresponds. The shaded area encloses moves in the same information set.

1.2.4 Example. The game of pure strategy, or GOPS, is played by giving each of two players an entire suit of cards (13 cards), A third suit is shuffled, and the cards of this third suit are then turned up, one by one. Each time one has been turned up, each player turns up one of his cards at will: the one who turns up the larger card "wins" the third card. (If both turn up a card of the same denomination, neither wins.) This continues until the three suits are exhausted. At this point, each player totals the number of spots on the cards he has "won"; the "score" is the difference between what the two players have.

With 13-card suits, this game's tree is too large to give here; however, we can give part of the tree of an analogous game using three-card suits (Figure 1.2).

There is a single chance move, the shuffle, which orders the cards in one of the six possible ways, each having a probability of $\frac{1}{6}$. After this the moves correspond to the two players, I and II, until the game ends. We have drawn parts of the game tree, including the initial point, several branches, and four of the terminal points. The remaining branches are similar to those we have already drawn. With respect to information, we have

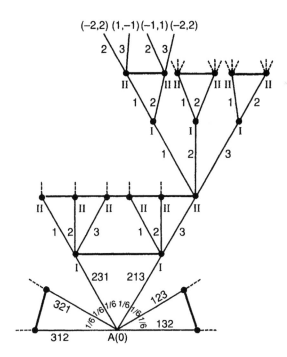

Figure 1.2.

1.2.5 Definition. Player i is said to have *perfect information* in Γ if his information sets S_i^j each consist of one element. The game Γ is said to have perfect information if each player has perfect information in Γ.

For example, chess and checkers have perfect information, whereas bridge and poker do not.

1.3.   Strategies: The Normal Form

The intuitive meaning of a strategy is that of a plan for playing a game. We may think of a player as saying to himself, "If such and such happens, I'll act in such and such a manner." Thus, we have

1.3.1 Definition. By a *strategy* for player i is meant a function which assigns, to each of player i's information sets S_i^j, one of the arcs which follows a representative vertex of S_i^j.

The set of all strategies for player i will be called Σ_i.

In general, we are accustomed to the idea that a player decides his move in a game only a few moves, at best, in advance, and quite usually only at the moment he must make it. In practice this must be so, for in a game such as chess or poker the number of possible moves is so great that no one can plan for every contingency very much in advance. From a purely theoretic point of view, however, we can overlook this practical limitation, and assume that, even before the game starts, each player has already decided what he will do in each case. Thus, we are actually assuming that each player chooses a strategy before the game starts.

Since this is so, it only remains to carry out the chance moves. Moreover, the chance moves may all be combined into a single move, whose result, together with the strategies chosen, determines the outcome of the game.

Actually, what we are interested in, and what the players are interested in, is deciding which one of the strategies is best, from the point of view of maximizing the player's share of the payoff (i.e., player i will want to maximize the ith component of the payoff function). As, however, no one knows, except probabilistically, what the results of the chance moves will be, it becomes natural to take the mathematical expectation of the payoff function, given that the players are using a given n-tuple of strategies. Therefore, we shall use the notation

$$\pi(\sigma_1, \sigma_2, \ldots, \sigma_n) = (\pi_1(\sigma_1, \ldots, \sigma_n), \pi_2(\ldots), \ldots, \pi_n(\sigma_1, \ldots, \sigma_n))$$

to represent the mathematical expectation of the payoff function, given that player i is using strategy $\sigma_i \in \Sigma_i$.

From this, it becomes possible to tabulate the function $\pi(\sigma_1, \ldots, \sigma_n)$ for all possible values of $\sigma_1, \ldots, \sigma_n$, either in the form of a relation, or by setting up an n-dimensional array of n-vectors. (In case $n = 2$, this reduces to a matrix whose elements are pairs of real numbers.) This n-dimensional array is called the *normal form* of the game Γ.

1.3.2 Example. In the game of matching pennies (see Example 1.2.3) each player has the two strategies, "heads" and "tails." The normal form of this game is the matrix

	H	T
H	$(-1, 1)$	$(1, -1)$
T	$(1, -1)$	$(-1, 1)$

(where each row represents a strategy of player I, and each column a strategy of player II).

1.3.3 Example. Consider the following game: An integer z is chosen at random, with possible values 1, 2, 3, 4 (each with probability $\frac{1}{4}$). Player I, without knowing the results of this move, chooses an integer x. Player II, knowing neither the result of the chance move nor I's choice, chooses an integer y. The payoff is

$$(|y - z| - |x - z|, |x - z| - |y - z|)$$

i.e., the point is to guess close to z.

In this game each player has four strategies: 1, 2, 3, 4, since other integers are of little use. If, for instance, I chooses 1 and II chooses 3, then the payoff will be $(2, -2)$ with probability $\frac{1}{4}$, $(0,0)$ with probability $\frac{1}{4}$, and $(-2, 2)$ with probability $\frac{1}{2}$. The expected payoff, then, is $\pi(1, 3) = (-\frac{1}{2}, \frac{1}{2})$. Calculating all the values of $\pi(\sigma_1, \sigma_2)$, we obtain

	1	2	3	4
1	$(0, 0)$	$(-\frac{1}{2}, \frac{1}{2})$	$(-\frac{1}{2}, \frac{1}{2})$	$(0, 0)$
2	$(\frac{1}{2}, -\frac{1}{2})$	$(0, 0)$	$(0, 0)$	$(\frac{1}{2}, -\frac{1}{2})$
3	$(\frac{1}{2}, -\frac{1}{2})$	$(0, 0)$	$(0, 0)$	$(\frac{1}{2}, -\frac{1}{2})$
4	$(0, 0)$	$(-\frac{1}{2}, \frac{1}{2})$	$(-\frac{1}{2}, \frac{1}{2})$	$(0, 0)$

1.3.4 Definition. A game is said to be *finite* if its tree contains only finitely many vertices.

Under this definition, most of our parlor games are finite. Chess, for instance, is finite, thanks to the laws which end the game after certain sequences of moves.

It should be seen that, in a finite game, each player has only a finite number of strategies.

1.4. Equilibrium *n*-Tuples

1.4.1 Definition. Given a game Γ, a strategy n-tuple $(\sigma_1^*, \sigma_2^*, \ldots, \sigma_n^*)$ is said to be *in equilibrium*, or an *equilibrium n-tuple*, if and only if, for any $i = 1, \ldots, n$, and $\hat{\sigma}_i \in \Sigma_i$.

$$\pi_i(\sigma_1^*, \ldots, \sigma_{i-1}^*, \hat{\sigma}_i, \sigma_{i+1}^*, \ldots, \sigma_n^*) \leq \pi_i(\sigma_1^*, \ldots, \sigma_n^*)$$

In other words, an *n*-tuple of strategies is said to be in equilibrium if no player has any positive reason for changing his strategy, assuming that none of the other players is going to change strategies. If, in such a case, each player knows what the others will play, then he has reason to play the strategy which will give such an equilibrium *n*-tuple, and the game becomes very stable.

1.4.2 Example. In the game with normal form

	β_1	β_2
α_1	(2, 1)	(0, 0)
α_2	(0, 0)	(1, 2)

both (α_1, β_1) and (α_2, β_2) are equilibrium pairs.

Unfortunately, not every game has equilibrium *n*-tuples. As an example, the game of matching pennies (Example 1.3.2) has no equilibrium pairs.

In general, if a game has no equilibrium *n*-tuples, we usually see the several players trying to outguess each other, keeping their strategies secret. This suggests (and it is indeed true) that in games of perfect information, equilibrium *n*-tuples exist.

To prove this statement, we must study the question of decomposition of a game.

A game Γ will be said to *decompose at a vertex* X if there are no information sets which include vertices from both of (a) X, and all its followers, and (b) the remainder of the game tree. In this case, we can distinguish the subgame, Γ_X, consisting of X, and all its followers, and the quotient game, Γ/X, which consists of all the remaining vertices, plus X. For the quotient game, X will be a terminal vertex; the payoff here can be considered to be Γ_X: i.e., the payoff at this vertex is a play of the subgame Γ_X.

Now, as we have seen, a strategy for i is a function whose domain consists of the information sets of player i. If we decompose a game at X, then we can also decompose the strategy σ into two parts: $\sigma_{|\Gamma/X}$, obtained by restricting σ to information sets in Γ/X, and $\sigma_{|\Gamma_X}$, obtained by restricting σ to Γ_X. Conversely, a strategy for Γ/X and a strategy for Γ_X can be combined in the obvious way to yield a strategy for the larger game Γ.

1.4.3 Theorem. Let Γ decompose at X. For $\sigma_i \in \Sigma_i$, assign to X (considered as a terminal vertex of Γ/X) the payoff

$$\pi_X(\sigma_{1|\Gamma_X}, \sigma_{2|\Gamma_X}, \ldots, \sigma_{n|\Gamma_X})$$

In this case

$$\pi(\sigma_1, \ldots, \sigma_n) = \pi_{\Gamma/X}(\sigma_{1|\Gamma/X}, \ldots, \sigma_{n|\Gamma/X})$$

The proof of this theorem is clear and can be left as an exercise to the reader. Briefly, it is only necessary to verify that, for each possible outcome of the chance moves, the same terminal vertex is eventually reached either in the original or in the decomposed game.

With this, we can prove.

1.4.4 Theorem. Let Γ decompose at X, and let $\sigma_i \in \Sigma_i$ be such that (a) $(\sigma_{1|\Gamma_X}, \ldots, \sigma_{n|\Gamma_X})$ is an equilibrium n-tuple for Γ_X, and (b) $(\sigma_{1|\Gamma/X}, \ldots, \sigma_{n|\Gamma/X})$ is an equilibrium n-tuple for Γ/X, with the payoff $\pi(\sigma_{1|\Gamma_X}, \ldots, \sigma_{n|\Gamma_X})$ assigned to the terminal vector X. Then $(\sigma_1, \ldots, \sigma_n)$ is an equilibrium n-tuple for Γ.

Proof. Let $\hat{\sigma}_i \in \Sigma_i$. Because $(\sigma_{1|\Gamma_X}, \ldots, \sigma_{n|\Gamma_X})$ is an equilibrium n-tuple for Γ_X, it follows that

$$\pi_i(\sigma_{1|\Gamma_X}, \ldots, \hat{\sigma}_{i|\Gamma_X}, \ldots, \sigma_{n|\Gamma_X}) \leq \pi_i(\sigma_{1|\Gamma_X}, \ldots, \sigma_{n|\Gamma_X})$$

On the other hand, by (b) we know that, if we assign the payoff $\pi(\sigma_{1|\Gamma_X}, \ldots, \sigma_{n|\Gamma_X})$ to the vector X, then

$$\pi_i(\sigma_{1|\Gamma/X}, \ldots, \hat{\sigma}_{i|\Gamma/X}, \ldots, \sigma_{n|\Gamma/X}) \leq \pi_i(\sigma_{1|\Gamma/X}, \ldots, \sigma_{n|\Gamma/X})$$

Now, the payoff (for a given set of strategies) is a weighted average of the payoffs at some of the terminal vertices of a tree. Hence, if the payoff to player i at a given terminal vertex (X, in this case) is decreased, his expected payoff, for any choice of strategies, will either remain equal or be decreased. Thus, applying Theorem 1.4.3, we find that

$$\pi_i(\sigma_1, \ldots, \hat{\sigma}_i, \ldots, \sigma_n) \leq \pi_i(\sigma_1, \ldots, \sigma_n)$$

and so $(\sigma_1, \ldots, \sigma_n)$ is an equilibrium n-tuple.

This is all we need to prove.

1.4.5 Theorem. Every finite n-person game with perfect information has an equilibrium n-tuple of strategies.

Proof. We shall define the length of a game as the largest possible number of edges that can be passed before reaching a terminal vertex, i.e., the largest possible number of moves before the game ends. Clearly a finite game has finite length. The proof is by induction on the length of the game.

If Γ has length 0, the theorem is trivially true. If it has length 1, then at most one player gets to move, and he obtains equilibrium by choosing his best alternative. If Γ has length m, then it decomposes (having perfect information) into several subgames of length less than m. By the induction hypothesis, each of these subgames has an equilibrium n-tuple; by Theorem 1.4.4 these form an equilibrium n-tuple for Γ.

1.5. The Monty Hall Game

We conclude this chapter with an example which, in the past, raised some controversy. Based on a well-known television show (*Let's Make a Deal*), the problem caused serious discussion among mathematicians and others who treated it as a problem in probability theory. (See items 3, 6, 8 in the bibliography for background on the problem and the controversy.) We will, rather, treat it as a game.

In this game, a Contestant (player I) is shown three screens. Behind one of these screens is an expensive, brand-new automobile. There is a goat behind each of the other two screens. It is understood that each screen has probability $\frac{1}{3}$ of being the "good" screen (i.e., the one with the automobile).

Once chance has chosen the good screen, player I, in ignorance of the results of the chance move, chooses one of the screens (call this screen a). At this point Monty Hall, the host of the show (player II), opens another one of the screens (call this screen b), to reveal a goat. In doing this, Monty knows the outcome of the chance move.

Player I may then *insist* on screen a, or opt to *switch* to screen c. In either case, he will receive whatever is behind his final choice (a or c). It is assumed that the automobile is much more valuable than the goat. What is the player's best strategy?

If we treat this as a game, the first thing we notice is that the game is not well defined. In fact, we are told how the Contestant may act, but we are told neither how Monty may act, nor what Monty's payoff will be. There would seem three possibilities.

Case 1. Monty the Dummy

Monty has no latitude in his actions. If the Contestant's choice, screen a, is the good screen, then Monty must open either b or c, each with probability $\frac{1}{2}$. If screen a is one of the two bad screens, then Monty must open the other bad screen (whether it be b or c).

Now, Contestant has two strategies: Insist (on his original choice) and Switch (from his original choice, a, to the remaining unopened screen, c).

Consider his first strategy, Insist. In this case, Contestant will win if (and only if) a was indeed the good screen. As mentioned above, this has probability $\frac{1}{3}$.

Consider next the strategy Switch. Then, if a is the good screen, Contestant will lose. If, on the other hand, a is a bad screen, then Monty will open the other bad screen. Contestant, by switching to the remaining screen, will win. This has probability $\frac{2}{3}$.

Thus, in this case, Contestant will do better by Switching.

Case 2. The Friendly Monty

Monty has some freedom. After player I has chosen screen a, Monty *may* (but need not) open a bad screen–either b or c. If Monty does not open a screen, then Contestant will be left with his original choice, a. Moreover, Monty wants Contestant to *win*.

In this case, Contestant, with Monty's help, is sure to win. In fact, if the original choice, a, is good, then Monty will not open any screen, and Contestant will have to open his choice, a. If, on the other hand, a is a bad screen, then Monty will open the other bad screen. Contestant, by switching, will win.

Thus, the strategy Switch will win with probability 1.

Case 3. The Unfriendly Monty

Monty has some freedom. After player I has chosen screen a, Monty *may* (but need not) open a bad screen–either b or c. If Monty does not open a screen, then Contestant will be left with his original choice, a. This time, however, Monty wants Contestant to *lose*.

In this case, Contestant, by Insisting, will win with probability $\frac{1}{3}$.

Suppose, however, Contestant uses his other strategy, Switch. If a is the good screen, then Monty's best strategy is to open one of the other screens–either b or c (both of which are bad). Contestant, by switching to the other bad screen, will lose. If a is a bad screen, then Monty's best strategy is not to open any screen, and thus Contestant will be left with his original choice. Thus Contestant will certainly lose.

In this case, then, Contestant will win with probability $\frac{1}{3}$ by Insisting, and probability 0 by Switching.

So what, then, is the best strategy for Contestant? Clearly it depends on the rules of the game, and also on Monty's objective: Contestant should Switch if Monty is a dummy, and also if Monty wants him to win; he should Insist if Monty wants him to lose. Unfortunately, Contestant is apparently not told the rules of the game, and probably does not know what Monty wants. Perhaps Monty's sponsors want Contestant to win–it will be good publicity for their products. Perhaps they want Contestant to lose–they have spent too much money already on prizes.

Clearly Contestant needs more information. This is an example of a *game with incomplete information*. We look at possible approaches to these games in Chapter 6 of this book.

Problems

1. An infinite game, even with perfect information, need not have an equilibrium n-tuple.
   (a) Consider a two-person game in which the two players alternate, and, at each move, each player chooses one of the two digits 0 and 1. If the digit x_i is chosen at the ith move, then each play of the game corresponds to a number

$$x = \sum_{i=1}^{\infty} x_i 2^{-i}$$

   in the interval $[0, 1]$. Then, player I wins one unit from player II if $x \in S$, and loses one unit if $x \notin S$, where S is some subset of $[0, 1]$.
   (b) Each player has exactly $2^{\aleph_0}$ strategies, which can therefore be indexed σ_β, τ_β respectively for $\beta < \alpha$, where α is the smallest ordinal preceded by at least $2^{\aleph_0}$ ordinals.
   (c) Let $\langle \sigma, \tau \rangle$ denote the play (or number x) obtained if the players choose strategies σ and τ, respectively. For each strategy σ of I, player II has $2^{\aleph_0}$ strategies τ which will give different values of $\langle \sigma, \tau \rangle$ (and similarly for each strategy τ of II).

(d) A set S can be constructed (using the axiom of choice) such that, for each σ_β, there is a τ_β with $\langle\sigma_\beta, \tau_\beta\rangle \notin S$, but for each τ_γ there is a σ_γ with $\langle\sigma_\gamma, \tau_\gamma\rangle \in S$.

2. Construct the normal form for the game whose tree is given in Figure 1.3. The game starts at the vertex 0 (a random move); each of three players has one information set containing two vertices, with two alternatives (indexed a and b, c and d, e and f, respectively) at each vertex.

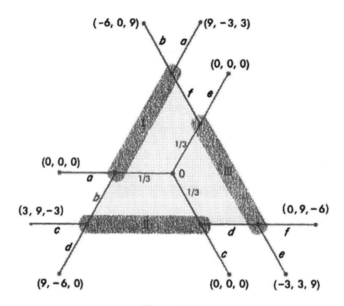

Figure 1.3.

For exercises 3 and 4, below,

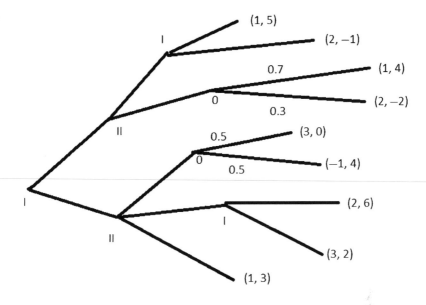

(a) State how many strategies each player has.
(b) Give the normal form of the game.
(c) Give equilibrium points.

For exercises 5–7, below, the shaded areas should be taken to represent information sets for the corresponding player.

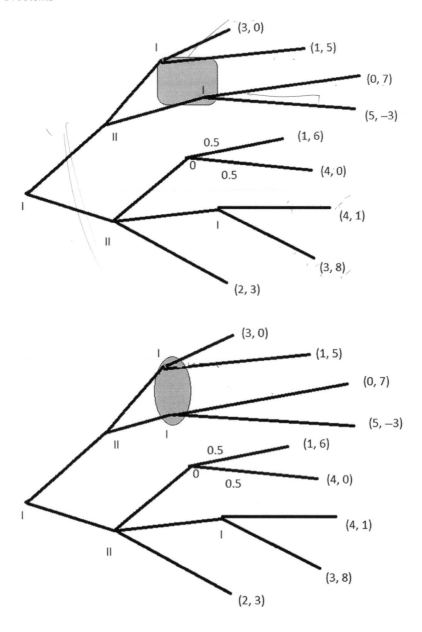

(a) State how many strategies each player has.
(b) Give the normal form of the game.

Chapter 2

Two-Person Zero-Sum Games

2.1. Zero-Sum Games

2.1.1 Definition. A game Γ is said to be *zero-sum* if and only if, at each terminal vertex, the payoff function $(p_1, \ldots, p_n)$ satisfies

$$\sum_{i=1}^{n} p_i = 0 \tag{2.1}$$

In general, a zero-sum game represents a closed system: everything that someone wins must be lost by someone else. Most parlor games are of the zero-sum type. Two-person zero-sum games are sometimes called strictly competitive games.

Because of condition (2.1), the nth component of a payoff vector is determined by the remaining $n - 1$ components. In the case of a two-person zero-sum game, we can simply give the first component of the payoff vector; the second component is necessarily the negative of the first. In this case, we call the first component simply the *payoff*, and it is understood that the second player gives this amount to the first player.

It will be seen that a two-person zero-sum game differs from other games in that there is no reason for any negotiation between the players: in fact, whoever wins, the other loses. The meaningfulness of this will be seen from:

2.1.2 Theorem. In a two-person zero-sum game, let (σ_1, σ_2) and (τ_1, τ_2) be two equilibrium pairs. Then

(i) (σ_1, τ_2) and (τ_1, σ_2) are also equilibrium pairs, and

(ii) $\pi(\sigma_1, \sigma_2) = \pi(\tau_1, \tau_2) = \pi(\sigma_1, \tau_2) = \pi(\tau_1, \sigma_2)$ $\tag{2.2}$

Proof. (σ_1, σ_2) is in equilibrium. Hence

$$\pi(\sigma_1, \sigma_2) \geq \pi(\tau_1, \sigma_2)$$

On the other hand, (τ_1, τ_2) is in equilibrium. Therefore

$$\pi(\tau_1, \sigma_2) \geq \pi(\tau_1, \tau_2)$$

Thus

$$\pi(\sigma_1, \sigma_2) \geq \pi(\tau_1, \sigma_2) \geq \pi(\tau_1, \tau_2)$$

But, similarly

$$\pi(\tau_1, \tau_2) \geq \pi(\sigma_1, \tau_2) \geq \pi(\sigma_1, \sigma_2)$$

and these two sets of inequalities prove (2.2). Now, for any $\hat{\sigma}_1$

$$\pi(\hat{\sigma}_1, \sigma_2) \le \pi(\sigma_1, \sigma_2) = \pi(\tau_1, \sigma_2)$$

and, for any $\hat{\sigma}_2$

$$\pi(\tau_1, \hat{\sigma}_2) \ge \pi(\tau_1, \tau_2) = \pi(\tau_1, \sigma_2)$$

Thus (τ_1, σ_2) is in equilibrium; similarly, (σ_1, τ_2) is in equilibrium.

This theorem is not true for other games: in the game of Example 1.4.2, for instance, (α_1, β_1) and (α_2, β_2) are equilibrium pairs, but the payoffs are different, and, moreover, neither (α_1, β_2) nor (α_2, β_1) is an equilibrium pair.

2.2. The Normal Form

As we have seen, the normal form of a finite zero-sum two-person game reduces to a matrix, A, with as many rows as player I has strategies and as many columns as player II has strategies. The expected payoff, assuming I chooses his ith strategy and II chooses his jth strategy, is the element a_{ij} in the ith row and jth column of the matrix.

It will be clear that a strategy pair will be in equilibrium if, and only if, the element a_{ij} corresponding to it is both the largest in its column and the smallest in its row. Such an element, if it exists, is called a *saddle point* (by analogy with the surface of a saddle, which curves upward in one direction and downward in the other direction).

2.2.1 Example. The game matrix

$$\begin{pmatrix} 5 & 1 & 3 \\ 3 & 2 & 4 \\ -3 & 0 & 1 \end{pmatrix}$$

has a saddle point in the second row and second column.

2.2.2 Example. The game matrix

$$\begin{pmatrix} -1 & 1 \\ 1 & -1 \end{pmatrix}$$

has no saddle points.

Let us suppose that two players are actually engaged in a matrix game. In choosing strategies, player I chooses, in effect, a row, i, while player II chooses a column, j. The payoff will be the entry a_{ij}. As this is an amount which I receives from II, it is clear that player I will be trying to maximize a_{ij}, while II will be trying to minimize it. Unfortunately, neither knows for certain what his opponent's strategy will be, and this knowledge is usually of importance in deciding what strategies to use.

For instance, in the game of matching pennies (see Example 1.3.2), we may picture player I as thinking, "People usually choose heads; hence II will expect me to choose heads and choose heads himself, and so I should choose tails. But perhaps II is reasoning along the same line: he'll expect me to choose tails, and so I'd better choose heads. But perhaps that is II's reasoning, so ..." In this case, player I could never reach a decision in which he could feel confident, and the outcome would be uncertain.

Suppose, however, that we consider the game of Example 2.2.1 above. In this case, player I would probably decide to play either his first or his second strategy, and we might hear him reasoning in this manner: "I should use either my first or my second strategy, but if II figures this, he will use his second strategy, so I had better use my second strategy." Even if II guesses I's strategy, I still does well to stick to it! Thus there is a tremendous difference between these two games. In one, secrecy is important; in the other, there is no point to it. The reason, of course, is that the second game has a saddle point.

2.3. Mixed Strategies

The preceding analysis, though it tells us how to play matrix games with saddle points, does not shed any light on a player's best choice for the vast majority of games, which contain no saddle points.

Let us suppose, now, that we are playing a game without saddle points, such as

$$\begin{pmatrix} 4 & 2 \\ 1 & 3 \end{pmatrix}$$

We cannot tell what will happen. However, let us suppose that our opponent is not only unpredictable, but omniscient: he will guess correctly whatever we decide. In this case, if we are player I, we will certainly use our first strategy, with which we cannot win less than two units, whereas with our other strategy we will win only one unit. This certain win of at least two units is our "gain-floor" and we shall denote it by v'_I:

$$v'_I = \max_i \{\min_j a_{ij}\} \tag{2.3}$$

If we were player II, we should choose, under the same conditions, our second strategy, which would give us a "loss-ceiling" of three units. This loss-ceiling will be denoted by v'_{II}:

$$v'_{II} = \min_j \{\max_i a_{ij}\} \tag{2.4}$$

Thus, for matrix games, we have the idea of a gain-floor and a loss-ceiling. It would be absurd to expect I's gain-floor to exceed II's loss-ceiling. In fact, we can easily prove

$$v'_I \leq v'_{II} \tag{2.5}$$

The proof of this is elementary and may be left to the reader. If equality holds in (2.5), we have a saddle point; if not, we have a game without saddle points. For such a game, we are not certain what will happen, but we can make one statement: *player* I *should not win less than* v'_I; *player* II *should not lose more than* v'_{II}.

In a game without saddle points, if we allow the opponent to know our strategy, the best we can hope for is the gain-floor v'_I, or the loss-ceiling v'_{II}, depending on whether we are player I or II. If we hope to do better, it must be by refusing to let our opponent know our strategy.

This is difficult to do if we choose our strategy rationally, since there is nothing to keep our opponent from reconstructing our reasoning. We seem to be saying that we should choose our strategy irrationally. But in that case what good is all our analysis?

The answer is that the strategy should be chosen at random — hence irrationally — but the randomization scheme should be chosen rationally. This is the idea behind mixed strategies.

2.3.1 Definition. A *mixed strategy* for a player is a probability distribution on the set of his pure strategies.

In case the player has only a finite number, m, of pure strategies, a mixed strategy reduces to an m-vector, $x = (x_1, \ldots, x_m)$, satisfying

$$x_i \geq 0 \tag{2.6}$$

$$\sum_{i=1}^{m} x_i = 1 \tag{2.7}$$

We shall use X to denote the set of all mixed strategies for player I, and let Y represent the set of player II's mixed strategies.

Let us suppose that players I and II are playing the matrix game A. If I chooses the mixed strategy x, and II chooses y, then the *expected payoff* will be

$$A(x, y) = \sum_{i=1}^{m} \sum_{j=1}^{n} x_i a_{ij} y_j \tag{2.8}$$

or, in matrix notation

$$A(x, y) = x A y^T \tag{2.9}$$

As before, player I must fear that II will discover his choice of a strategy. If this should happen, then II will certainly choose y so as to minimize $A(x, y)$; i.e., I's expected gain-floor, assuming he uses x, will be

$$v(x) = \min_{y \in Y} x A y^T \tag{2.10}$$

Now, $x A y^T$ can be thought of as a weighted average of the expected payoffs for I if he uses x against II's pure strategies. Thus the minimum will be attained by a pure strategy j:

$$v(x) = \min_j xA_{\cdot j} \tag{2.11}$$

(where $A_{\cdot j}$ is the jth column of the matrix A).

Hence player I should choose x so as to maximize $v(x)$ i.e., so as to obtain

$$v_{\mathrm{I}} = \max_{x \in X} \min_j xA_{\cdot j} \tag{2.12}$$

(One should prove that a maximum exists; however, since X is compact and the function $v(x)$ is continuous, there is no problem.) Such an x is I's *maximin* strategy.

Similarly, if II chooses y he will obtain the expected loss-ceiling

$$v(y) = \max_i A_{i \cdot} y^{\mathrm{T}} \tag{2.13}$$

(where $A_{i \cdot}$ is the ith row of A) and he should choose y so as to obtain

$$v_{\mathrm{II}} = \min_{y \in Y} \max_i A_{i \cdot} y^{\mathrm{T}} \tag{2.14}$$

Such a y is II's *minimax* strategy.

Thus we obtain the two numbers v_{I} and v_{II}. These numbers are called the *values* of the game to I and II, respectively.

2.4. The Minimax Theorem

It is easily proved that, for any function $F(x, y)$ defined on any Cartesian product $X \times Y$

$$\max_{x \in X} \min_{y \in Y} F(x, y) \leq \min_{y \in Y} \max_{x \in X} F(x, y) \tag{2.15}$$

Hence we have

$$v_{\mathrm{I}} \leq v_{\mathrm{II}}$$

This is analogous to (2.5), and indeed, in this case as before, it is only natural that I's gain-floor cannot exceed II's loss-ceiling. In the previous case, however, we saw that equality only held in exceptional cases. In this case, we shall actually prove

2.4.1 Theorem. (The Minimax Theorem)

$$v_{\mathrm{I}} = v_{\mathrm{II}}$$

This theorem, the most important of game theory, has been proved in many ways. Here we shall give the proof that was given by von Neumann and Morgenstern. We start with two lemmas:

2.4.2 Lemma. (Theorem of the Supporting Hyperplane) Let B be a closed convex set of points in n-dimensional Euclidean space, and let $x = (x_1, \ldots, x_n)$ be a point not in B. Then there exist numbers $p_1, \ldots, p_n, p_{n+1}$ such that

$$\sum_{i=1}^{n} p_i x_i = p_{n+1} \tag{2.16}$$

and

$$\sum_{i=1}^{n} p_i y_i > p_{n+1}, \quad \text{for all} \quad y \in B \tag{2.17}$$

(Geometrically, this means that we can pass a hyperplane through x such that B lies entirely "above" the hyperplane.)

Proof. Let z be the point in B whose distance from x is a minimum. (Such a point exists because B is closed.) Now, let

$$p_i = z_i - x_i \quad i = 1, \ldots, n,$$

$$p_{n+1} = \sum_{i=1}^{n} z_i x_i - \sum_{i=1}^{n} x_i^2$$

Clearly, (2.16) holds. We must show (2.17) also holds. Now

$$\sum_{i=1}^{n} p_i z_i = \sum_{i=1}^{n} z_i^2 - \sum_{i=1}^{n} z_i x_i$$

and hence

$$\sum p_i z_i - p_{n+1} = \sum_{i=1}^{n} z_i^2 - 2\sum_{i=1}^{n} z_i x_i + \sum_{i=1}^{n} x_i^2 = \sum_{i=1}^{n} (z_i - x_i)^2 > 0$$

Therefore

$$\sum p_i z_i > p_{n+1}$$

Suppose now that there exists $y \in B$ such that

$$\sum_{i=1}^{n} p_i y_i \leq p_{n+1}$$

Because B is convex, the line joining y to z must be entirely contained in B, i.e., for all $0 \leq r \leq 1, w_r = ry + (1 - r)z \in B$. Now the square of the distance from x to w_r is given by

$$\rho^2(x, w_r) = \sum_{i=1}^{n} (x_i - ry_i - (1 - r)z_i)^2$$

Therefore

$$\frac{\partial \rho^2}{\partial r} = 2 \sum_{i=1}^{n} (z_i - y_i)(x_i - r y_i - (1 - r) z_i)$$

$$= 2 \sum (z_i - x_i) y_i - 2 \sum (z_i - x_i) z_i + 2 \sum r (z_i - y_i)^2$$

$$= 2 \sum p_i y_i - 2 \sum p_i z_i + 2r \sum (z_i - y_i)^2$$

If we evaluate this at $r = 0$ (i.e., $w_r = z$)

$$\left. \frac{\partial \rho^2}{\partial \rho} \right|_{r=0} = 2 \sum p_i y_i - 2 \sum p_i z_i$$

But the first term on the right is assumed less than or equal to $2p_{n+1}$ while the second is greater than $2p_{n+1}$. Thus

$$\left. \frac{\partial \rho^2}{\partial r} \right|_{r=0} < 0$$

It follows that, for r close enough to zero

$$\rho(x, w_r) < \rho(x, z)$$

But this contradicts the manner in which z was chosen. Therefore, for all $y \in B$, (2.17) must hold.

2.4.3 Lemma. (Theorem of the Alternative for Matrices) Let $A = (a_{ij})$ be an $m \times n$ matrix. Then *either* (i) *or* (ii), below, must hold:

(i) The point 0 (in m-space) is contained in the convex hull of the $m + n$ points

$$a_1 = (a_{11}, \ldots, a_{m1})$$

$$\vdots$$

$$a_n = (a_{1n}, \ldots, a_{mn})$$

and

$$e_1 = (1, 0, \ldots, 0)$$
$$e_2 = (0, 1, 0, \ldots, 0)$$

$$\vdots$$

$$e_m = (0, 0, \ldots, 1)$$

(ii) There exist numbers $x_1, \ldots, x_m$ satisfying

$$x_i > 0$$

$$\sum_{i=1}^{m} x_i = 1$$

$$\sum_{i=1}^{m} a_{ij} x_i > 0 \qquad \text{for} \qquad j = 1, \ldots, n$$

Proof. Suppose (i) does not hold. If we apply Lemma 2.4.2, there exist numbers $p_1, \ldots, p_{m+1}$, such that

$$\sum_{j=1}^{m} 0 \cdot p_j = p_{m+1}$$

(this means, of course, that $p_{m+1} = 0$) and

$$\sum_{j=1}^{m} p_j y_j > 0$$

for all y in the convex set. In particular, this will hold if y is one of the $m + n$ vectors a_i, e_j. Therefore

$$\sum a_{ij} p_i > 0 \quad \text{for all} \quad j$$
$$p_i > 0 \quad \text{for all} \quad i$$

As $p_i > 0$, it follows that $\sum p_i > 0$, and we can let

$$x_i = \frac{p_i}{\Sigma p_i}$$

Hence

$$\sum a_{ij} x_i > 0$$
$$x_i > 0$$
$$\sum x_i = 1$$

With these two lemmas, we are able to prove our theorem:

Proof of Minimax Theorem. Let A be a matrix game. By Lemma 2.4.3, either (i) or (ii), above, must hold.

If (i) holds, 0 is a convex linear combination of the $m + n$ vectors. Thus, there exist $s_1, \ldots, s_{m+n}$ such that

$$\sum_{j=1}^{n} s_j a_{ij} + s_{n+i} = 0, \qquad i = 1, \ldots, m$$
$$s_j \geq 0 \quad j = 1, \ldots, m+n$$
$$\sum_{j=1}^{m+n} s_j = 1$$

Now, if all the numbers $s_1, \ldots, s_n$ were equal to 0, it would follow that 0 would be a convex linear combination of the m unit vectors $e_1, \ldots, e_m$, which is obviously not possible as these are independent vectors. Hence at least one of $s_1, \ldots, s_n$ is positive, and $\sum_{j=1}^{n} s_j > 0$. We can then write

$$y_j = \frac{s_j}{\sum_{j=1}^n s_j}$$

and we have

$$y_j \geq 0$$

$$\sum_{j=1}^n y_i = 1$$

$$\sum_{j=1}^n a_{ij} y_j = \frac{-s_{n+i}}{\sum_{j=1}^n s_j} \leq 0 \quad \text{for all} \quad i$$

Thus $v(y) \leq 0$, and $v_{\mathrm{II}} \leq 0$.

Suppose, instead, that (ii) holds. Then $v(x) > 0$, and so $v_{\mathrm{I}} > 0$.

We know therefore that it is not possible to have $v_{\mathrm{I}} \leq 0 \leq v_{\mathrm{II}}$. Let us suppose that we change the game A: we replace it by $B = (b_{ij})$, where

$$b_{ij} = a_{ij} + k$$

It should be clear that for any x, y,

$$xBy^{\mathrm{T}} = xAy^{\mathrm{T}} + k$$

Hence

$$v_{\mathrm{I}}(B) = v_{\mathrm{I}}(A) + k$$

$$v_{\mathrm{II}}(B) = v_{\mathrm{II}}(A) + k$$

As it is not possible that

$$v_{\mathrm{I}}(B) < 0 < v_{\mathrm{II}}(B)$$

it is also not possible that

$$v_{\mathrm{I}}(A) < -k < v_{\mathrm{II}}(A)$$

But k was arbitrary. Thus we cannot have $v_{\mathrm{I}} < v_{\mathrm{II}}$. But we have already seen that $v_{\mathrm{I}} \leq v_{\mathrm{II}}$. Therefore

$$v_{\mathrm{I}} = v_{\mathrm{II}}$$

Q.E.D.

Thus, if we deal with mixed strategies, we find that I's gain-floor is precisely equal to II's loss-ceiling. The common value, v, of these two numbers is called the *value* of the game. We see that a strategy x which satisfies

$$\sum_{i=1}^m x_i a_{ij} \geq v, \quad j = 1, \ldots, n \tag{2.18}$$

is *optimal* for the first player, in the sense that there is no strategy which will give him a higher expectation than v against every strategy of II. Conversely, if y satisfies

$$\sum_{j=1}^{n} a_{ij} y_j \leq v, \quad i = 1, \ldots, m \tag{2.19}$$

then y is *optimal* for the second player in the same sense. Now, clearly

$$xAy^T = v$$

for, if the left side in this equation were smaller, this would contradict (2.18); if it were larger, this would contradict (2.19). Hence x and y, besides being optimal, are also optimal against each other and against any other optimal strategies. We shall call any pair of optimal strategies, (x, y), a *solution* to the game.

The following theorem (a somewhat stronger form of the minimax theorem) will be useful later:

2.4.4 Theorem. In an $m \times n$ matrix game A, either player II has an optimal strategy y with $y_n > 0$, or player I has an optimal strategy x with

$$\sum_{i=1}^{m} a_{in} x_i > v$$

To prove 2.4.4, we give first this definition:

2.4.5 Definition. Let $r^k = (r_1^k, \ldots, r_n^k)$, $k = 1, \ldots, p$, be p n-vectors. Then by the *convex cone* generated by $r^1, \ldots, r^p$, we mean the set of all vectors x such that

$$x = \sum_{k=1}^{p} \lambda_k r^k$$

for some nonnegative $\lambda_1, \ldots, \lambda_p$.

It is easily checked that such a cone is indeed convex.

2.4.6 Lemma. (Farkas) Suppose $r^k = (r_1^k, \ldots, r_n^k)$, $k = 1, \ldots, p+1$, are n-vectors such that any $(q_1, \ldots, q_n)$ satisfying

$$\sum_{j=1}^{n} q_j r_j^k \geq 0, \quad k = 1, \ldots, p \tag{2.20}$$

will also satisfy

$$\sum_{j=1}^{n} q_j r_j^{p+1} \geq 0 \tag{2.21}$$

Then r^{p+1} is in the convex cone C generated by $r^1, \ldots, r^p$.

Proof. Suppose $r^{p+1} \notin C$. By Lemma 2.4.2, there exist numbers such $q_1, \ldots, q_{n+1}$ such that

$$\sum_{j=1}^{n} q_j r_j^{p+1} = q_{n+1}$$

and

$$\sum_{j=1}^{n} q_j s_j > q_{n+1} \quad \text{for all} \quad s \in C$$

Now, $0 \in C$; hence $q_{n+1} < 0$. Suppose, moreover, that $\sum_{j=1}^{n} q_j s_j$ were negative for any $s \in C$. For any positive number α, $\alpha s \in C$. But by making α large enough, the form $\Sigma q_j \alpha s_j = \alpha \Sigma q_j s_j$ can be made smaller than q_{n+1}. It follows, then, that

$$\sum_{j=1}^{n} q_j r_j^{p+1} < 0$$

but

$$\sum_{j=1}^{n} q_j s_j \geq 0, \quad \text{for all} \quad s \in C$$

and in particular for $r = r^1, \ldots, r^p$. Hence if the conclusion of 2.4.6 is false, so is its hypothesis.

Let us now proceed to prove 2.4.4. Assume first that $v(A) = 0$. Consider the $m + n$ m-vectors

$$e_1 = (1, 0, \ldots, 0)$$
$$e_2 = (0, 1, \ldots, 0)$$
$$\vdots$$
$$e_m = (0, 0, \ldots, 1)$$
$$a_1 = (a_{11}, a_{21}, \ldots, a_{m1})$$
$$\vdots$$
$$a_{n-1} = (a_{1,n-1}, a_{2,n-1}, \ldots, a_{m,n-1})$$
$$-a_n = (-a_{1n}, -a_{2n}, \ldots, -a_{mn})$$

Either $-a_n$ is in the convex cone C generated by the other $m + n - 1$ points, or it is not. Suppose $-a_n \in C$. Then there exist nonnegative numbers $\mu_1, \ldots, \mu_m, \lambda_1, \ldots, \lambda_{n-1}$ such that

$$-a_{in} = \sum_{j=1}^{m} \mu_j e_{ij} + \sum_{j=1}^{n-1} \lambda_j a_{ij} \quad \text{for all} \quad i$$

and this means

$$\sum_{j=1}^{n-1} \lambda_j a_{ij} + a_{in} = -\mu_i \leq 0 \quad \text{for all} \quad i$$

We now set

$$y_i = \frac{\lambda_j}{1 + \sum \lambda_i}, \quad j = 1, \ldots, n - 1$$

$$y_n = \frac{1}{1 + \sum \lambda_i}$$

It is clear that $y = (y_1, \ldots, y_n)$ is an optimal strategy for II such that $y_n > 0$.

Suppose, on the other hand, $-a_n \notin C$. There exist numbers $q_1, \ldots, q_m$ such that

$$\sum_{i=1}^{m} q_i e_{ij} \geq 0 \quad \text{for all} \quad j = 1, \ldots, m \tag{2.22}$$

$$\sum_{i=1}^{m} q_i a_{ij} \geq 0 \quad \text{for} \quad j = 1, \ldots, n - 1 \tag{2.23}$$

and

$$\sum_{i=1}^{m} q_i(-a_{in}) < 0 \tag{2.24}$$

By (2.22), $q_i \geq 0$, and by (2.24), the q_i are not all zero. Hence we can let

$$x_i = \frac{q_i}{\sum_{j=1}^{m} q_j} \quad i = 1, \ldots, m$$

and it follows directly from (2.23) and (2.24) that x is an optimal strategy for I such that $\Sigma x_i a_{in} > 0$.

Suppose, now, that $v(A) = k \neq 0$. As before, we construct a matrix $B = (b_{ij})$ where $b_{ij} = a_{ij} - k$; the proof proceeds exactly as in Theorem 2.4.1.

2.5. Computation of Optimal Strategies

Theorem 2.4.1 (the minimax theorem) assures us that every two-person zero-sum game will have optimal strategies. Unfortunately, the proof given is an existence proof and does not explain how these optimal strategies are to be computed.

Here we give methods for solving the easiest games; Chapter 3 will deal with a more general method.

2.5.1 Saddle Points. The simplest case of all occurs if a saddle point exists, i.e., if there exists an entry a_{ij} which is both the maximum entry in its column and the minimum entry in its row. In this case, the pure strategies i and j, or equivalently, the mixed strategies x and y with $x_i = 1, y_j = 1$ and all other components equal to zero, will be optimal strategies for players I and II, respectively.

2.5.2 Domination. In a matrix A, we say the ith row *dominates* the kth row if

$$a_{ij} \geq a_{kj} \quad \text{for all} \quad j \tag{2.25}$$

and

$$a_{ij} > a_{kj} \quad \text{for at least one} \quad j \tag{2.26}$$

Similarly, we say the jth column dominates the lth column if

$$a_{ij} \leq a_{il} \quad \text{for all} \quad i \tag{2.27}$$

and

$$a_{ij} < a_{il} \quad \text{for at least one} \quad i \tag{2.28}$$

Briefly, a pure strategy (represented by its row or column) is said to dominate a second pure strategy if the choice of the first (dominating) strategy is at least as good as the choice of the second (dominated) strategy, and in some cases better. It follows that a player can always dispense with any dominated strategies and use only the undominated strategies. The following theorem, which is given without proof, expresses this:

2.5.3 Theorem. Let A be a matrix game, and assume that rows $i_1, i_2, \ldots, i_k$ of A are dominated. Then player I has an optimal strategy x such that $x_{i_1} = x_{i_2} = \cdots = x_{i_k} = 0$; moreover, any optimal strategy for the game obtained by removing the dominated rows will also be an optimal strategy for the original game.

A similar theorem will hold for column domination; the general effect of these theorems is that any dominated rows and columns may be discarded, allowing us to work with a smaller matrix.

2.5.4 Example. Consider the game with matrix

$$\begin{pmatrix} 2 & 0 & 1 & 4 \\ 1 & 2 & 5 & 3 \\ 4 & 1 & 3 & 2 \end{pmatrix}$$

It is seen that the second column dominates the fourth column. It follows that player II will never use his fourth strategy; we can therefore disregard it and are left with the matrix

$$\begin{pmatrix} 2 & 0 & 1 & 4 \\ 1 & 2 & 5 & 3 \\ 4 & 1 & 3 & 2 \end{pmatrix}$$

In this matrix, we find the third row dominates the first row. Removing this, we have

$$\begin{pmatrix} 2 & 0 & 1 & 4 \\ 1 & 2 & 5 & 3 \\ 4 & 1 & 3 & 2 \end{pmatrix}$$

and, in this matrix, we see that the third column is dominated by the second. Hence our matrix reduces to

$$\begin{pmatrix} 2 & 0 & 1 & 4 \\ 1 & 2 & 5 & 3 \\ 4 & 1 & 3 & 2 \end{pmatrix}$$

and we need only look for optimal strategies to the small 2×2 matrix game. The next section discusses such games.

2.5.5 2×2 Games. Suppose we are given the 2×2 matrix game

$$\begin{pmatrix} a_{11} & a_{12} \\ a_{21} & a_{22} \end{pmatrix}$$

It may be that this game has a saddle point; if so, there is no problem. Suppose, however, that the game has no saddle point. It follows that the optimal strategies, $x = (x_1, x_2)$ and $y = (y_1, y_2)$ must have positive components. Now, if the value of the game is v, we have

$$a_{11}x_1y_1 + a_{12}x_1y_2 + a_{21}x_2y_2 + a_{22}x_2y_2 = v$$

or

$$x_1(a_{11}y_1 + a_{12}y_2) + x_2(a_{21}y_1 + a_{22}y_2) = v \tag{2.29}$$

Now, the two terms in parentheses on the left side of (2.29) are both less than or equal to v, since y is by hypothesis an optimal strategy. Suppose one of them were less than v; i.e., suppose

$$a_{11}y_1 + a_{12}y_2 < v \tag{2.30}$$

$$a_{21}y_1 + a_{22}y_2 \leq v \tag{2.31}$$

Then, since $x_i > 0$ and $x_1 + x_2 = 1$, the left-hand term in (2.29) would be strictly smaller than v. It follows that both the terms in parentheses in (2.29) must equal v. Hence

$$a_{11}y_1 + a_{12}y_2 = v \tag{2.32}$$

$$a_{21}y_1 + a_{22}y_2 = v \tag{2.33}$$

Similarly, it can be seen that

$$a_{11}x_1 + a_{21}x_2 = v \tag{2.34}$$

$$a_{12}x_1 + a_{22}x_2 = v \tag{2.35}$$

or, in matrix form

$$Ay^{\mathrm{T}} = \begin{pmatrix} v \\ v \end{pmatrix} \tag{2.36}$$

$$xA = (v, v) \tag{2.37}$$

These equations, together with the equations

$$x_1 + x_2 = 1 \tag{2.38}$$

$$y_1 + y_2 = 1 \tag{2.39}$$

allow us to solve for x, y, and v. In fact, if A is nonsingular, we will have

$$x = vJA^{-1}$$

where J is the vector $(1,1)$. We now eliminate v by noticing that the sum of the components of x, i.e., xJ^{T}, must be equal to 1; this gives

$$vJA^{-1}J^{\mathrm{T}} = 1$$

or

$$v = \frac{1}{JA^{-1}J^{\mathrm{T}}} \tag{2.40}$$

and

$$x = \frac{JA^{-1}}{JA^{-1}J^{\mathrm{T}}} \tag{2.41}$$

Similarly, we find

$$y = \frac{A^{-1}J^{\mathrm{T}}}{JA^{-1}J^{\mathrm{T}}} \tag{2.42}$$

In case A is singular, this is of course meaningless; it is easy then to see that

$$x = \frac{JA^{*}}{JA^{*}J^{\mathrm{T}}} \tag{2.43}$$

and

$$y = \frac{A^{*}J^{\mathrm{T}}}{JA^{*}J^{\mathrm{T}}} \tag{2.44}$$

(where A^{*} is the adjoint of A) will give optimal strategies; note that (2.43) and (2.44) coincide with (2.41) and (2.42) when A is nonsingular. We also see that

$$v = \frac{|A|}{JA^{*}J^{\mathrm{T}}} \tag{2.45}$$

where $|A|$ is the determinant of A, will give the value of the game, whether A is singular or not. We recapitulate these laws in the following theorem:

2.5.6 Theorem. Let A be a 2×2 matrix game. Then if A does not have a saddle point, its unique optimal strategies and value will be given by

$$x = \frac{JA^*}{JA^*J^{\mathrm{T}}} \tag{2.46}$$

$$y = \frac{A^*J^{\mathrm{T}}}{JA^*J^{\mathrm{T}}} \tag{2.47}$$

$$v = \frac{|A|}{JA^*J^{\mathrm{T}}} \tag{2.48}$$

where A^* is the adjoint of A, $|A|$ the determinant of A, and J the vector $(1,1)$.

2.5.7 Example. Solve the matrix game

$$\begin{pmatrix} 1 & 0 \\ -1 & 2 \end{pmatrix}$$

It is easily checked that the game has no saddle point. Now, the adjoint of A is

$$A^* = \begin{pmatrix} 2 & 0 \\ 1 & 1 \end{pmatrix}$$

and $|A| = 2$, $JA^* = (3, 1)$; $A^*J^{\mathrm{T}} = (2, 2)$ and $JA^*J^{\mathrm{T}} = 4$. Thus we have

$$x = \left(\frac{3}{4}, \frac{1}{4}\right), \quad y = \left(\frac{1}{2}, \frac{1}{2}\right) \quad v = \frac{1}{2}$$

It may be checked that these strategies x and y do indeed guarantee the payoff value of $\frac{1}{2}$.

2.5.8 $2 \times n$ and $m \times 2$ Games. The simplest games to solve, after the 2×2 games, are those games in which one of the players has only two strategies ($2 \times n$ or $m \times 2$ games). We shall consider $2 \times n$ games here; a similar analysis may be used for $m \times 2$ games.

The first player's problem is to maximize

$$v(x) = \min_j \{a_{1j}x_1 + a_{2j}x_2\}$$

Now, $x_1 = 1 - x_2$ and so we have

$$v(x) = \min_j \{(a_{2j} - a_{1j})x_2 + a_{1j}\}$$

Thus $v(x)$ is the minimum of n linear functions of the single variable x_1; these can all be plotted and their minimum $v(x)$ is then maximized by graphic methods.

2.5.9 Example. Consider the game with matrix

$$\begin{pmatrix} 2 & 3 & 1 & 5 \\ 4 & 1 & 6 & 0 \end{pmatrix}$$

The several functions $(a_{2j} - a_{1j})x_2 + a_{1j}$ are easily plotted if we notice that they must pass through the points $(0, a_{1j})$ and $(1, a_{2j})$ (Figure 2.1).

The heavy irregular line represents the function $v(x)$. The highest point on this line, X, is seen to be at the intersection of the lines corresponding to the second and third columns. The abscissa at this point is $\frac{2}{7}$, the ordinate $\frac{17}{7}$. Thus we have $x = (\frac{5}{7}, \frac{2}{7})$ and $v = \frac{17}{7}$. These numbers could also be calculated by application of the formulas (2.46) and (2.48) to the 2×2 matrix consisting of the second and third columns for our game matrix. Application of (2.47) to this matrix also yields $y = (0, \frac{5}{7}, \frac{2}{7}, 0)$; it is easily checked that these are indeed the optimal strategies and value of the game.

This covers the simplest type of games; generalizations of these methods for other games, while possible, are generally too laborious to be of practical use. For more general games, several schemes are in use. The easiest one to explain is given below. We do not give a proof of the scheme's validity; proofs of it may be found in the literature.

2.5.10 Solution by Fictitious Play. Let us suppose two players, perhaps with no knowledge of game theory, decide to play a game many times. While they might know nothing about game theory, they are nevertheless statistically inclined, and so keep track of the pure strategies used by their opponent at the several plays of the game. At each play, each player acts in such a way as to maximize his expected return against his opponent's observed empirical probability distribution: if player II has used his jth strategy q_j times, player I will choose i so as to maximize $\Sigma_j a_{ij} q_j$. Similarly, if player I has played his ith strategy p_i times, player II will choose j so as to minimize $\Sigma_i a_{ij} p_i$.

Now, naïve as this method of play may seem, the amazing fact is that the empirical distributions will converge to optimal strategies. More precisely, let p_i^N be the number of times that player I uses his ith strategy during the first N plays of the game. If we let $x_i^N = P_i^N / N$, it is obvious that x^N is a mixed strategy. This sequence of strategies, x^N, need not converge; however, lying in the compact set of mixed strategies, it must

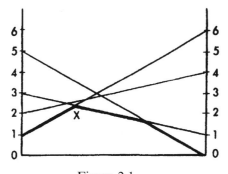

Figure 2.1.

have a convergent subsequence. The fact is that *the limit of any convergent subsequence will be an optimal strategy.*

We omit the proof of this theorem, as it is much too involved for our present purposes. Briefly, it is based on the fact that, if x^N and y^N are the strategies obtained as explained above for players I and II, respectively, the equation

$$\lim_{N \to \infty} \{v(y^N) - v(x^N)\} = 0 \tag{2.49}$$

will hold. This method, incidentally, is most useful in the case of very large games (i.e., games with large numbers of pure strategies), which are, because of their size, difficult to handle in any other way. (See Problem 2.9 for an outline of the proof of this theorem.)

2.6. Symmetric Games

Let us conclude this chapter with a brief discussion of a special type of game.

2.6.1 Definition. A square matrix $A = (a_{ij})$ is said to be *skew-symmetric* if $a_{ij} = -a_{ji}$ for all i, j. A matrix game is said to be *symmetric* if its matrix is skew-symmetric.

2.6.2 Theorem. The value of a symmetric game is zero. Moreover, if x is an optimal strategy for I, then it is also optimal for II.

Proof. Let A be the game matrix, and let x be any strategy. It is easy to see that $A = -A^T$. Hence

$$x A x^T = -x A^T x^T = -(x A x^T)^T = -x A x^T$$

Therefore, $x A x^T = 0$. It follows that, for any x

$$\min_{y} x A y^T \leq 0$$

so that the value is nonpositive, while

$$\max_{y} y A x^T \geq 0$$

so that the value is nonnegative. Hence the value is zero. If, now, x is an optimal strategy for I, we have

$$x A \geq 0$$

But this means

$$x(-A^T) \geq 0$$

and so

$$x A^T \leq 0$$

or

$$Ax^{\mathrm{T}} \leq 0$$

Hence x is also optimal for II.

An iterative algorithm, somewhat related to that given in 2.34 above, has been devised for solving symmetric games. Unlike the method of 2.34, which gives a discrete sequence of strategies, this one relies on a continuous modification of the given strategy, which, in the limit, will approach optimality. The method, that of von Neumann and Brown, is, however, of very limited value (other than as a constructive existence proof of the minimax theorem) because the differential equations which it involves generally cannot be solved analytically.

2.6.3 Example. Consider the game with matrix

$$\begin{pmatrix} 0 & 1 & -2 \\ -1 & 0 & 3 \\ 2 & -3 & 0 \end{pmatrix}$$

It can be considered a generalization of the well-known children's game "scissors, stone, and paper." As the matrix is skew-symmetric, we know the value of the game must be zero. Now, it is clear that the game has no saddle point. Nor can an optimal strategy use only two of the pure strategies, for if, say $x_1 > 0, x_2 > 0$, and $x_3 = 0$, it is easily seen that this mixed strategy of I gives a negative expectation against II's first pure strategy. Thus the optimal strategy x must have all positive components. This is also optimal for player II, and an argument similar to that given in 2.29 above shows that x must satisfy the system of linear equations:

$$-x_2 + 2x_3 = 0$$
$$x_1 - 3x_3 = 0$$
$$-2x_1 + 3x_2 = 0$$
$$x_1 + x_2 + x_3 = 1$$

The solution to these equations is easily found: it is $(\frac{1}{2}, \frac{1}{3}, \frac{1}{6})$. This is the unique optimal strategy for both players.

Problems

1. Show that, if C is any convex set, and x is any point on the boundary of C, then there exists a vector $p \neq 0$ such that $\Sigma x_i p_i = p_{m+1}$, and, for all $y \in C$, $\Sigma y_i p_i \geq p_{m+1}$. (This is a somewhat stronger version of the theorem of the supporting hyperplane. Use this theorem, together with the fact that, as x is on the boundary of C, there is a sequence of points $x^k \notin \overline{C}$ which converges to x.)

2. Show that the set of optimal strategies is always closed, convex, and bounded, and hence the convex hull of its extreme points.

3. Show that, if we treat an $m \times n$ matrix game as a point in mn-dimensional Euclidean space, the value of the game is a continuous function of the game.

4. Show that the set of optimal strategies of a game is a lower semicontinuous function of the game matrix.

5. Show that the set of $m \times n$ matrix games with unique optimal strategies is dense and open.

6. Let x and y be extreme points of the sets of optimal strategies of a matrix game, M, with nonzero value. Show that M has a square nonsingular submatrix $\dot{M}$ such that

$$\dot{x} = \frac{\dot{j}\dot{M}^{-1}}{\dot{j}\dot{M}^{-1}\dot{j}^{\mathrm{T}}}$$

$$\dot{y}^{\mathrm{T}} = \frac{\dot{M}^{-1}\dot{j}^{\mathrm{T}}}{\dot{j}M^{-1}\dot{j}^{\mathrm{T}}}$$

$$v = \frac{1}{\dot{j}\dot{M}^{-1}\dot{j}^{\mathrm{T}}}$$

where $\dot{j}$ is the vector, all of whose components are 1, of desired size:
(a) The product of x with any column of M that belongs to $\dot{M}$ must be v.
(b) The product of y with any row of M that belongs to $\dot{M}$ must be v.
(c) The ith row and the jth column of M must belong to $\dot{M}$ if $x_i > 0$ and $y_j > 0$.
(d) Rows and columns can be added to M, subject to (a) and (b) above, so long as they are linearly independent.
(e) If the matrix formed in this manner has linearly dependent rows, then x is not extreme. If the columns are linearly dependent, then y is not extreme.

7. Let $A = (a_{ij})$ be a skew-symmetric game matrix. Let $x = (x_1, \ldots, x_m)$ be a mixed strategy, and use the notation

$$u_i = \sum_{j=1}^{m} a_{ij} x_j$$

$$\varphi(u_i) = \max(0, u_i)$$

$$\Phi(x) = \sum_{i=1}^{m} \varphi(u_i)$$

$$\psi(x) = \sum_{i=1}^{m} \varphi^2(u_i)$$

Show that, for any "initial" strategy x^0, the solution to the system of differential equations

$$\frac{dx_i}{dt} = \varphi(u_i) - x_i \Phi(x)$$

$$x(0) = x^0$$

converges to an optimal strategy, in the sense that it must have accumulation points, and any accumulation point is optimal:

(a) If x^0 is a strategy, then $x(t)$ will be a strategy for all t.

(b) The function ψ satisfies

$$\frac{d\psi}{dt} = -2\Phi(x)\psi(x)$$

(c) $\psi(x) \leq \Phi^2(x)$

(d) From (b) and (c), we must have

$$\psi(x) \leq \frac{\psi(x_0)}{\left(1 + t\sqrt{\psi(x_0)}\right)^2}$$

8. Let $A = (a_{ij})$ be any $m \times n$ matrix, and consider the $mn \times mn$ matrix $B = (b_{pq})$ defined by

$$b_{(i-1)n+j,(k-1)n+1} = a_{il} - a_{kj}$$

Show that B is skew-symmetric and hence represents a symmetric game. Moreover, show that if $\lambda = (\lambda_1, \ldots, \lambda_{mn})$ is an optimal strategy for B, then the vectors x and y, defined by

$$x_i = \sum_{j=1}^{n} \lambda_{(i-1)n+j}$$

and

$$y_j = \sum_{i=1}^{m} \lambda_{(i-1)n+j}$$

are optimal strategies for players I and II, respectively, in A.

9. (Solutions of games by fictitious play.) Consider the following procedure for a matrix game $A = (a_{ij})$. Two players play the game a large number of times. At the first play, they may use any of their strategies. Suppose, now, that in the first k plays, I has used his ith strategy X_i^k times $(i = 1, \ldots, m)$ while II has used his jth strategy Y_j^k times $(j = 1, \ldots, n)$. Then, at the $(k + 1)$th play, I will use his i_{k+1}th strategy, and II his j_{k+1}th strategy, where

$$\sum_j a_{i_{k+1}j} Y_j^k = \max_i \sum_j a_{ij} Y_j^k = U_k$$

and

$$\sum_i a_{ij_{k+1}} X_i^k = \min_j \sum_i a_{ij} X_i^k = V_k$$

(Thus, at each play, each plays whatever is best against his opponent's cumulative past choices.) Show that any accumulation point of the sequence X^k/k, Y^k/k is a solution of the game A.

(a) We say the i_0th row is eligible in the interval $[k, k']$ if there is a $k_1, k \le k_1 \le k'$, such that

$$\sum_j a_{i_0 j} Y_j^{k_1} = U_{k_1}$$

and similarly, the j_0th column is eligible in $[k, k']$ if there is a $k_2, k \le k_2 \le k'$, such that

$$\sum_i a_{i j_0} X_i^{k_2} = V_{k_2}$$

Then, if all rows are eligible for $[k, l]$, we have

$$V_l - U_l \le 4a(k - l)$$

where $a = \max |a_{ij}|$.

(b) For any $\varepsilon > 0$, there exists k_0 such that $V_k - U_k \le k\varepsilon$ for all $k \ge k_0$. Prove by induction on the size of the matrix: If A is a 1×1 matrix, this is obvious. Assume true for all submatrices $\tilde{A}$ of A. Choose k^* so that $V_{k^*} - U_{k^*} \le k\varepsilon/2$ for all $\tilde{A}$. Then $k_0 \ge 8ak^*/\varepsilon$ will suffice. (Show that if some row or column is ineligible for an interval, it may effectively be discarded for that interval, treating only the remaining submatrix; consider intervals of the form $(k, k + k^*)$ with $k + k^* \le k_0$.)

10. Solve the following matrix games:

(a) $\begin{pmatrix} 3 & 8 \\ 5 & 1 \end{pmatrix}$

(b) $\begin{pmatrix} 4 & 6 \\ 2 & 7 \end{pmatrix}$

(c) $\begin{pmatrix} 1 & 4 & 8 \\ 9 & 3 & 2 \end{pmatrix}$

(d) $\begin{pmatrix} 2 & 10 \\ 5 & 8 \\ 6 & 3 \end{pmatrix}$

(e) $\begin{pmatrix} 0 & 7 & -3 \\ -7 & 0 & 4 \\ 3 & -4 & 0 \end{pmatrix}$

(f) $\begin{pmatrix} 1 & -2 & 0 \\ -4 & 0 & 2 \\ 0 & 4 & -1 \end{pmatrix}$

Chapter 3

Linear Programming

3.1. Introduction

A linear program is, briefly, the problem of maximizing (or minimizing) a linear function (called the *objective function*) subject to linear constraints. In the most common case, the constraints are loose inequalities, and the variables are restricted to nonnegative values. Thus the most common form will be: find $(x_1, \ldots, x_m)$ so as to

$$\text{Maximize} \sum_{i=1}^{m} c_i x_i \tag{3.1}$$

subject to

$$\sum a_{ij} x_i \le b_j \quad j = 1, \ldots, n \tag{3.2}$$

$$x_i \ge 0 \tag{3.3}$$

or, in matrix notation

$$\text{Maximize} \quad x C^{\mathrm{T}} \tag{3.4}$$

subject to

$$x A \le B \tag{3.5}$$

$$x \ge 0 \tag{3.6}$$

It should be pointed out that this is not the most general type of linear program. In fact, we may be asked to minimize the objective function, rather than to maximize it. Some of the constraints may be equations, and some of the variables may be allowed to take on negative values. However, it can also be pointed out that minimizing a function is the same as maximizing its negative, that an equation can be replaced by two opposite inequalities, and that an unrestricted variable can be expressed as the difference of two nonnegative variables. Thus there is no loss in generality, strictly speaking, in considering programs such as (3.1)–(3.6).

3.1.1 Definition. The set of all points satisfying the constraints (3.2) and (3.3) shall be called the *constraint set* of program (3.1)–(3.3). The maximum of (3.1) is called the *value* of the program.

It may be seen that the solution of a matrix game can be reduced to a linear program. In fact, if player I uses the strategy $(x_1, \ldots, x_m)$ he can assure himself of an expectation at least equal to λ, where λ is any number such that

$$\sum a_{ij} x_i \geq \lambda \quad \text{for} \quad j = 1, \ldots, n$$

Thus the problem of obtaining an optimal strategy for I reduces to the program

$$\text{Maximize} \quad \lambda \tag{3.7}$$

subject to

$$-\sum_{i=1}^{m} a_{ij} x_i + \lambda \leq 0, \quad j = 1, \ldots, n \tag{3.8}$$

$$\sum_{i=1}^{m} x_i = 1 \tag{3.9}$$

$$x_i \geq 0, \quad i = 1, \ldots, m \tag{3.10}$$

Similarly, the problem for II reduces to the program

$$\text{Minimize} \quad \mu \tag{3.11}$$

subject to

$$-\sum_{j=1}^{n} a_{ij} y_j + \mu \geq 0, \quad i = 1, \ldots, m \tag{3.12}$$

$$\sum_{j=1}^{n} y_j = 1 \tag{3.13}$$

$$y_i \geq 0, \quad j = 1, \ldots, n$$

3.2. Duality

Consider the two linear programs

$$\text{Maximize} \quad xC^{\mathrm{T}} \tag{3.14}$$

subject to

$$xA \leq B \tag{3.15}$$

$$x \geq 0 \tag{3.16}$$

and

$$\text{Minimize} \quad By^{\mathrm{T}} \tag{3.17}$$

subject to

$$yA^{\mathrm{T}} \geq C \tag{3.18}$$

$$y \geq 0 \tag{3.19}$$

The program (3.17)–(3.19) is said to be the *dual* of (3.14)–(3.16). Moreover, if we change the signs of A, B, and C, (3.17)–(3.19) can be rewritten as a maximization program

$$\text{Maximize} \quad y(-B)^{\mathrm{T}} \tag{3.20}$$

subject to

$$y(-A^{\mathrm{T}}) \leq -C \tag{3.21}$$

$$y \geq 0 \tag{3.22}$$

The dual of (3.20)–(3.22), as defined above, must be

$$\text{Minimize} \quad -Cx^{\mathrm{T}} \tag{3.23}$$

subject to

$$x(-A) \geq -B \tag{3.24}$$

$$x \geq 0 \tag{3.25}$$

But it is clear that (3.23)–(3.25) is equivalent to (3.14)–(3.16). Hence we may state: the dual of (3.17)–(3.19) is (3.14)–(3.16). Duality is, then, a reciprocal relation: the two programs (3.14)–(3.16) and (3.17)–(3.19) are dual to each other.

The following theorems explain the relation between dual linear programs.

3.2.1 Theorem. Let x and y be in the constraint sets of programs (3.14)–(3.16) and (3.17)–(3.19), respectively. Then

$$xC^{\mathrm{T}} \leq By^{\mathrm{T}} \tag{3.26}$$

Proof. We have

$$
\begin{aligned}
xC^{\mathrm{T}} &= \sum_{i=1}^{m} x_i c_i \\
&\leq \sum_{i=1}^{m} x_i (\Sigma a_{ij} y_i) \qquad \text{[by (3.16) and (3.18)]} \\
&= \sum_{j=1}^{n} \left(\sum_{i=1}^{m} x_i a_{ij} \right) y_j \\
&\leq \sum_{j=1}^{n} b_j y_j = By^{\mathrm{T}} \qquad \text{[by (3.15) and (3.19)]}
\end{aligned}
$$

3.2.2 Corollary. Suppose the vectors x^* and y^* are in the constraint sets of (3.14)–(3.16) and (3.17)–(3.19), respectively, and $x^*C^T = By^{*T}$. Then x^* and y^* are the solutions of the two programs.

In general, a linear program need not have a solution. One of two things may go wrong. It may be that the constraints are inconsistent so that the constraint set is empty, in which case the program is said to be *infeasible*. Or it may be that the desired maximum (or minimum) does not exist even if the program is *feasible*, since the function may take on arbitrarily large (or small, in the case of a minimization problem) values in the constraint set. For example, the program

Maximize $2x$

subject to

$$-x \leq 1$$
$$x \geq 0$$

does not have a solution, as the objective function can be made arbitrarily large. Such a program is said to be *unbounded*.

Let us consider the following corollary:

3.2.3 Corollary. (to 3.2.1) If the linear program (3.14)–(3.16) is unbounded, then the dual program (3.17)–(3.19) is infeasible. Similarly, if (3.17)–(3.19) is unbounded, then (3.14)–(3.16) is infeasible.

Proof. Suppose (3.17)–(3.19) is feasible, i.e., suppose y is in the constraint set of (3.17)–(3.19). Then the objective function xC^T is bounded above by By^T. Hence (3.14)–(3.16) is not unbounded.

The converse of Corollary 3.2.3 is not true. That is, if a program is infeasible, it does not follow that its dual is unbounded. It may be that both programs are infeasible.

3.2.4 Example. The dual programs

Maximize $5x_1 + x_2$

subject to

$$x_1 - x_2 \leq 1$$
$$-x_1 + x_2 \leq -2$$
$$x_1, x_2 \geq 0$$

and

Minimize $y_1 - 2y_2$

subject to

$$y_1 - y_2 \geq 5$$
$$-y_1 + y_2 \geq 1$$
$$y_1, y_2 \geq 0$$

are both infeasible.

Thus, if a program is infeasible its dual may be unbounded, or it may also be infeasible. We shall prove next that these are the only possibilities.

3.2.5 Theorem. Let the program (3.14)–(3.16) be feasible, and suppose the dual program (3.17)–(3.19) is infeasible. Then (3.14)–(3.16) is unbounded.

Proof. Suppose (3.17)–(3.19) is infeasible. We consider the game with the $m \times (n+1)$ matrix

$$(-A \ \vdots \ C^{\mathrm{T}}) \tag{3.27}$$

which is obtained by adjoining the column vector C^{T} to the matrix $-A$. Suppose that the value of this game is negative. Then, if $s = (s_1, \ldots, s_n, s_{n+1})$ is an optimal strategy for II, we find that

$$-\sum_{j=1}^{n} a_{ij}s_j + c_i s_{n+1} < 0, \quad i = 1, \ldots, m$$

Now, if $s_{n+1} = 0$, we have

$$\sum_{j=1}^{n} a_{ij}s_j > 0, \quad i = 1, \ldots, m$$

and so, for sufficiently large k, we will have

$$\sum_{j=1}^{n} a_{ij}(ks_j) > c_i, \quad i = 1, \ldots, m$$

$$ks_j \geq 0$$

i.e., the vector ks satisfies (3.18) and (3.19), so that (3.17)–(3.19) is feasible. If, on the other hand, $s_{n+1} > 0$, we write $y_j = s_j/s_{n+1}$, and it follows that y satisfies (3.18) and (3.19). Thus *the value of the game must be zero or positive.*

Suppose now that the value of the game is zero; suppose also that II has an optimal strategy s with $s_{n+1} > 0$. Once again, if we set $y_j = s_j/s_{n+1}$, we find that y satisfies (3.18) and (3.19), contradicting the hypothesis that (3.17)–(3.19) was infeasible.

Hence the value of the matrix game (3.27) is nonnegative. Moreover, if the value is zero, all optimal strategies s for player II must have $s_{n+1} = 0$.

By Lemma 2.4.3 and Theorem 2.5.3, it follows that player I must have a strategy $t = (t_1, \ldots, t_m)$ such that

$$\sum_{i=1}^{m}(-a_{ij})t_i \geq 0, \quad j = 1, \ldots, n$$

$$\sum_{i=1}^{m} c_i t_i > 0$$

By hypothesis, (3.14)–(3.16) is feasible. Suppose that $x = (x_1, \ldots, x_m)$ lies in its constraint set. Then, for every positive number α, the vector $x + \alpha t$ will also lie in the constraint set. Moreover,

$$(x + \alpha t)C^{\mathrm{T}} = xC^{\mathrm{T}} + \alpha tC^{\mathrm{T}}$$

and the right side of this equation can be made as large as desired simply by increasing α. Thus (3.14)–(3.16) is unbounded.

We see thus that there are four possibilities for the dual pair of programs (3.14)–(3.16) and (3.17)–(3.19). They can both be infeasible, or one can be infeasible while the other is unbounded, or both may be feasible and bounded. In this last case, the fact that the constraint set is closed assures us that the programs will both have solutions. The following very important theorem explains the relationship between them in this last case.

3.2.6 Theorem. Let the two dual programs (3.14)–(3.16) and (3.17)–(3.19) both be feasible. Then both will have solutions, x^* and y^*, respectively, and, moreover

$$x^* C^{\mathrm{T}} = By^{*\mathrm{T}}$$

i.e., both programs have the same value.

Proof. Consider the game with matrix

$$M = \begin{pmatrix} 0 & A^{\mathrm{T}} & -B^{\mathrm{T}} \\ -A & 0 & C^{\mathrm{T}} \\ B & -C & 0 \end{pmatrix}$$

where the 0's represent matrices of suitable size, all of whose entries are zero. The matrix M is skew-symmetric and therefore its value is zero.

Suppose now that $s = (s_1, \ldots, s_{m+n+1})$ is an optimal strategy for player I. We write:

$$y = (s_1, \ldots, s_n)$$
$$x = (s_{n+1}, \ldots, s_{n+m})$$
$$w = x_{n+m+1}$$

Because of the optimality of s, and since the game has value zero, we have

$$-xA + wB \geq 0$$
$$yA^{\mathrm{T}} - wC \geq 0$$
$$-yB^{\mathrm{T}} + xC^{\mathrm{T}} \geq 0$$

Let us suppose that $w > 0$. If we set $x^* = x/w$ and $y^* = y/w$, it is clear that

$$x^* A \leq B$$
$$y^* A^T \geq C$$
$$x^*, y^* \geq 0$$

and

$$x^* C^T \geq y^* B^T$$

Now this means that x^* and y^* lie in the constraint sets of (3.14)–(3.16) and (3.17)–(3.19). By Theorem 3.2.1 and its Corollary 3.2.2 it follows that x^* and y^* are the solutions to the two programs and, moreover

$$x^* C^T = y^* B^T$$

Suppose, on the other hand, that none of the optimal strategies s for player I has $s_{m+n+1} > 0$. In this case, by Theorem 2.5.3, player II will have an optimal strategy, $r = (r_1, \ldots, r_{m+n+1})$, whose inner product with the last row of M is negative. Because the two players have identical sets of optimal strategies, we must have $r_{n+m+1} = 0$. If we now set

$$y = (r_1, \ldots, r_n)$$
$$x = (r_{n+1}, \ldots, r_{m+n})$$

we shall have

$$-xA \geq 0$$
$$yA^T \geq 0$$
$$-yB^T + xC^T > 0$$

or, equivalently

$$xA \leq 0$$
$$yA^T \geq 0 \qquad\qquad (3.28)$$
$$xC^T > yB^T$$

Now either the left side of (3.28) is positive or its right side is negative. Suppose that the left side is positive. By hypothesis (3.14)–(3.16) is feasible; suppose x' lies in its constraint set. It is easily seen that for any positive $\alpha, x' + \alpha x$ also lies in the constraint set, while $(x' + \alpha x) C^T$ can be made as large as desired simply by increasing α. Thus (3.14)–(3.16) will be unbounded. But this means (3.17)–(3.19) is infeasible. Similarly, if the right side of (3.28) is negative, (3.17)–(3.19) is unbounded and (3.14)–(3.16) is infeasible. This contradiction proves the theorem.

3.3. Solution of Linear Programs

At first glance it might seem that a linear program may be solved by the methods of the calculus, setting the partial derivatives equal to zero. In fact, however, the objective function is linear, so that all its partial derivatives are constant; the solution to a linear program will generally be found on the boundary of the constraint set. The method which is generally used for solving linear programs is based on the following observations:

(3.3.1) The constraint set is the intersection of several half-spaces; since each half-space is convex, the constraint set is a convex hyperpolyhedron.

(3.3.2) Because the constraint set is convex and the objective function is linear, any local extremum of the function will be the global extremum.

(3.3.3) Because the objective function is linear, the extremum will be obtained at one of the extreme points (vertices) of the constraint set (hyperpolyhedron).

Geometrically, the simplex method may be described as follows: One of the vertices of the hyperpolyhedron is found (this may be done analytically by solving a system of n equations given by the constraints). All the edges which meet at this vertex are considered. If the objective function cannot be improved by moving along any one of these edges, it follows that the given vertex is a local extremum, and hence (by Remark (3.3.2)) the global extremum. If, on the other hand, the objective function is improved along one of the edges, we follow this edge to the vertex which lies at its other end and repeat the procedure. Since the objective function is improved at each step, it follows that we cannot go through the same vertex twice. Since there are only a finite number of vertices, this procedure will bring us to the solution in a finite number of steps.

3.3.1 Example. We illustrate this notion with this example:

Maximize $w = 2x + y$

subject to

$$x \le 1$$
$$y \le 1$$
$$2x + 2y \le 3$$
$$x, y \ge 0$$

The constraint set is shaded in Figure 3.1. The arrow shows the gradient of the objective function. If we start from the origin, we see that w is increased along either of the edges that start there. We take, say, the edge that goes to $(1,0)$. At $(1,0)$, we find that w is increased along the edge to $(1,\frac{1}{2})$. At this point, we find that w decreases along both edges that meet here. Thus we can conclude that $(1,\frac{1}{2})$ is the solution of our program.

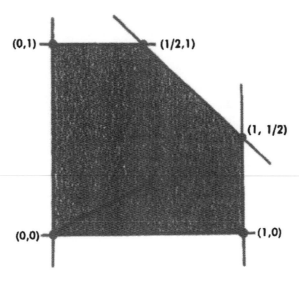

Figure 3.1.

3.4. The Simplex Algorithm

Let us describe algebraically the method given above.

We introduce a schematic notation, the purpose of which is simply to achieve a small degree of economy in symbols. Let us suppose we are given the system of inequalities.

$$
\begin{aligned}
a_{11}x_1 + a_{21}x_2 + \cdots + a_{m1}x_m &\leq b_1 \\
a_{12}x_1 + a_{22}x_2 + \cdots + a_{m2}x_m &\leq b_2 \\
&\;\;\vdots \\
a_{1n}x_1 + a_{2n}x_2 + \cdots + a_{mn}x_m &\leq b_n
\end{aligned}
\tag{3.29}
$$

which, together with the usual nonnegativity constraints, characterizes the constraint set of some linear program. We can rewrite these in the form

$$
\begin{aligned}
a_{11}x_1 + a_{21}x_2 + \cdots + a_{m1}x_m - b_1 &= -u_1 \\
a_{12}x_1 + a_{22}x_2 + \cdots + a_{m2}x_m - b_2 &= -u_2 \\
&\;\;\vdots \\
a_{1n}x_1 + a_{2n}x_2 + \cdots + a_{mn}x_m - b_n &= -u_n
\end{aligned}
\tag{3.30}
$$

where the u_j, $j = 1, \ldots, n$, are so-called *slack* variables, restricted to take nonnegative values. The constraints of (3.29) reduce now to this system of equations, together with the nonnegativity constraints $x_i \geq 0$, $u_j \geq 0$. We propose to identify the system of equations (3.30) with the schema

$$
\begin{array}{|cccc|l}
x_1 & x_2 & \cdots & x_m & 1 & \\
\hline
a_{11} & a_{21} & \cdots & a_{m1} & -b_1 & = -u_1 \\
a_{12} & a_{22} & \cdots & a_{m2} & -b_2 & = -u_2 \\
& & & & & \vdots \\
a_{1n} & a_{2n} & \cdots & a_{mn} & -b_n & = -u_n
\end{array}
\qquad (3.31)
$$

Thus each row inside the box of such a schema represents the linear equation obtained by multiplying each entry in the row by the term in the top row (outside the box) and corresponding column, and setting their sum equal to the term outside the box, on the right. The principal advantage of such a scheme, as explained previously, is that it avoids repetition of the variables $x_1, \ldots, x_m$, and thus saves much space. It is also possible to introduce the objective function into the schema (3.31) by calling it, say, w, and adjoining the row

$$
\begin{array}{|cccc|l}
c_1 & c_2 & \cdots & c_m & 0 & = w
\end{array}
\qquad (3.32)
$$

at the bottom of the schema.

If we do this, it may be seen that the schema which gives the linear program (3.14)–(3.16) can also be used to give the dual program (3.17)–(3.19). In fact, we can use the columns to represent the equations of system (3.18), obtaining the double schema

$$
\begin{array}{c|cccc|c|l}
 & x_1 & x_2 & \cdots & x_m & 1 & \\
\hline
y_1 & a_{11} & a_{21} & \cdots & a_{m1} & -b_1 & = -u_1 \\
y_2 & a_{12} & a_{22} & \cdots & a_{m2} & -b_2 & = -u_2 \\
 & & & & & & \vdots \\
y_n & a_{1n} & a_{2n} & \cdots & a_{mn} & -b_n & = -u_n \\
-1 & c_1 & c_2 & \cdots & c_m & 0 & = w \\
\hline
 & \| & \| & & \| & \| & \\
 & v_1 & v_2 & \cdots & v_m & -w' &
\end{array}
\qquad (3.33)
$$

which gives the two dual programs (3.14)–(3.16) and (3.17)–(3.19) simultaneously.

Consider next the maximizing problem (3.14)–(3.16) as represented by the rows of schema (3.33). We have a system of $n+1$ equations in the $m+n+1$ unknowns $x_1, \ldots, x_m, u_1, \ldots, u_n$, and w. We can, in general, solve for $n+1$ of these unknowns in terms of the remaining m. The schema, of course, represents a solution for $u_1, \ldots, u_n$, and w in terms of the x_i, but, except in degenerate cases, which involve the inversion of a singular matrix, we can solve for any $n+1$ in terms of the other m.

Suppose that we have solved for the $n+1$ variables $s_1, s_2, \ldots, s_n, w$, in terms of the variables $r_1, r_2, \ldots, r_m$, and obtain the new schema (simplex tableau)

r_1	r_2	$\cdots$	r_m	1		
a'_{11}	a'_{21}	$\cdots$	a'_{m1}	$-b'_1$	$= -s_1$	
a'_{12}	a'_{22}	$\cdots$	a'_{m2}	$-b'_2$	$= -s_2$	(3.34)
					$\vdots$	
a'_{1n}	a'_{2n}	$\cdots$	a'_{mn}	$-b'_n$	$= -s_n$	
c'_1	c'_2	$\cdots$	c'_m	δ	$= w$	

where every entry in the bottom row (except possibly the right-hand entry δ) and every entry in the right-hand column (again excepting the bottom entry) is nonpositive. If this happens, then the tableau (3.34) gives us the solution to program (3.14)–(3.16). In fact, if we let all the variables $r_1, \ldots, r_m$ be equal to zero, each of the variables s_j will be equal to the corresponding b. But the entries $-b_j$ are nonpositive. Hence all the s_j will be nonnegative. Since the r_i are all equal to zero, this point lies in the constraint set of (3.14)–(3.16). Moreover, the bottom row gives us the equation

$$w = c'_1 r_1 + \cdots + c'_m r_m + \delta$$

where all the c'_i are nonpositive. It is clear that the largest possible value of w will be obtained by setting all the r_i equal to zero (since they cannot be made smaller). Thus the solution of (3.14)–(3.16) is obtained from (3.34) by setting

$$r_i = 0, \quad i = 1, \ldots, m \tag{3.35}$$

$$s_j = b'_j, \quad j = 1, \ldots, n \tag{3.36}$$

$$w = \delta \tag{3.37}$$

We are looking for a method which will allow us to rewrite our system (3.33) in the form of (3.34), with nonpositive entries along the right-hand column and bottom row (except possibly the bottom right-hand entry δ). We give first some definitions:

3.4.1 Definition. In the simplex tableau (schema) (3.34) the variables s_j are called the *basic* variables, while the variables r_i are called the *nonbasic* variables.

3.4.2 Definition. A simplex tableau will also be called a *basic point*. If all the entries in the right-hand column (except possibly the bottom entry) are nonpositive, it will be called a *basic feasible point* (b.f.p.). If all the entries in the bottom row (except possibly the right-hand entry) are nonpositive, it will be called a *basic dual feasible point* (b.d.f.p.).

As already noted, when the right-hand entries are nonpositive, we obtain a point in the constraint set of (3.14)–(3.16) by letting the nonbasic variables vanish. This is what is meant by a b.f.p. If all the bottom entries are nonpositive, we see that we get a point in the constraint set of the dual program (3.17)–(3.19). This is a b.d.f.p. Finally, if a b.f.p. is also a b.d.f.p., it is the solution to the program (as was noted above).

Geometrically, a basic feasible point represents a vertex of the convex hyperpolyhedron which is the constraint set of (3.14)–(3.16). As explained before, we are interested in looking at the edges which pass through this vertex, or, equivalently, at the vertices which lie at the other ends of these edges. A vertex is obtained by setting m of the variables (the nonbasic ones) equal to zero; an edge is obtained by setting $m - 1$ of the variables equal to zero. An edge passes through a vertex if these $m - 1$ variables are among the m which correspond to the vertex. Thus two vertices lie on an edge if they have $m - 1$ of their nonbasic variables in common, which brings us to the following definition:

3.4.3 Definition. Two basic points are said to be *adjacent* if they represent different points in space and their sets of basic variables differ by at most one.

It should be noticed that it is not impossible for two simplex tableaux to represent the same point even though their basic variables might be different. In fact, there is nothing to prevent the entries in the right-hand column from vanishing. If some such b'_j does vanish, the corresponding s_j will vanish when the nonbasic variables vanish. Thus we could place it among the nonbasic variables without changing the point represented by the tableau.

3.4.4 Pivot Steps. A pivot step is a procedure for changing the sets of basic variables by one; i.e., it is a procedure for solving the $m + 1$ equations of (3.34) for $n - 1$ of the s_j, one of the r_i, and w in terms of the remaining m variables. Let us see how this is done.

Suppose we wish to make s_j a nonbasic variable, while making r_i a basic variable. We take the equation

$$a_{1j}r_1 + a_{2j}r_2 + \cdots + a_{mj}r_m - b_j = -s_j$$

and solve for $-r_i$, obtaining

$$\frac{a_{1j}}{a_{ij}}r_1 + \cdots + \frac{a_{i-1,j}}{a_{ij}}r_{i-1} + \frac{1}{a_{ij}}s_j + \cdots + \frac{a_{mj}}{a_{ij}}r_m - \frac{b_j}{a_{ij}} = -r_i$$

(assuming, of course, that $a_{ij} \neq 0$). This new equation will be the jth row of our modified simplex tableau. To get the remaining rows, it is necessary to substitute this value of r_i. Thus (assuming that $j \neq 1$) the first equation will now be

$$\left(a_{11} - \frac{a_{i1}a_{1j}}{a_{ij}}\right)r_1 + \cdots + \left(a_{i-1,1} - \frac{a_{i1}a_{i-1,j}}{a_{ij}}\right)r_{i-1} - \frac{a_{1j}}{a_{ij}}s_j + \cdots - \left(b_1 - \frac{a_{i1}b_j}{a_{ij}}\right) = -s_1$$

and the other equations (except, of course, the jth equation) will be similarly changed. To get the new tableau, we simply modify the coefficients accordingly, and interchange the position of the two variables r_i and s_j.

The entry a_{ij}, which cannot be zero, is called the *pivot* of the transformation. The transformation itself is called a *pivot step*. We can summarize the transformation most easily according to the diagram

$$\begin{bmatrix} p & q \\ r & s \end{bmatrix} \rightarrow \begin{bmatrix} \dfrac{1}{p} & \dfrac{q}{p} \\ -\dfrac{r}{p} & s - \dfrac{qr}{p} \end{bmatrix} \tag{3.38}$$

In (3.38), p represents the pivot, q is any other entry in the pivot row, r is any other entry in the pivot column, and s is the entry in the row of r and the column of q. Thus, the pivot is replaced by its reciprocal, $1/p$. The remaining entries in the pivot row are simply divided by p, while the remaining entries in the pivot column are divided by p and change in sign. Finally, the other entries in the tableau are decreased by the product of the entries in the same row and pivot column, and in the same column and pivot row, divided by p.

3.4.5 Example. Given the simplex tableau

r_1	r_2	r_3	1	
4	1	3	2	$= -s_1$
-1^*	2	-1	4	$= -s_2$
1	2	2	0	$= -s_3$
0	1	4	1	$= -s_4$

$$\tag{3.39}$$

which represents a system of equations solved for s_1, s_2, s_3, s_4 in terms of r_1, r_2, r_3, we wish to solve the system for s_1, r_1, s_3, s_4 in terms of s_2, r_2, r_3. The pivot, then, will be the starred entry which lies below r_1 and to the left of s_2. Performing the transformation given by (3.38) and interchanging the position of r_1 and s_2, we obtain the tableau

s_2	r_2	r_3	1	
4	9	-1	18	$= -s_1$
-1	-2	1	-4	$= -r_1$
1	4	1	4	$= -s_3$
0	1	4	1	$= -s_4$

$$\tag{3.40}$$

It may be checked that the system of equations (3.40) is equivalent to the system (3.39).

3.4.6 The Dual Problem. In 3.4.4, we determined the form of the transformation which must be effected on the simplex tableau in order that the system of equations represented by the rows of the new tableau be equivalent to that represented by the old tableau. On the other hand, we had mentioned above that the tableau can be used to represent a pair of dual programs simultaneously. It remains to be verified, then, that the equations represented by the columns of the tableau will also be equivalent to the old system.

The *i*th column of tableau (3.33) represents the linear equation

$$\sum a_{ij} y_j - c_i = v_i$$

Here, of course, it is the v_i which are basic variables, while the y_i are nonbasic. Now, let us suppose we wish to interchange the role of v_i and y_j; i.e., we wish to make y_j a basic variable, while v_i is made nonbasic. As before, we solve this equation for y_j in terms of v_i and the remaining y's, thus obtaining

$$-\frac{a_{i1}}{a_{ij}} y_1 + \cdots - \frac{a_{i,j-1}}{a_{ij}} y_{j-1} + \frac{1}{a_{ij}} v_i + \cdots + \frac{c_i}{a_{ij}} = y_j$$

If we substitute this value of y_j in the remaining equations, we obtain for the first column (assuming of course $i \neq 1$)

$$\left(a_{11} - \frac{a_{i1} a_{1j}}{a_{ij}} \right) y_1 + \cdots + \frac{a_{1j}}{a_{ij}} v_i + \cdots - \left(c_1 - \frac{c_i a_{1j}}{a_{ij}} \right) = v_1$$

and so on for the other columns. If we represent these changes schematically, as in (3.38), we obtain the schema

$$\begin{bmatrix} p & q \\ r & s \end{bmatrix} \rightarrow \begin{bmatrix} \dfrac{1}{p} & \dfrac{q}{p} \\ -\dfrac{r}{p} & s - \dfrac{qr}{p} \end{bmatrix}$$

But this is precisely the schema (3.38). Thus the same transformation which preserves the (primal) problem (3.14)–(3.16) will preserve the dual problem (3.17)–(3.19). That is, each simplex tableau obtained by means of pivot steps will represent the pair of dual problems simultaneously.

3.5. The Simplex Algorithm (Continued)

We have just explained the fundamental pivot step which is the basic tool of the simplex algorithm. What we need now, however, is a rule to decide how to use this tool. In fact, the simplex tableau will generally have a large number of nonzero entries which can be used as pivots; we need a rule to tell us which of these entries should be used as pivot.

When we refer to the geometric analogue, we note that each of our $m + n$ variables $x_1, \ldots, x_m, u_1, \ldots, u_n$ can be thought of as representing a bounding hyperplane of our constraint set, i.e., the hyperplane in which that variable vanishes. Now the point of intersection of m of these hyperplanes (assuming these are independent) is a basic point of our constraint set. If all the other variables are also nonnegative, the basic point is actually a b.f.p.; i.e., it will lie in the constraint set.

Our choice of pivot steps will have two objectives. First, starting from a basic point, we must move toward a b.f.p. Once we have a b.f.p., we must try to improve (increase or decrease) our objective function, being careful to keep at all times to b.f.p.s., until a solution (i.e., a point where no further improvement is possible) has been reached. We thus give two sets of rules: The first, assuming nothing, will give us a b.f.p.; the second, assuming we start from a b.f.p., will give us a solution. Let us call these Case I and Case II, respectively.

3.5.1. Rules for Choosing a Pivot

(3.5.1) Case I. Let $-b_k$ be the lowest positive entry in the right-hand column (excepting possibly the bottom entry). Choose any negative entry $a_{i_0 k}$ in the kth row. The pivot will be in the i_0th column. Now, for all $j \geq k$ for which $-b_j/a_{i_0 j}$ is negative, let j_0 be that value of j for which this quotient is largest (i.e., closest to zero). Then $a_{i_0 j_0}$ will be our pivot. (If there is a tie, any one of those entries which give the maximum may be chosen.)

(3.5.2) Case II. Let c_{i_0} be any positive entry in the bottom row (again, excepting possibly the right-hand entry). The pivot will be in the i_0th column. For all j for which $-b_j/a_{i_0 j}$ is negative, let j_0 be that value of j for which this quotient is greatest (i.e., closest to zero). Then $a_{i_0 j_0}$ will be our pivot. (Again, in case of ties, we may choose as above.)

3.5.2 Proof of Convergence. We prove now that our algorithm will actually converge in a finite number of steps. We shall make the assumption of nondegeneracy, which means that no zeros will appear in either the bottom row or the right-hand column (again, excepting the right-hand bottom entry).

(3.5.3) In Case I, we shall prove that all the entries in the right-hand column below the kth remain negative, while the kth entry itself decreases.

We know that, for each $j \neq j_0$, the entry $-b_j$ will be replaced by

$$-b'_j = -b_j + \frac{a_{i_0 j} b_{j_0}}{a_{i_0 j_0}}$$

We also know that $-b_{j_0}/a_{i_0 j_0}$ is negative. Hence, if $a_{i_0 j} < 0$, it is clear that $-b'_j < -b_j$.

If, on the other hand, $a_{i_0 j} > 0$, we know (by our rule) that, for $j \geq k$

$$-\frac{b_{j_0}}{a_{i_0 j_0}} \geq -\frac{b_j}{a_{i_0 j}}$$

Now

$$-b'_j = a_{i_0 j}\left(-\frac{b_j}{a_{i_0 j}} + \frac{b_{j_0}}{a_{i_0 j_0}}\right)$$

The term in parentheses will be nonpositive. Hence $-b'_j < 0$.

Finally, we notice that $-b_{j_0}$ will be replaced by $-b_{j_0}/a_{i_0 j_0}$, which is, by hypothesis, negative.

Thus for $j > k$, we see $b'_j < 0$. We also see $-b'_k < -b_k$. As there are only a finite number of possible simplex tableaux, we see that, if we continue to pivot according to our rule, we shall eventually reach a tableau in which, for $j \geq k, -b_j < 0$. We then work on the next higher row, etc., until all entries in the right-hand column are nonpositive.

(3.5.4) In Case II, we see that, by the same argument as above, each $-b_j$ will be replaced by $-b'_j$, where $-b'_j \leq 0$. The bottom right-hand entry, δ, will be replaced by

$$\delta' = \delta + \frac{c_{i_0} b_{j_0}}{a_{i_0 j_0}}$$

We know that $c_{i_0} > 0$, while $-b_{j_0}/a_{i_0} < 0$. Hence $\delta' > \delta$. Now, δ and δ' are the values of the objective function at the two given b.f.p.s. Hence we see that the objective function is increased. As there are only a finite number of b.f.p.s, it follows that we shall eventually reach the solution.

We have thus seen that (under the assumption of nondegeneracy), if we can continue to pivot according to the rules given in 3.5.2, the simplex algorithm will give us the solution to the linear program in a finite number of steps. It may be, however, that the two rules (3.5.3) and (3.5.4) will fail us. Two things could conceivably go wrong. In Case I, it may be that a row, ending in a positive entry, has only nonnegative entries. In Case II, it may be that a column which has a positive entry in the bottom row has only nonpositive entries otherwise. In either situation, it becomes impossible to follow the rules 3.5.1. We shall see that such a situation implies the insolubility (either because of infeasibility or because of unboundedness) of the program.

Suppose, then, in Case I, that a row which ends in a positive entry has only nonnegative entries. The row represents the

$$\sum_{i=1}^{n} a_{ij} r_i - b_j = -s_j \tag{3.41}$$

Now, the a_{ij} and r_i are all, by hypothesis, nonnegative, while $-b_j$ is positive. But this means s_j must be negative. Thus the constraints are inconsistent; i.e., the program is infeasible.

In Case II, suppose we have a column with a positive entry in the bottom row, but all nonpositive entries otherwise. Now, if the coefficient a_{ij} is nonpositive, it follows from equation (3.41) that r_i can be increased indefinitely without decreasing s_j. As all the entries in the given column (except the bottom entry) are nonpositive, it follows that we can increase r_i, indefinitely, keeping all the other r's fixed at zero, without leaving the constraint set. But, as the bottom entry in this column is positive, this will increase the objective function indefinitely. Hence the program is unbounded.

3.5.3 Degeneracy. In the discussion above, an explicit assumption of nondegeneracy was made; i.e., we assumed that no zeros would ever appear in the right-hand column (except possibly for the bottom entry). Now, if zeros do appear, it is clear that certain strict inequalities, which were given above, must be replaced by loose inequalities. For example, the inequality $\delta' > \delta$, which characterizes the steps of Case II when nondegeneracy is assumed, must be replaced by the weaker $\delta' \geq \delta$. Now, $\delta' > \delta$ means that the objective function is increased at each step, and hence it is clear that we cannot pass through the same b.f.p. twice. This guarantees termination. The weaker $\delta' \geq \delta$ allows for the possibility that we might pivot several times, and, after all these pivots, reach the same tableau again.

Degeneracy of the program, as characterized by the appearance of zeros in the right-hand column, occurs when more than m of the variables vanish simultaneously, i.e., when more than m of the bounding hyperplanes of the constraint set pass through one point. Thus nondegeneracy will hold if the $m + n$ bounding hyperplanes are in general position in space. But this means that, in a degenerate program, an arbitrarily small perturbation of the constraints will remove the degeneracy.

A technique which is often used, then, is a perturbation technique. If one of the rows ends in a zero, that is, if it looks like this:

$$|a_{1j} \quad a_{2j} \quad \cdots \quad a_{mj} \quad 0| = -s_j$$

then we replace the variable s_j by the variable $s'_j = s_j + \varepsilon$, to obtain the row

$$|a_{1j} \quad a_{2j} \quad \cdots \quad a_{mj} \quad -\varepsilon| = -s'_j$$

where ε is assumed to be positive but very small. We proceed to solve the modified program by the simplex algorithm. Because these perturbations will remove any degeneracy, it follows that the program, as modified by this finite number of perturbations, will be solved (if it has a solution) in this manner. The solution will be a function of the various perturbations; i.e., the value of the $x_1, \ldots, x_m$ will depend on the particular values given to the ε's introduced in these perturbations. Generally, however, they depend linearly on these ε's; it is merely a question of setting these equal to zero to obtain a solution to the original (unperturbed) problem.

In fact the perturbations, which replace the constraints $s_j \geq 0$ by the weaker constraints $s'_j \geq 0$, tend to enlarge the constraint set. We are interested in the extreme points of the constraint set; it is clear that the only ones affected by this perturbation are those for which $s_j = 0$. At each of these (and there are only a finite number) the perturbation will increase (or decrease) the objective function by some constant times ε. Suppose now that the solution to the program is at an extreme point P_1. There is some positive number η, such that, if P_2 is any other extreme point (not also a solution to the program), then $w(P_2) \leq w(P_1) - \eta$ (where $w(P)$ is the value of the objective function at the point P). It is clear that if all the ε's are sufficiently small, the objective function cannot be increased by more than η at any extreme point. It follows that P_2 cannot be the solution to the perturbed problem because it was not the solution to the unperturbed problem. Thus the solution to the perturbed problem will always give us a solution to the unperturbed problem.

In general the perturbation method, while useful (especially in proving the finiteness of the simplex algorithm), is tedious, especially since one never knows how small the ε's must be to guarantee that the solution to the perturbed problem will solve the original problem. In computer work, it has been found easier to instruct the computer to work in a random manner whenever it has a choice of pivots. This guarantees, with probability 1, that the cycling which complicates the algorithm in such cases will be avoided (since in fact such cycling occurs only when there is a systematic method of choosing the pivot).

3.6.   Examples

Let us now consider several examples of the solution of linear programs by the simplex method.

3.6.1 Example

Maximize $5x_1 + 2x_2 + x_3$

subject to

$$x_1 + 3x_2 - x_3 \le 6$$
$$x_2 + x_3 \le 4$$
$$3x_1 + x_2 \le 7$$
$$x_1, x_2, x_3 \ge 0$$

We form the tableau

x_1	x_2	x_3	1	
1	3	-1	-6	$= -u_1$
0	1	1	-4	$= -u_2$
3^*	1	0	-7	$= -u_3$
5	2	1	0	$= w$

$$(3.42)$$

Because all the columns have positive entries in the bottom row, we can choose any column for our pivot. We may, for instance, choose the first column; the pivot will then be the starred entry. We pivot on it to obtain the tableau

u_3	x_2	x_3	1	
$-\frac{1}{3}$	$\frac{8}{3}$	-1	$-\frac{11}{3}$	$= -u_1$
0	1	1^*	-4	$= -u_2$
$\frac{1}{3}$	$\frac{1}{3}$	0	$-\frac{7}{3}$	$= -x_1$
$-\frac{5}{3}$	$\frac{1}{3}$	1	$\frac{35}{3}$	$= w$

$$(3.43)$$

Again, we pivot on the starred entry; this gives us

u_3	x_2	u_2	1	
$-\frac{1}{3}$	$\frac{11}{3}$	1	$-\frac{23}{3}$	$= -u_1$
0	1	1	-4	$= -x_3$
$\frac{1}{3}$	$\frac{1}{3}$	0	$-\frac{7}{3}$	$= -x_1$
$-\frac{5}{3}$	$-\frac{2}{3}$	-1	$\frac{47}{3}$	$= w$

$$(3.44)$$

Here it is clear that we have our solution: $(\frac{7}{3}, 0, 4)$. For this choice of variables, the value of w is $\frac{47}{3}$.

3.6.2 Example

Minimize $6y_1 + 4y_2 + 7y_3$

subject to

$$y_1 + 3y_3 \geq 5$$
$$3y_1 + y_2 + y_3 \geq 2$$
$$-y_1 + y_2 \geq 1$$
$$y_1, y_2, y_3 \geq 0$$

It is easily seen that this problem is the dual of 3.6.1. If we had written tableau (3.42) so as to incorporate both problems, and continued as above, the last tableau would be

	u_3	x_2	u_2	1	
y_1	$-\frac{1}{3}$	$\frac{11}{3}$	1	$-\frac{23}{3}$	$= -u_1$
v_3	0	1	1	-4	$= -x_3$
v_1	$\frac{1}{3}$	$\frac{1}{3}$	0	$-\frac{7}{3}$	$= -x_1$
-1	$-\frac{5}{3}$	$-\frac{2}{3}$	-1	$\frac{47}{3}$	$= w$
	$\parallel$	$\parallel$	$\parallel$	$\parallel$	
	y_3	v_2	y_2	w'	

$$(3.45)$$

Hence, the solution to 3.6.2 is $(0, 1, \frac{5}{3})$. The value of this program is the same as 3.6.1, i.e., $\frac{47}{3}$.

3.6.3 Example. Solve the matrix game

$$\begin{pmatrix} 3 & 6 & 1 & 4 \\ 5 & 2 & 4 & 2 \\ 1 & 4 & 3 & 5 \end{pmatrix}$$

Solution. We write the maximizing problem in the form

Maximize λ

subject to

$$3x_1 + 5x_2 + x_3 \geq \lambda$$
$$6x_1 + 2x_2 + 4x_3 \geq \lambda$$
$$x_1 + 4x_2 + 3x_3 \geq \lambda$$
$$4x_1 + 2x_2 + 5x_3 \geq \lambda$$
$$x_1 + x_2 + x_3 = 1$$
$$x_1, x_2, x_3 \geq 0$$

This gives us the tableau

x_1	x_2	x_3	λ	
-3	-5	-1	1	$= -u_1$
-6	-2	-4	1	$= -u_2$
-1	-4	-3	1	$= -u_3$
-4	-2	-5	1^*	$= -u_4$
-1	-1	-1	0	$= -1$

(3.46)

Here, we must pivot first in such a way as to put λ among the basic variables, and 1 among the nonbasic. Unfortunately, the entry below λ and on the same row as 1 turns out to be a zero; we need, therefore, two pivot steps rather than one to do this.

In the following sequence, asterisks are used to denote the pivot at each step:

x_1	x_2	x_3	u_4	
1	-3	4	-1	$= -u_1$
-2	0	1	-1	$= -u_2$
3	-2	2	-1	$= -u_3$
-4	-2	-5	1	$= -\lambda$
-1	-1	-1^*	0	$= -1$

(3.47)

x_1	x_2	1	u_4	
-3	-7	4	-1	$= -u_1$
-3	-1	1	-1	$= -u_2$
1	-4	2	-1	$= -u_3$
1	3	-5	1	$= -\lambda$
1	1	-1	0	$= -x_3$

(3.48)

We now permute rows and columns so as to make the right-hand column correspond to the constant term, and the bottom row correspond to the objective function, λ. We also change all signs in the bottom row, obtaining the following tableau:

x_1	x_2	u_4	1	
-3	-7	-1	4	$= -u_1$
-3	-1	-1	1	$= -u_2$
1	-4	-1^*	2	$= -u_3$
1	1	0	-1	$= -x_3$
-1	-3	-1	5	$= \lambda$

$$(3.49)$$

We proceed now according to the rules (3.5.1), until we reach a b.f.p.:

x_1	x_2	u_3	1	
-4	-3	-1^*	2	$= -u_1$
-4	3	-1	-1	$= -u_2$
-1	4	-1	-2	$= -u_4$
1	1	0	-1	$= -x_3$
-2	1	-1	3	$= \lambda$

$$(3.50)$$

x_1	x_2	u_1	1	
4	3	-1	-2	$= -u_3$
0	6^*	-1	-3	$= -u_2$
3	7	-1	-4	$= -u_4$
1	1	0	-1	$= -x_3$
2	4	-1	1	$= \lambda$

$$(3.51)$$

As we now have a b.f.p., we proceed according to the rules (3.5.2):

x_1	u_2	u_1	1	
4^*	$-\frac{1}{2}$	$-\frac{1}{2}$	$-\frac{1}{2}$	$= -u_3$
0	$\frac{1}{6}$	$-\frac{1}{6}$	$-\frac{1}{2}$	$= -x_2$
3	$-\frac{7}{6}$	$\frac{1}{6}$	$-\frac{1}{2}$	$= -u_4$
1	$-\frac{1}{6}$	$\frac{1}{6}$	$-\frac{1}{2}$	$= -x_3$
2	$-\frac{2}{3}$	$-\frac{1}{3}$	3	$= \lambda$

$$(3.52)$$

	u_3	u_2	u_1	1	
v_1	$\frac{1}{4}$	$-\frac{1}{8}$	$-\frac{1}{8}$	$-\frac{1}{8}$	$= -x_1$
v_2	0	$\frac{1}{6}$	$-\frac{1}{6}$	$-\frac{1}{2}$	$= -x_2$
y_4	$-\frac{3}{4}$	$-\frac{19}{24}$	$\frac{13}{24}$	$-\frac{1}{8}$	$= -u_4$
v_3	$-\frac{1}{4}$	$-\frac{1}{24}$	$\frac{7}{24}$	$-\frac{3}{8}$	$= -x_3$
-1	$-\frac{1}{2}$	$-\frac{5}{12}$	$-\frac{1}{12}$	$\frac{13}{4}$	$= \lambda$
	$\parallel$	$\parallel$	$\parallel$	$\parallel$	
	y_3	y_2	y_1	μ	

$$(3.53)$$

Tableau (3.53) gives us the solution: I's optimal strategy is $(\frac{1}{8},\frac{1}{2},\frac{3}{8})$. The dual variables, which we had omitted in the intermediate tableaux, tell us that II's optimal strategy is $(\frac{1}{12},\frac{5}{12},\frac{1}{2},0)$. The value of the game is $\frac{13}{4}$.

Several observations might be made by referring to this example. The first is that, in pivoting as we did in tableau (3.49), the two variables, u_4 and u_3, were interchanged. But the next step interchanged u_3 and u_1. We could have saved a step by interchanging u_4 and u_1 to begin with — this would have taken us from tableau (3.49) to (3.51) in a single step. Thus we see that the rules give in (3.5.1) will not always give the quickest results; modifications may be made which will cut down the number of operations (and hence the time) needed by a significant amount.

A second observation is this: Tableau (3.49) does not give us a b.f.p. It does, however, give us a b.d.f.p. A modification of the rules (3.5.1) may be given which will allow us to proceed, not through b.f.p.s., but through b.d.f.p.s., until the solution is found. In other words, our method starts with the b.d.f.p. of (3.49), but does not make use of the fact that this is a b.d.f.p. Instead, it proceeds to give us a b.f.p., after which it looks for the solution. If, in looking for a b.f.p., we are careful to keep to tableaux which are also b.d.f.p.s, then it is clear that the first b.f.p. we reach will be the solution.

The third observation is that we need not keep track of the dual variables y_j and v_i. In fact, each y_j will necessarily be located opposite the corresponding u_j of the maximizing problem; the same holds for the v_i and x_i. Indeed, it is clear that each pivot step will preserve these relative positions.

3.6.4 Example

Maximize $x_1 + x_2$

subject to

$$-2x_1 + x_2 \leq -3$$
$$x_1 - 2x_2 \leq 4$$
$$x_1, x_2 \geq 0$$

Solution. We obtain the tableau

x_1	x_2	1	
-2	1	-3	$=-u_1$
1	-2^*	4	$=-u_2$
1	1	0	$=w$

(3.54)

Pivoting on the starred entry, we obtain the tableaux

x_1	u_2	1	
$-\frac{3}{4}^*$	$\frac{1}{2}$	5	$=-u_1$
$-\frac{1}{2}$	$-\frac{1}{2}$	-2	$=-x_2$
$\frac{3}{2}$	$\frac{1}{2}$	2	$=w$

(3.55)

u_1	u_2	1	
$-\frac{4}{3}$	$-\frac{2}{3}$	$-\frac{20}{3}$	$= -x_1$
$-\frac{2}{3}$	$-\frac{5}{6}$	$-\frac{16}{3}$	$= -x_2$
2	$\frac{3}{2}$	12	$= w$

$$(3.56)$$

It is clear from (3.56) that the problem is unbounded.

3.7. Constrained Games

The theory of linear programming has many applications other than to the theory of games. While many of these are very interesting in their own right, it is not possible, within the scope of this book, to treat them in any detail. We give here an application which does pertain to games.

Let us consider a game in which not all mixed strategies are permitted. There is generally some practical reason why this should be. Now, we assume that the mixed strategies x and y, respectively, must be chosen from some convex hyperpolyhedron, i.e., from constraint sets determined by linear inequalities and equations.

If the game matrix is $A = (a_{ij})$, then player I's problem is to find

$$\max_{x \in X} \left\{ \min_{y \in Y} x A y^T \right\} \tag{3.57}$$

where the two sets, X and Y, are determined by

$$xB \leq C \tag{3.58}$$

$$x \geq 0 \tag{3.59}$$

and

$$yE^T \geq F \tag{3.60}$$

$$y \geq 0 \tag{3.61}$$

respectively. Player II's problem would, similarly, be to find

$$\min_{y \in Y} \left\{ \max_{x \in X} x A y^T \right\} \tag{3.62}$$

Consider then the expression (3.57). The expression inside the brackets, clearly, is a function of x. More exactly, it is the value of a linear program whose objective function has coefficients depending on x. By the duality theorems given in 3.2 we know that if this program is feasible and bounded (which generally does not depend on x), then the two programs

$$\text{Minimize} \quad (xA)y^{\mathrm{T}} \tag{3.63}$$

subject to

$$yE^{\mathrm{T}} \geq F \tag{3.64}$$

$$y \geq 0 \tag{3.65}$$

and

$$\text{Maximize} \quad zF^{\mathrm{T}} \tag{3.66}$$

subject to

$$zE \leq xA \tag{3.67}$$

$$z \geq 0 \tag{3.68}$$

will have the same value. Thus, player I's problem reduces to the pure maximization problem

$$\text{Maximize} \quad zF^{\mathrm{T}} \tag{3.69}$$

subject to

$$zE - A \leq 0 \tag{3.70}$$

$$xB \leq C \tag{3.71}$$

$$z, x \geq 0 \tag{3.72}$$

This problem, of course, can be solved by the usual linear programming techniques (generally, the simplex algorithm).

In a similar way, it can be seen that player II's problem (3.62) reduces to the pure minimization problem

$$\text{Minimize} \quad Cs^{\mathrm{T}} \tag{3.73}$$

subject to

$$sB^{\mathrm{T}} - yA^{\mathrm{T}} \geq 0 \tag{3.74}$$

$$yE^{\mathrm{T}} \geq F \tag{3.75}$$

$$s, y \geq 0 \tag{3.76}$$

It can be seen that (3.73)–(3.76) is the dual of (3.69)–(3.72). Therefore, if both of these programs are feasible the two expressions (3.57) and (3.62) will be equal, and thus the constrained game will have a saddle point in mixed strategies.

Problems

1. Give the duals of the following linear programs:
   (a) Maximize $x_1 + 3x_2 - x_3$
subject to

$$3x_1 + x_2 \leq 5$$
$$x_1 - 3x_2 + x_3 \geq -6$$
$$x_2 + 3x_3 \leq 4$$
$$x_i \geq 0, \quad i = 1, 2, 3$$

   (b) Minimize $2x_1 + 5x_2 + 6x_3$
subject to

$$x_1 + 3x_3 + x_3 \geq 5$$
$$3x_2 + 2x_3 \geq 7$$
$$x_1 + x_3 \geq 3$$
$$x_i \geq 0, \quad i = 1, 2, 3$$

2. Show that the program
   Minimize $2x_1 - 3x_2 + 4x_4$
subject to

$$2x_2 - x_3 \geq -2$$
$$-2x_1 + 3x_3 \geq +3$$
$$x_1 - 3x_2 \geq -4$$
$$x_i \geq 0, \quad i = 1, 2, 3$$

has value zero. (*Hint*: show that it is feasible; then find its dual.)

3. Let $A(x, y)$ be a continuous function of x and y in the unit square $[0, 1] \times [0, 1]$. Let $b(y)$ and $c(x)$ be defined and continuous on $[0, 1]$. Show that, if $f(x)$ and $g(y)$ are continuous nonnegative functions on $[0, 1]$, such that

$$\int_0^1 A(x, y)f(x)dx \leq b(y)$$

and

$$\int_0^1 A(x, y)g(y)dy \geq c(x)$$

then

$$\int_0^1 c(x)f(x)dx \leq \int_0^1 b(y)g(y)dy$$

4. Let A be a matrix game, and let B be a matrix of the same size as A. Letting $v(A)$ be the value of A define

$$\frac{\partial v(A)}{\partial B} = \lim_{a \to 0+} \frac{v(A + \alpha B) - v(A)}{\alpha}$$

Show that

$$\frac{\partial v(A)}{\partial B} = \max_{x \in X^*} \min_{y \in Y^*} xBy^{\mathsf{T}}$$

where X^* and Y^* are the sets of optimal strategies for I and II, respectively, in A.

5. A *flow problem* consists of a system of nodes (vertices), numbered $0, 1, 2, \ldots, n+1$, connected by arcs, $A_{ij}(i \neq j)$. The nodes numbered 0 and $n+1$ are distinguished, being called, respectively, the *source* and the *sink*. The arcs are undirected, hence A_{ij} and A_{ji} are the same. Each arc has a *capacity* C_{ij}, while each node also has a capacity C_{ii} (Some of the arcs A_{ij} may be missing, in which case we say that $C_{ij} = 0$.) A flow is a vector $x = (x_{ij})$ in which the component x_{ij} represents the amount "flowing" from node i to note j. The flow must, of course, satisfy the obvious constraints

$$x_{ij} \leq C_{ij}, \quad i, j = 1, \ldots, n$$

$$x_{ii} = \sum_{j \neq i} x_{ij} = \sum_{j \neq i} x_{ji}, \quad i = 1, \ldots, n$$

$$x_{00} = \sum_{j=1}^{n+1} x_{0j}$$

$$x_{n+1, n+1} = \sum_{j=0}^{n} x_{j, n+1}$$

The *value* of the flow is the number x_{00}. Show that $x_{00} = x_{n+1, n+1}$.

By a *cut* is meant a collection of arcs and nodes which intersects every chain of arcs and nodes from the source to the sink. The value of a cut is equal to the sum of the capacities of the arcs and nodes in the collection. Prove the max-flow min-cut theorem: The value of the maximal flow is equal to the value of the minimal cut.

6. Maximize $2x_1 + 5x_2$
subject to

$$x_1 + 3x_2 \leq 7$$
$$2x_1 + x_3 \leq 6$$
$$x_1, x_2 \geq 0$$

7. Minimize $3x_1 + 6x_2 - x_3$
subject to

$$x_1 + 4x_2 + 5x_3 \geq 7$$
$$3x_2 + x_3 \geq 5$$
$$3x_1 - x_3 \geq 8$$
$$x_1, x_2, x_3 \geq 0$$

8. Maximize $x_1 + 4x_2 - 6x_3$
subject to

$$3x_1 + 2x_2 - x_3 \leq 5$$
$$x_2 + 4x_3 \leq 7$$
$$-x_1 + 4x_2 + x_3 \leq 8$$
$$x_1, x_2, x_3 \geq 0$$

9. Minimize $x_1 + x_2 + x_3 + x_4$
subject to

$$3x_1 + 2x_2 + x_3 + x_3 \geq 1$$
$$x_2 + x_3 - x_4 \geq 6$$
$$x_1 + x_3 + 2x_4 = 8$$
$$x_1, x_2, x_3, x_4 \geq 0$$

Find a pair of optimal strategies, and the value, for the following matrix games:

10. $\begin{pmatrix} 4 & 3 & 1 & 4 \\ 2 & 5 & 6 & 3 \\ 1 & 0 & 7 & 0 \end{pmatrix}$

11. $\begin{pmatrix} 0 & 5 & -2 \\ -3 & 0 & 4 \\ 6 & -4 & 0 \end{pmatrix}$ *rewrite each as mutually dual programs*

12. $\begin{pmatrix} 5 & 8 & 3 & 1 & 6 \\ 4 & 2 & 6 & 3 & 5 \\ 2 & 4 & 6 & 4 & 1 \\ 1 & 3 & 2 & 5 & 3 \end{pmatrix}$

For each of the following games, the mixed strategies are constrained as noted.
  (a) Express each of these as a pair of dual linear programs.
  (b) Obtain optimal strategies, and the value.

13. $\begin{pmatrix} 7 & -3 \\ 4 & 6 \end{pmatrix}$

subject to the constraints

$$-x_1 + x_2 \leq 0.4$$

$$y_1 - y_2 \geq 0.1$$

14. $\begin{pmatrix} 7 & 5 & 2 \\ 2 & 1 & 6 \end{pmatrix}$

subject to the constraints

$$x_1 - 2x_2 \geq 0$$

$$y_2 \leq 0.1$$

Chapter 4

Infinite Games

4.1. Games with Countably many Strategies

Thus far we have dealt mainly with finite games, i.e., games in which each player has a finite number of strategies. Let us consider now a generalization to infinite games. We consider first of all games with countably many pure strategies. These will, generally, be labeled with the positive integers. As in the case of finite games, we shall let a_{ij} represent the expected payoff from II to I, assuming that I chooses his ith pure strategy, while II chooses his jth pure strategy.

A mixed strategy for I, in this case, will be a sequence $(x_1, x_2, \ldots)$ satisfying

$$\sum_{i=1}^{\infty} x_i = 1 \tag{4.1}$$

$$x_i \geq 0 \tag{4.2}$$

A mixed strategy for II will be a sequence $(y_1, y_2, \ldots)$ defined similarly. The payoff function for the mixed strategies (x, y) will be defined as

$$A(x, y) = \sum_{x, y=1}^{\infty} x_i a_{ij} y_j \tag{4.3}$$

assuming that this series converges absolutely.

Games with countably many strategies exhibit several undesirable properties which are not found among finite games. The difficulties are as follows: First, the series (4.3) need not converge, and when this does not happen, it may be that the two iterated sums

$$\sum_{i=1}^{\infty} \sum_{j=1}^{\infty} x_i a_{ij} y_j \tag{4.4}$$

and

$$\sum_{j=1}^{\infty} \sum_{i=1}^{\infty} x_i a_{ij} y_j \tag{4.5}$$

will exist and be different. Second, the sets of mixed strategies are not compact, and thus maxima and minima will not exist. We give two examples to demonstrate these difficulties; it should be added that there is very little other interest in games with countably many strategies.

4.1.1 Example. Consider the game with payoff $a_{ij} = \text{sgn}(i - j)$. (In essence, this games reduces to the following: each player chooses a number; the one choosing the smaller number pays his opponent one unit.) The game as such is ridiculous; its behavior is even worse.

Indeed, suppose x is a mixed strategy for *I*. We know that, given any $\varepsilon > 0$, there will exist some N such that

$$\sum_{i=1}^{N-1} x_i > 1 - \varepsilon$$

and so

$$\sum_{i=N}^{\infty} x_i < \varepsilon \qquad (4.6)$$

If I chooses the mixed strategy x, then let II choose his Nth pure strategy. It will be seen that the expected payoff then is

$$\sum a_{iN} x_i < -1 + 2\varepsilon$$

and so

$$\inf_{N} \sum a_{iN} x_i = -1$$

But x was arbitrary; hence

$$\sup_{x} \inf_{N} \sum a_{iN} x_i = -1 \qquad (4.7)$$

Similarly, it may be seen that

$$\inf_{y} \sup_{N} \sum a_{Nj} y_j = 1 \qquad (4.8)$$

and we see that the minimax theorem (even generalized by putting inf and sup instead of min and max) does not hold. Note also that every row and column is dominated.

4.1.2 Example. Consider the game with payoff $a_{ij} = i - j$. As in Example 4.1.1, the object is to choose a large number; this is further complicated, however, by the fact that the payoff function is not bounded.

Consider now the strategy x, defined by

$$x_i = \begin{cases} 1/(2i) & \text{if } i = 2^k, \quad k \text{ integer} \\ 0 & \text{otherwise} \end{cases} \qquad (4.9)$$

It is easily checked that this is indeed a strategy. Now for any j

$$\sum_{i=1}^{\infty} x_i a_{ij} = \sum_{i=1}^{\infty} i x_i - \sum_{i=1}^{\infty} j x_i$$

$$= \sum_{i=1}^{\infty} i x_i - j$$

Therefore

$$\sum_{i=1}^{\infty} x_i a_{ij} = +\infty \tag{4.10}$$

and so it is seen that this mixed strategy x gives an infinite expectation for player I against any pure strategy for II. That is, player I can assure himself of an infinite expectation, whatever II does. But the game is symmetric, and it follows that player II can also assure himself of an infinite expectation. Thus, we have a pathological game. Indeed, Example 4.1.1 merely contradicts the minimax theorem; this example seems to contradict common sense.

4.2. Games on the Square

An important type of infinite game is that in which each player has a continuum of pure strategies, generally represented as points in the interval [0,1]. A pure strategy for each player, then, is a real number in this interval. The payoff is a real-valued function $A(x, y)$ defined on the unit square $[0,1] \times [0,1]$.

A mixed strategy, as before, is a probability distribution over the set of pure strategies. In this case, a mixed strategy can be represented as a cumulative distribution function, i.e., as a function F, defined on [0,1], satisfying

$$F(0) = 0 \tag{4.11}$$

$$F(1) = 1 \tag{4.12}$$

$$\text{if} \quad x > x', \quad F(x) \geq F(x') \tag{4.13}$$

$$\text{if} \quad x \neq 0, \quad F(x) = F(x+) \tag{4.14}$$

If player I uses the pure strategy x, while II uses the mixed strategy G, the expected payoff is the Stieltjes integral

$$E(x, G) = \int_0^1 A(x, y) \, dG(y) \tag{4.15}$$

but, if II uses the pure strategy y, and I uses the mixed strategy F, the expected payoff is

$$E(F, y) = \int_0^1 A(x, y) \, dF(x) \tag{4.16}$$

Finally, if I uses F and II uses G, we have

$$E(F, G) = \int_0^1 \int_0^1 A(x, y) \, dF(x) \, dG(y) \tag{4.17}$$

assuming, in each case, that these integrals exist. If they exist, of course, we have

$$E(F, G) = \int_0^1 E(F, y) \, dG(y) = \int_0^1 E(x, G) \, dF(x) \tag{4.18}$$

4.2.1 Optimal Strategies and Value. As with finite games, we can define the two numbers

$$v_{\mathrm{I}} = \sup_F \inf_y E(F, y) \tag{4.19}$$

and

$$v_{\mathrm{II}} = \inf_G \sup_x E(x, G) \tag{4.20}$$

Two questions arise now: (1) Will $v_{\mathrm{I}} = v_{\mathrm{II}}$? (2) Can sup inf and inf sup be replaced by max min and min max, respectively? If the answer to both of these questions is in the affirmative, optimal mixed strategies will exist; if these can be found, the game is as well determined as the finite games are. If only the first question can be answered in the affirmative, the game has a value (the common value of v_{I} and v_{II}) but no optimal strategies. It will have ε-optimal strategies, however, i.e., given any $\varepsilon > 0$, there exist mixed strategies, F and G for I and II, respectively, such that

$$E(F, y) > v - \varepsilon \tag{4.21}$$

and

$$E(x, G) < v + \varepsilon \tag{4.22}$$

for any x or y in [0,1]. Thus, although the game is not as well determined as the finite games, the game seems to show a type of stability.

4.3. Games with Continuous Kernel

The most obvious candidates for inspection, among games on the square, are those in which the payoff function, $A(x, y)$ (usually called the *kernel*), is continuous. For these, optimal strategies will exist, as we now prove.

4.3.1 Theorem. If the kernel $A(x, y)$ is a continuous function, the forms sup inf and inf sup may be replaced by max min and min max, respectively.

Proof. We know that $A(x, y)$ is continuous. Therefore, for any F, the function

$$E(F, y) = \int_0^1 A(x, y) \; dF(x)$$

is a continuous function of y. Since the interval $[0,1]$ is compact, it follows that $E(F, y)$ will attain a minimum value in that interval. Thus, sup inf may be replaced by sup min.

By the definition of v_{I}, given any n, there exists a distribution F_n such that

$$\min_y E(F_n, y) > v_{\mathrm{I}} - (1/n)$$

Now the set of all functions mapping $[0,1]$ into itself is compact in the topology of pointwise convergence; hence the sequence $(F_1, F_2, \ldots)$ will contain a convergent subsequence. Let the limit of this subsequence be F_0. Since each of the functions F_n satisfies (4.11)–(4.13), it is clear that F_0 will also satisfy these three conditions. On the other hand, F_0 need not satisfy (4.14), as this is not the type of property which is preserved in the limit. Nevertheless, we define the function $\overline{F}_0$ by

$$\overline{F}_0(x) = \begin{cases} 0 & \text{if} \quad x = 0 \\ F_0(x+) & \text{if} \quad 0 < x < 1 \\ 1 & \text{if} \quad x = 1 \end{cases} \tag{4.23}$$

and it is easily seen that $\overline{F}_0$ is a strategy. F_0 and $\overline{F}_0$ differ only at their points of discontinuity but, as they are monotone functions, their points of discontinuity form a countable set, and so, for all y

$$\int_0^1 A(x, y) \; d\overline{F}_0(x) = \int_0^1 A(x, y) \; dF_0(x)$$

The function $A(x, y)$ is continuous and hence uniformly continuous. It follows that, as F_0 is the limit of a subsequence of the F_n

$$\int_0^1 A(x, y) \; dF_0(x) = \lim_n \int_0^1 A(x, y) \; dF_n(x)$$

But this limit is at least v_{I}, for each value of y. Thus

$$\min_y E(\overline{F}_0, y) \geq v_{\mathrm{I}} \tag{4.24}$$

and $\overline{F}_0$ gives the desired maximum.

In a similar way, it can be shown that inf sup can be replaced by min max.

4.3.2 Theorem. If the kernel $A(x, y)$ is continuous, then $v_\mathrm{I} = v_\mathrm{II}$.

Proof. For any integer n, consider the $(n+1) \times (n+1)$ matrix $A_n = (a_{ij}^n)$, where

$$a_{ij}^n = A(i/n, \ j/n), \quad i, j = 0, 1, \ldots, n \tag{4.25}$$

The game with matrix A_n has a value w_n and optimal strategies $r_n = (r_0^n, \ldots, r_n^n)$ and $s_n = (s_0^n, \ldots, s_n^n)$, for I and II, respectively.

The function $A(x, y)$ is continuous. Since the square is compact, it follows that A is uniformly continuous. Thus, given $\varepsilon > 0$, there exists a $\delta > 0$ such that, if

$$\sqrt{(x - x')^2 + (y - y')^2} < \delta$$

then

$$|A(x, y) - A(x', y')| < \varepsilon$$

Let us take n large enough so that $1/n < \delta$. Form the strategy F_n defined by

$$F_n(x) = \sum_{i=1}^{[nx]} r_i^n \tag{4.26}$$

(where $[nx]$ is the greatest integer in nx).

For any y, let $j = [ny]$. It is clear that

$$E(F_n, \ j/n) = \sum_{i=0}^{n} a_{ij}^n r_i^n > w_n \tag{4.27}$$

and, since $|y - (j/n)| < \delta$

$$|A(x, y) - A(x, \ j/n)| < \varepsilon$$

and so

$$|E(F_n, y) - E(F_n, \ j/n)| < \varepsilon$$

Hence, for any y

$$E(F_n, y) > w_n - \varepsilon$$

and so

$$v_\mathrm{I} > w_n - \varepsilon \tag{4.28}$$

In a similar way, we can show that

$$v_\mathrm{II} < w_n + \varepsilon \tag{4.29}$$

These two inequalities give us

$$v_\mathrm{I} > v_\mathrm{II} - 2\varepsilon$$

But, as $v_I \leq v_{II}$, and ε is arbitrary, it follows that

$$v_I = v_{II} \tag{4.30}$$

It should be remarked from the proof of Theorem 4.3.2 first, that

$$v = \lim_n w_n \tag{4.31}$$

and second, that the strategies F_n will approximate the optimal strategy as closely as desired. In this sense, the matrix games A_n approximate the continuous game $A(x, y)$. Now, if the kernel is very smooth or flat, it may be that a very small value of n will be sufficient for a good approximation. On the other hand, if the kernel is very irregular, a good approximation will require a larger value of n. The techniques most generally used for solving matrix games (viz, fictitious play and the simplex method) are not generally quick enough for handling games of size, say, 100×100. Thus, it would be preferable to solve these games by analytic methods; unfortunately, none are available. *Development of such a method is one of the outstanding problems today.* It would also be of interest to develop a method to determine whether the optimal strategy F can be a step function, i.e., whether a finite number of pure strategies will suffice, or whether a continuous distribution should be used. The following sections show some progress that has been made in these directions.

4.4. Concave–Convex Games

4.4.1 Definition. A game on the square will be said to be concave–convex if its kernel $A(x, y)$ is concave in x for each value of y and convex in y for each value of x.

Essentially, a concave–convex game must have a saddle-shaped kernel. Because it is saddle shaped, it is to be expected that it will have a saddle point in pure strategies. That this is so, under the additional assumption of continuity, is proved as follows:

4.4.2 Theorem. Let the concave–convex game $A(x, y)$ be continuous. Then it will have optimal pure strategies.

Proof. Because the game is continuous, it will have optimal strategies; let these be F and G, for I and II, respectively. Now, let

$$x_0 = \int_0^1 x \ dF(x) \tag{4.32}$$

$$y_0 = \int_0^1 y \ dG(y) \tag{4.33}$$

By the concavity of A, given any y, there exists some α such that the function

$$B_y(x) = A(x, y) - \alpha x$$

attains its maximum value (keeping y fixed) at x_0. Thus we have

$$E(F, y) = \int_0^1 (B_y(x) + \alpha x) \, dF(x)$$

$$E(F, y) = \int_0^1 B_y(x) \, dF(x) + \alpha \int_0^1 x \, dF(x)$$

(4.34)

Now, the function B_y has its maximum at x_0, hence the first integral in (4.34) is at most equal to $B_y(x_0)$. Thus

$$E(F, y) \leq B_y(x_0) + \alpha x_0 = A(x_0, y)$$

(4.35)

and it follows that x_0 is at least as good as F against any y. Similarly, y_0 is at least as good as G against any x. Thus, x_0 and y_0 are optimal pure strategies.

In a sense the proof of 4.4.2 is overly complicated. In fact, Theorem 4.4.2 is concerned with pure strategies; it is hardly to be desired that the proof should involve mixed strategies. We give, therefore, a proof which uses only pure strategies, under the somewhat stronger hypothesis of strict concavity and strict convexity.

Alternative Proof of 4.4.2. (Assuming Strict Concavity and Convexity): For any value of x there exists a unique value of y (because of strict convexity) which minimizes $A(x, y)$. Let this value of y be called $\varphi(x)$. Thus we have

$$A(x, \varphi(x)) = \min_y A(x, y)$$

(4.36)

Suppose that the function φ is not continuous. Say, then, that φ is discontinuous at x_0, and let $\varphi(x_0) = y_0$. There exists a sequence of numbers, $(x_1, x_2, \ldots)$ such that x_0 is the limit of x_n, but y_0 is not the limit of $\varphi(x_n)$. By compactness, there exists a subsequence of $\{x_n\}$, say $\{x_{n_k}\}$, such that $\varphi(x_{n_k})$ converges, say, to y', where $y' \neq y_0$.

Now

$$A(x_{n_k}, \varphi(x_{n_k})) < A(x_{n_k}, y_0)$$

and so, taking limits, and using the continuity of A

$$A(x_0, y') \leq A(x_0, y_0)$$

But y_0 was the *unique* value of y which minimized $A(x_0, y)$. This contradiction proves that φ must be a continuous function.

Similarly, we can define the function $\psi(y)$ to be that value of x which maximizes $A(x, y)$. That is

$$A(\psi(y), y) = \max_x A(x, y)$$

(4.37)

Consider now the composite function, $\psi \circ \varphi$. It is clearly continuous, mapping $[0, 1]$ into itself. By the Brouwer fixed point theorem, there must exist some $\bar{x}$ such that

$$\bar{x} = \psi \circ \varphi(\bar{x})$$

If we let $\bar{y} = \varphi(\bar{x})$, this means

$$\bar{x} = \psi(\bar{y}) \qquad (4.38)$$

$$\bar{y} = \varphi(\bar{x}) \qquad (4.39)$$

But this means $\bar{x}$ and $\bar{y}$ are in equilibrium, i.e., they are optimal pure strategies.

4.4.3 Example. Consider the game with kernel

$$A(x, y) = -2x^2 + y^2 + 3xy - x - 2y$$

It is seen that $A_{xx} = -4$ and $A_{yy} = 2$, so the game is indeed concave–convex. We see that

$$A_x = -4x + 3y - 1$$

and, setting this equal to zero, we have

$$x = (3y - 1)/4$$

This value of x will maximize A however, it does not always lie in the unit interval, being negative when $y < \frac{1}{3}$. In such cases the maximum is obtained by letting $x = 0$. Thus we have

$$\psi(y) = \begin{cases} 0 & \text{if } y \le \frac{1}{3} \\ (3y - 1)/4 & \text{if } y \ge \frac{1}{3} \end{cases} \qquad (4.40)$$

Similarly we find that

$$A_y = 2y + 3x - 2$$

$$\varphi(x) = \begin{cases} (2 - 3x)/2 & \text{if } x \le \frac{2}{3} \\ 0 & \text{if } x \ge \frac{2}{3} \end{cases} \qquad (4.41)$$

It is then easy to find the optimal strategies and the value

$$\bar{x} = \tfrac{4}{17} \quad \bar{y} = \tfrac{11}{17} \quad v = -\tfrac{13}{17}$$

4.5. Games of Timing

It was mentioned before that, although continuous games always have optimal strategies, there is no analytic method of computing these. On the other hand, while we cannot be certain that games with discontinuous kernels will have optimal strategies, there are certain cases in which the discontinuities will enable us to find these optimal strategies (if they exist) by analytic methods.

We are interested here in a type of game on the square called a *game of timing*. The prototype of this game would be a game in which each player can act only once, within a given interval of time. The order in which the two players act is most important; this fact will cause a discontinuity along the diagonal $x = y$ of the square.

We consider, then, a game with kernel $A(x, y)$, of the form

$$A(x, y) = \begin{cases} K(x, y) & \text{if} \quad x < y \\ \varphi(x) & \text{if} \quad x = y \\ L(x, y) & \text{if} \quad x > y \end{cases} \qquad (4.42)$$

where the function K is assumed to be defined and continuous on the set $0 \le x \le y \le 1$, L is assumed to be defined and continuous on $0 \le y \le x \le 1$, and φ to be continuous on $[0, 1]$.

There is no guarantee that optimal strategies will exist. However, we can determine certain properties of the optimal strategies, *assuming that they do exist*.

Let F be a mixed strategy for I. For $y \in [0, 1]$ we have

$$E(F, y) = \int_0^y K(x, y) \, dF(x) + \varphi(y)[F(y) - F(y-)] \\ + \int_y^1 L(x, y) \, dF(x) \qquad (4.43)$$

If F is a continuous distribution we can, of course, omit the middle term here and have

$$E(F, y) = \int_0^y K(x, y) \, dF(x) + \int_y^1 L(x, y) \, dF(x) \qquad (4.44)$$

Let us suppose that F and G are optimal strategies for I and II, respectively, and that they are both continuous distributions. By the optimality of F and G, we know that, if y_0 is any point such that

$$G'(y_0) > 0 \qquad (4.45)$$

then

$$E(F, y_0) = v \qquad (4.46)$$

where v is the value of the game. Now, if G' is positive at y, it will also be positive for a neighborhood of y_0, and so, for all y close enough to y_0, we will have $E(F, y) = v$. But this means that

$$\frac{\partial E(F, y)}{\partial y} = 0 \qquad (4.47)$$

and (4.47) can be rewritten in the form

$$[L(y,y) - K(y,y)]F'(y)$$

$$= \int_0^y K_2(x,y)F'(x)\,dx + \int_y^1 L_2(x,y)F'(x)\,dx \tag{4.48}$$

(an integral equation in F' which is sometimes reasonably easy to solve).

Suppose, then, that we are given a game with kernel of the form (4.42). We have no guarantee that optimal strategies will exist; we don't even know whether these will be continuous, discrete, or partly continuous and partly discrete distributions. It seems reasonable to assume that the optimal strategies F and G will be continuous distributions, with derivatives F' and G' which are positive in a certain pair of intervals (a,b) and (c,d) respectively, and vanish outside these intervals. We then have several relations which must hold among the functions F, G and the numbers a, b, c, d, v. These are

$$E(F,y) = v \quad \text{for} \quad y \in (c,d) \tag{4.49}$$

$$E(F,y) \geq v \quad \text{for all} \quad y \tag{4.50}$$

$$E(x,G) = v \quad \text{for} \quad x \in (a,b) \tag{4.51}$$

$$E(x,G) \leq v \quad \text{for all} \quad x \tag{4.52}$$

$$[L(y,y) - K(y,y)]F'(y)$$

$$= \int_0^y K_2(x,y)F'(x)\,dx + \int_y^1 L_2(x,y)F'(x)\,dx$$

$$\text{for} \quad y \in (c,d) \tag{4.53}$$

$$[K(x,x) - L(x,x)]G'(x)$$

$$= \int_0^x L_1(x,y)G'(y)\,dy + \int_x^1 K_1(x,y)G'(y)\,dy$$

$$\text{for} \quad x \in (a,b) \tag{4.54}$$

To these six relations, we must, of course, add the constraints that F and G must be strategies and that $0 \leq a < b \leq 1$, $0 \leq c < d \leq 1$. If the complete system admits a solution, this solution will give us the optimal strategies as well as the value of the game. If there is no solution we conclude that the game either does not have optimal strategies or, if it does, these are not of the type assumed above.

Sometimes the payoff kernel is skew-symmetric; i.e., we will have

$$L(x,y) = -K(y,x) \tag{4.55}$$

$$\varphi(x) = 0 \tag{4.56}$$

If so, an analysis similar to that in 2.6 shows that the value, if it exists, must be zero, and the optimal strategies, if they exist, must be the same for both players. Thus, in the formulation of (4.49)–(4.54) we will have $F = G$, $a = c$, $b = d$, and $v = 0$. Thus, the relations given above will reduce to

$$E(F, y) = 0 \quad \text{if} \quad y \in (a, b) \tag{4.57}$$

$$E(F, y) \geq 0 \quad \text{for all} \quad y \tag{4.58}$$

$$[L(y, y) - K(y, y)]F'(y)$$
$$= \int_a^y K_2(x, y)F'(x)\, dx + \int_y^b L_2(x, y)F'(x)\, dx$$
$$\text{for} \quad y \in (a, b) \tag{4.59}$$

This system, together with the usual constraints, will give us the solution of the game, if such a solution exists.

4.5.1 Example. Consider the following duel. Two men, starting at time $t = 0$, walk toward each other; they will reach each other (if nothing intervenes) at time $t = 1$. Each man has a pistol with exactly one shot, and may fire at any time he wishes. If one hits the other, the duel is immediately over, and the one who fired successfully is the winner. If neither hits the other, the duel is a standoff. If both fire simultaneously and hit each other, the duel is also considered a standoff.

Let us make two assumptions: first, that the accuracy of a shot increases as the players come nearer together, so that, if either player shoots at time t, he has a probability t of hitting his opponent; second, the duel is silent, i.e., a player does not know that his opponent has fired (unless, of course, he is hit).

We compute, now, the kernel. If I chooses time x, and II chooses time y, with $x < y$, then player I has a probability x of hitting his opponent (in which case his payoff is $+1$). If he misses (which has probability $1 - x$) then there is probability y that he will be hit (giving him a payoff of -1). Thus we have

$$K(x, y) = x - y + xy \tag{4.60}$$

The game is obviously symmetric; we will have

$$L(x, y) = x - y - xy \tag{4.61}$$

and

$$\varphi(x) = 0$$

We also have

$$K_2(x, y) = -1 + x \tag{4.62}$$

$$L_2(x, y) = -1 - x \tag{4.63}$$

and

$$L(y, y) - K(y, y) = -2y^2 \tag{4.64}$$

Assume, now, that the optimal strategy is a continuous distribution F, with positive derivative in an interval (a, b). We will have

$$-2y^2 F'(y) = \int_a^y (-1 + x)F'(x)\, dx + \int_y^b (-1 - x)F'(x)\, dx \tag{4.65}$$

This integral equation can be transformed into a differential equation by differentiating both sides; it gives us

$$-4yF' - 2y^2 F'' = (y - 1)F' + (y + 1)F'$$

or, simplifying

$$yF'' = -3F' \tag{4.66}$$

which has the solution

$$F'(y) = ky^{-3} \tag{4.67}$$

We must now find a, b, and k. Suppose $b < 1$. We know that, for all $y \in (a, b)$, we have

$$E(F, y) = 0$$

But $E(F, y)$ is continuous in y; it follows that

$$E(F, b) = 0$$

We thus have

$$\int_a^b (x - b + bx)\, dF(x) = 0$$

But, if $b < 1$, this means

$$\int_a^b (x - 1 + x)\, dF(x) < 0$$

and so $E(F, 1) < 0$. This contradicts (4.58). Thus we must have

$$b = 1 \tag{4.68}$$

We saw that $b = 1$, hence $E(F, 1) = 0$. This gives us

$$k \int_a^1 \frac{2x - 1}{x^3}\, dx = 0$$

which will give us

$$3a^2 - 4a + 1 = 0$$

This equation has two solutions, $a=1$ and $a=\frac{1}{3}$. Clearly $a=1$ is not possible. Thus

$$a = \tfrac{1}{3} \qquad\qquad (4.69)$$

Now F is a strategy; hence

$$\int_{1/3}^{1} kx^{-3} = 1$$

and this gives us

$$k = \tfrac{1}{4} \qquad\qquad (4.70)$$

This gives us, then, the optimal strategy for either player. It is a continuous distribution F defined by

$$F'(x) = \begin{cases} 0 & \text{if } x<\tfrac{1}{3} \\ 1/(4x^3) & \text{if } x>\tfrac{1}{3} \end{cases} \qquad\qquad (4.71)$$

It may be checked directly that this is the solution.

4.5.2 Example. Let us consider again the duel given in the previous example, with the sole exception that this time the guns are noisy; i.e., a player knows whether his opponent shoots and misses. In such a case, he will naturally hold his fire until time $t=1$, when he is certain to hit. Therefore, if I chooses x and II chooses y, with $x<y$, we find that I has probability x of winning and probability $1-x$ of losing. Thus we have

$$K(x, y) = 2x - 1 \qquad\qquad (4.72)$$

$$L(x, y) = 1 - 2y \qquad\qquad (4.73)$$

$$\varphi(x) = 0 \qquad\qquad (4.74)$$

We could attempt the same type of analysis as in the previous problem. It is easily seen, however, that this game has a saddle point in pure strategies. In fact, we have

$$\begin{aligned} A(\tfrac{1}{2}, y) &= L(x, y) = 1 - 2y > 0 && \text{if } y < \tfrac{1}{2} \\ A(\tfrac{1}{2}, y) &= \varphi(\tfrac{1}{2}) = 0 && \text{if } y = \tfrac{1}{2} \\ A(\tfrac{1}{2}, y) &= K(\tfrac{1}{2}, y) = 0 && \text{if } y > \tfrac{1}{2} \end{aligned}$$

and so $\frac{1}{2}$ is an optimal pure strategy.

4.6.   Higher Dimensions

The last four sections have all dealt with games on the square, i.e., games in which the pure strategies for each player form a one-dimensional continuum. It is, of course, true that any set with the cardinality of the continuum can be mapped into the unit interval in a 1–1 manner. Thus, any game in which each player has a set of pure strategies with the cardinality of the continuum can be represented as a game on the square. The difficulty with this, however, is that in doing this many of the properties of the payoff function might be lost. We prefer therefore to maintain the most natural structure possible for the strategy sets.

As before, we will have a function $A(x, y)$, the kernel of the game, defined on the cartesian product $X \times Y$ of the two sets of pure strategies. A mixed strategy for I will be a measure μ defined on X such that

$$\mu(X) = 1 \tag{4.75}$$

and, similarly, a mixed strategy for II will be a measure v such that

$$v(Y) = 1 \tag{4.76}$$

The expected payoff is defined as

$$E(\mu, y) = \int_X A(x, y) \, d\mu(x) \tag{4.77}$$

$$E(x, v) = \int_Y A(x, y) \, dv(y) \tag{4.78}$$

$$E(\mu, v) = \int_{X \times Y} A(x, y) \, d(\mu \times v) \tag{4.79}$$

when these integrals exist. It should be noted, of course, that if (4.79) exists, it can be replaced by either of the corresponding iterated integrals.

Values and optimal strategies are defined for these games, exactly as for games on the square. We give the following theorems, easy generalizations of theorems given for games on the square.

4.6.1 Theorem. If the sets X and Y are compact, and the kernel $A(x, y)$ is continuous, then $v_I = v_{II}$, and, moreover, optimal mixed strategies will exist.

4.6.2 Theorem. If the sets X and Y are convex compact subsets of some linear spaces, and the kernel $A(x, y)$ is continuous, concave in x (for each y) and convex in y (for each x), then the game will have a saddle point in pure strategies.

The proofs of 4.6.1 and 4.6.2 are easy generalizations of the proofs of 4.3.1, 4.3.2, and 4.4.2. (Note that these proofs did not use the particular structure of the sets of pure strategies, but only their compactness and convexity.) We will not give them here.

As before, it is necessary to point out that techniques for solving continuous games are not available, but that discontinuities will very often give us the tool to solve such games analytically.

4.6.3 Example. Two generals, each with an equal number of forces, must fight to take three strategic battlefields. The manner of fighting is such that whichever side has more men at a given field will take that field. It is assumed that the two armies are infinitely divisible, and that the payoff is simply the number of battlefields captured.

The set X will be the set of all ordered triples $x = (x_1, x_2, x_3)$ such that

$$x_1 + x_2 + x_3 = 1 \tag{4.80}$$

$$x_i \geq 0, \quad i = 1, 2, 3 \tag{4.81}$$

The set Y is identical; the payoff is

$$A(x, y) = \operatorname{sgn}(x_1 - y_1) + \operatorname{sgn}(x_2 - y_2) + \operatorname{sgn}(x_3 - y_3) \tag{4.82}$$

Equivalently, each player chooses three nonnegative numbers whose sum is 1. A player wins if two of his numbers are larger than the corresponding two numbers chosen by his opponent. There will be a standoff if a pair of corresponding numbers are equal.

We can represent the set X by the equilateral triangle in Figure 4.1.

It may be seen that, if a player chooses a point such as P, in the diagram, he will defeat any point in the three areas D, E, and F, but be defeated by any point in the three areas A, B, and C. (There will be a standoff against points on the lines which separate these areas.)

Let us assume that optimal strategies do exist, and that they are continuous distributions which can be represented as the Lebesgue integral of some function f which is positive inside some connected set M and zero outside it.

Next, suppose that f is positive in the neighborhood of a point, P. If $P' = P + \Delta P$ is any other point in that neighborhood, we know (by the symmetry of the game) that

$$E(\mu, P) = E(\mu, P') = 0$$

Now, the fact that $E(\mu, P) = 0$ means that

$$\int_{A \cup B \cup C} f d\sigma = \tfrac{1}{2} \tag{4.83}$$

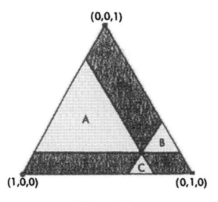

Figure 4.1.

(where σ is Lebesgue measure). Similarly

$$\int_{A'\cup B'\cup C'} f \, d\sigma = \tfrac{1}{2} \tag{4.84}$$

where A', B', C' are the sets of points which defeat P'.

It can be seen from Figure 4.2 that the difference between the two sets $A \cup B \cup C$ and $A' \cup B' \cup C'$ consists simply of the three strips between the lines L_1, L_2, L_3 through P parallel to the sides of the triangle (the solid lines) and the corresponding lines through P'. If $P = (x_1, x_2, x_3)$ and $\Delta P = (\Delta x_1, \Delta x_2, \Delta x_3)$, we can see that for small ΔP the difference between the left-hand sides of (4.78) and (4.79) must be approximately given by

$$\Delta x_1 \int_{L_1} f \, ds + \Delta x_2 \int_{L_2} f \, ds + \Delta x_3 \int_{L_3} f \, ds \tag{4.85}$$

(where s is Lebesgue linear measure). Now, the expression (4.85) must be equal to zero for all values of Δx_1, Δx_2, Δx_2 which satisfy

$$\Delta x_1 + \Delta x_2 + \Delta x_3 = 0 \tag{4.86}$$

This means that we must have

$$\int_{L_1} f \, ds = \int_{L_2} f \, ds = \int_{L_3} f \, ds = k \tag{4.87}$$

We can therefore state that the integral of f must be the same along all lines which are parallel to the sides of the triangle and which intersect M.

We must now determine M. Suppose first that the closure of M does not touch the line $x_1 = 0$. Let us take a point P, of minimal x_1, in the closure of M. Because it is in the closure of M, equation (4.78) must hold. Consider the point

$$P' = (x_1 - 2\varepsilon, x_2 + \varepsilon, x_3 + \varepsilon)$$

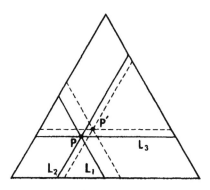

Figure 4.2.

We have, by (4.80)

$$\int_{A'\cup B'\cup C'} f d\sigma = \tfrac{1}{2} + 2\varepsilon \int_{L'_1} f \, ds - \varepsilon \int_{L'_2} f \, ds - \varepsilon \int_{L'_3} f \, ds$$

Now, the line L'_1 clearly does not intersect the set M, as on it, the first coordinate is smaller than the assumed minimal value of x. On the other hand, at least one of the two lines, L'_2 and L'_3, must intersect M (if ε is sufficiently small) because M is, by assumption, connected, and not all of M can lie between the two lines L_2 and L_3 (as the area between L_2 and L_3 is precisely the set F). It follows then that

$$\int_{A'\cup B'\cup C'} f d\sigma < \tfrac{1}{2}$$

and so $E(\mu, P') < 0$. But this would mean μ is not optimal. Thus, *the closure of the set M must touch the three sides of the triangle.*

We know therefore that the minimal values of x_1, x_2, and x_3 in the set M must be zero. To find the maximal values, we note first of all that, if M is connected, then every line $x_1 = a$, for a between 0 and the maximal value r_1 of x_1, intersects M. But this means that the integral of f over M is $r_1 k$. Hence $r_1 = 1/k$. Similarly, the maximal values of x_2 and x_3 (in M) will be $r = 1/k$.

We then find that the set M is inscribed in the hexagon determined by the inequalities

$$0 \le x_i \le r, \quad i = 1, 2, 3$$

which is shown in Figure 4.3.

Now let P be a point (in the closure of M) with $x_3 = r$. It can be seen that

$$E(\mu, P) = E(\mu, Q)$$

where $Q = (0, 1 - r, r)$, because this expression is constant along the line PQ. In the figure the dotted line through Q divides M into two parts. The part of M which lies

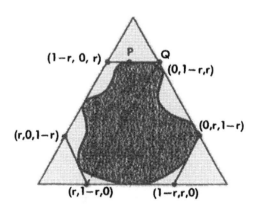

Figure 4.3.

above the line consists of points which are defeated by Q; that below the line, of points which defeat Q. Thus, the integral of f, over this part of M, must be equal to $\frac{1}{2}$. Now all the lines $x_2 = a$, with a in the interval $(0, 1 - r)$, pass through this part; the interval of f here, then, is $k(1 - r)$. Therefore

$$\frac{1 - r}{r} = \frac{1}{2}$$

and so

$$r = \tfrac{2}{3} \tag{4.88}$$

Thus we find that M must be a connected set, inscribed in the hexagon $0 \le x_i \le \frac{2}{3}$. The function f is zero outside the set M, positive in M, and such that, for every line parallel to one of the sides of the triangle and passing through M, the integral of f is $\frac{3}{2}$.

These observations do not determine f or M uniquely; generally, there are many optimal strategies. We might, for example, let M be a circle (the hexagon is clearly regular), and let f be a function only of the distance from the center of the circle. Then if f is chosen so that its integral is constant over all lines in one direction, it will clearly be the same over all lines which intersect M.

Consider, then, a circle, $x^2 + y^2 \le 1$. We want a function $f(R)$ such that, for any $x \in (-1, 1)$, the integral

$$\int_0^{\sqrt{1-s^2}} f(x^2 + y^2)\, dy \tag{4.89}$$

is constant. If we perform the change of variables

$$y = t\sqrt{1 - x^2}$$

we have

$$R = x^2 + y^2 = x^2 + t^2 - t^2 x^2$$

and the integral becomes

$$\int_0^1 \sqrt{1 - x^2} f(x^2 + t^2 - t^2 x^2)\, dt$$

This is constant in x; hence, we differentiate with respect to x, and set the derivative equal to zero. The derivative is, of course, an integral; it suffices to set the integrand equal to zero. We have thus the differential equation

$$xf(R) = 2x(1 - x^2)(1 - t^2)f'(R)$$

which reduces to

$$f(R) = 2(1 - R)f'(R)$$

and has the solution

$$f(R) = \frac{C}{\sqrt{1-R}} = \frac{C}{\sqrt{1-r^2}}$$

We put this back into our problem; we must remember, however, that the radius of the circle we are dealing with is only $\frac{1}{3}$ unit, thus we have

$$f(x_1, x_2, x_3) = \frac{C}{\sqrt{1-9r^2}}$$

where

$$r^2 = (x_1 - \tfrac{1}{3})^2 + (x_2 - \tfrac{1}{3})^2 + (x_3 - \tfrac{1}{3})^2 \qquad (4.90)$$

The constant C is evaluated by remembering that the integral along any line must be $\frac{3}{2}$; this gives us $C = \frac{9}{2}$. Thus, the optimal strategy is

$$f = \frac{9}{2\sqrt{1-9r^2}} \qquad (4.91)$$

for points satisfying $r \le \frac{1}{3}$, and 0 otherwise.

Problems

1. Let $K(x, y)$ be the kernel of a game over the unit square. Assume K has continuous partial derivatives up to the nth order, and that

$$\partial^n K / \partial y^n \ge 0$$

throughout the unit square. Show that player II has an optimal strategy which uses not more than $n/2$ points (with the stipulation that if an end-point of the interval is used, it shall count only as half a point).
(a) Assume first $\partial^n K / \partial y^n > 0$. Assume $F(x)$ is an optimal strategy for I, and show that any strategy optimal against F can use at most $n/2$ points.
(b) Assume now that $\partial^n K / \partial y^n \ge 0$. Let $L_\varepsilon(x, y) = K(x, y) + \varepsilon y^n$, where $\varepsilon > 0$. Let G_ε be an optimal strategy for II in L_ε then, if G_0 is any limit of the strategies G_ε, G_0 will be an optimal strategy for II in K. (Similarly, if F_ε is optimal for I in L_ε, and F_0 is a limit of F_ε as $\varepsilon \to 0$, then F_0 is optimal in K.)

2. Show that, if K is continuous and $K_{yy} \ge 0$ throughout the unit square, then player I has an optimal mixed strategy using at most 2 points.
(a) Show that II has an optimal pure strategy y_0.
(b) If F is an optimal strategy then, for any point x used in F, $K(x, y_0) = v$.
(c) If I does not have an optimal pure strategy, then he must have two pure strategies x_1 and x_2 such that $K_y(x_1, y_0) > 0$ and $K_y(x_2, y_0) < 0$.
(d) I has an optimal mixed strategy using only x_1 and x_2.

3. Let $P(x,y) = \sum_{i=0}^{m}\sum_{j=0}^{n} a_{ij}x^i y^j$ be the kernel of a game over the unit square.
   (a) If the two distributions $F_1(x)$ and $F_2(x)$ have the same moments, i.e., if for $i = 1, \ldots, m$

$$F_1^{(i)} \equiv \int_0^1 x^i \, dF_1(x) = \int_0^2 x^i \, dF_2(x) \equiv F_2^{(i)}$$

then F_1 is optimal if and only if F_2 is optimal. Similarly, if $G_1(y)$ and $G_2(y)$ have the same moments, then G_1 is optimal if and only if G_2 is optimal.
(b) The moments $F^{(i)}$ of any distribution are in the convex hull of the m-dimensional space curve given by the equations $r_i = r^i$.
(c) Players I and II have optimal strategies using at most m and n points, respectively.
(d) Give the payoff function, $E(F,G)$ in terms of the moments $F^{(i)}$ and $G^{(j)}$. Show that the optimal strategies for the two players can be computed as solutions of a set of algebraic equations.

4. The fact that a game on the square has a continuous, or even rational, kernel does not guarantee that players have optimal strategies using only a finite number of points.
   (a) If $K(x,y)$ is a continuous rational function, and I has an optimal strategy using at most m points, this optimal strategy can be obtained as the solution of a system of algebraic equations.
   (b) If $K(x,y)$ is rational, and v is transcendental, neither player has optimal strategies which use only a finite number of points.
   (c) The game with kernel

$$K(x,y) = \frac{(1+x)(1+y)(1-xy)}{(1+xy)^2}$$

has optimal strategies

$$F(x) = \frac{4}{\pi}\arctan\sqrt{x}$$

$$G(y) = \frac{4}{\pi}\arctan\sqrt{y}$$

and value $4/\pi$.

5. Even a game with a "very regular" payoff kernel need not have a value. Consider the kernel

$$K(x,y) = \begin{cases} -1 & \text{if} & x < y < x + \tfrac{1}{2} \\ 0 & \text{if} & x = y \text{ or } y = x + \tfrac{1}{2} \\ +1 & \text{otherwise} \end{cases}$$

(a) Whatever $F(x)$ may be, there is some y such that $E(F,y) \le \tfrac{1}{3}$. [Distinguish the two cases $F(\tfrac{1}{2} - 0) \le \tfrac{1}{3}$ and $F(\tfrac{1}{2} - 0) > \tfrac{1}{3}$.]

(b) Whatever $G(y)$ may be, there is some x such that $E(x, G) > \frac{3}{7}$. [Distinguish the three cases: (i) $G(1 - 0) \geq \frac{3}{7}$; (ii) $G(1 - 0) < \frac{3}{7}$ and $G(\frac{1}{2} - 0) < \frac{1}{7}$; (iii) $G(1 - 0) < \frac{3}{7}$ and $G(\frac{1}{2} - 0) \geq \frac{1}{7}$.]

(c) Thus, $v_I \leq \frac{1}{3}$ and $v_{II} \geq \frac{3}{7}$. Show that in fact the values $\frac{1}{3}$ and $\frac{3}{7}$ are attained by using strategies F^* and G^*, respectively, where F^* uses only 0, $\frac{1}{2}$, and 1, while G^* uses only $\frac{1}{4}$, $\frac{1}{2}$, and 1. Find the probabilities with which these points are used in F^* and G^*.

6. Find the value and optimal strategies of the game with kernel

$$A(x, y) = \begin{cases} 2x - y + xy & \text{if} \quad x < y \\ 0 & \text{if} \quad x = y \\ x - 2y - xy & \text{if} \quad x > y \end{cases}$$

7. Develop a method for solving a "duel game" in which each player has two bullets instead of one. Assume that the duel is silent so that neither player knows whether the other has shot unless he is hit.

8. Let c_1, c_2, c_3 be positive numbers satisfying the strict triangle in equality $c_i + c_j > c_k$. Consider the following game:
Player I choose nonnegative numbers (x_1, x_2, x_3) whose sum is 1. Player II similarly chooses (y_1, y_2, y_3). The payoff is $A(x, y) = c_1 \text{sgn}(x_1 - y_1) + c_2 \text{sgn}(x_2 - y_2) + c_3 \text{sgn}(x_3 - y_3)$. Find optimal strategies for this game.

9. Even a game with rational payoff may have a singular solution. Consider the Cantor function, $C(x)$, which satisfies the relations

$$C(\tfrac{x}{3}) = \tfrac{1}{2} C(x)$$

$$C(x) + C(1 - x) = 1$$

$$x_1 \geq x_2 \Rightarrow C(x_1) \geq C(x_2)$$

(a) For any integrable function f

$$\int_0^1 f(x) dC(x) = \int_0^{1/3} f(x) dC(x) + \int_{2/3}^1 f(x) dC(x)$$

(b) If f is any continuous function

$$\int_0^1 \left[2f(x) - f\left(1 - \frac{x}{3}\right) + f\left(\frac{x}{3}\right) \right] dC(x) = 0$$

(c) The function

$$K(x, y) = \sum_{n=0}^{\infty} \frac{1}{2^n} \left[2x^n - \left(1 - \frac{x}{3}\right)^n - \left(\frac{x}{3}\right)^n \right] \left[2y^n - \left(1 - \frac{y}{3}\right)^n - \left(\frac{y}{3}\right)^n \right]$$

is continuous and, in fact, rational, for $0 \leq x \leq 1$, $0 \leq y \leq 1$. (Show this by expanding the summand and distributing.) Moreover, $C(x)$ and $C(y)$ are optimal strategies for the two players, giving the value zero.

(d) Let $F(x)$ be any optimal strategy for player I. The function $E(f, y)$ is an analytic function of y. Hence, it either vanishes throughout [0, 1] or has only a finite number of zeros in [0, 1].

(e) If $E(F, y)$ does not vanish throughout [0,1], then $C(y)$ cannot be optimal for II.

(f) The fact that $E(F, y) = 0$ throughout [0, 1] whenever F is optimal allows us to compute its moments $F^{(i)}$. Hence, any two optimal strategies have the same moments. (By a well-known theorem, this means any two optimal strategies are identical.) Hence C is the unique optimal strategy.

10. Obtain optimal strategies, and the value, for the game on the square with payoff kernel

$$K(x, y) = 7x - 10xy + 6y^2 - 2xy^3$$

Hint: use the results of problem 2 above.

11. Modify Example 4.6.3, above, by assuming that battle field 3 is more valuable than the others. Specifically, let the payoff function be

$$A(x, y) = 2\operatorname{sgn}(x_1 - y_1) + 2\operatorname{sgn}(x_2 - y_2) + 3\operatorname{sgn}(x_3 - y_3)$$

Chapter 5

Multistage Games

5.1. Behavioral Strategies

Let us consider a game with many moves. It may be a game as simple, say, as tic-tac-toe. This game is simple enough so that even a child can learn to master it. Yet suppose we wish to count the number of strategies for the first player. We note (disregarding all symmetries) that he has nine choices for the first move. Then, for any of the eight possible replies, he will have seven choices on his second move. If we consider only the first player's first two moves, we find that he has $9 \cdot 7^8 = 51,883,209$ pure strategies. Any attempt to enumerate them, naturally, is out of the question. Even if we consider symmetries, we find that the number of pure strategies for this game is astronomical. And yet the game is quite trivial (in the practical sense of the word, as opposed to chess, which is theoretically trivial but practically very complex).

It is clear, then, that pure strategies leave something to be desired. Let us remember the definition of a pure strategy: it is a function, defined on the collection of a particular player's information sets, assigning to each information set a number between 1 and k (where k is the number of choices at the given information set). Thus if a player has N information sets, and k choices at each one, the total number of pure strategies is k^N, which can be very large.

Going back to the example of tic-tac-toe we find that no one plays the game by actually considering all possible pure strategies (i.e., all possible sequences of moves from the first to the last). Rather it is played by considering, at each move, all the possible choices *for that move only*, and deciding (from experience or otherwise) which is best.

This, then, is the essence of simplification: take the moves one at a time. It reduces one choice among $k_1 k_2 \cdots k_N$ possible strategies to N choices among the k_i possible moves at each information set. It leads to the following definition:

5.1.1 Definition. A *behavioral strategy* is a collection of N probability distributions, one each over the set of possible choices at each information set.

5.1.2 Example. A player is given a card from a deck of 52 cards; after seeing his card, he has a choice of either *passing* or *betting* a fixed amount. The total number of pure strategies is 2^{52}; thus, the set of mixed strategies will have dimension $2^{52} - 1$. On the other hand, a behavioral strategy simply gives the probability of betting (a number between 0 and 1) with each hand. Thus the dimension of the set of behavioral strategies is 52.

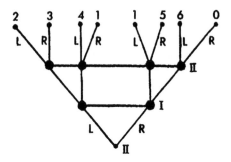

Figure 5.1.

In general, then, the set of behavioral strategies is of much smaller dimension than the set of mixed strategies. On the other hand, it should be pointed out that under certain conditions, not all mixed strategies can be attained by using behavioral strategies, as can be seen by the following example:

5.1.3 Example. In the game with tree in Figure 5.1, we can label player II's pure strategies as LL, LR, RL, RR. A mixed strategy is a vector (x_1, x_2, x_3, x_4) satisfying the usual constraints; a behavioral strategy is a pair of numbers (y_1, y_2) satisfying only $0 \le y_i \le 1$. To the behavioral strategy (y_1, y_2) there corresponds the mixed strategy

$$(y_1, y_2, y_1(1 - y_2), (1 - y_1)y_2, (1 - y_1)(1 - y_2)) \tag{5.1}$$

Now, the optimal strategy for player II is (see Example 2.5.9) the vector $(0, \frac{5}{7}, \frac{2}{7}, 0)$. Clearly, this is not of the form (5.1). Hence, the game does not admit a solution in terms of behavioral strategies.

In a sense, the difficulty with Example 5.1.3 is that player II does not know, at his second move, what he did at his first move. Such a game is said to have *imperfect recall*. This difficulty complicates matters: it is not certain whether all mixed strategies are permitted. In bridge, the two partners are treated as a single player who "forgets" some of his previous choices. As secret arrangements are forbidden, one partner cannot know whether the bid made by his partner was honest or a "psychic" bid, although he may know the probability of such a "psychic" bid.

5.2. Games of Exhaustion

We shall not enter into these difficulties here. The type of game which we propose to study is very nearly a game of perfect information: after a given number of moves, both players are (simultaneously) given perfect information (which will not be forgotten) so that in a sense the game can be recommenced from the given position. The general pattern will be: first, player I moves; then, player II moves (in ignorance of player I's previous move); next, a random move is made, after which both players are given perfect information. (This pattern may vary, but only slightly.) Each of these cycles will be called a *stage* of the game.

In a game of exhaustion, the two players have some finite resources, which decrease (by at least one unit) after each stage. The game will end when one player's resources are exhausted.

It is advisable to solve these games by working backward. The general idea is that each stage can be treated as a separate game. When the strategies for this stage are chosen, the payoff will be either a true payoff (in case the multistage game terminates) or an obligation to play a subsequent stage of the game.

Since we generally deal with expected values, it follows that we can replace the obligation to play a game by the value of that particular game.

5.2.1 Example. Consider the game with matrix

$$\begin{pmatrix} a_{11} & \Gamma_1 \\ \Gamma_2 & a_{12} \end{pmatrix} \tag{5.2}$$

where the entries Γ_1 and Γ_2 represent the obligation to play two other games with the respective matrices

$$\Gamma_1 = \begin{pmatrix} b_{11} & b_{12} \\ b_{21} & b_{22} \end{pmatrix}, \quad \Gamma_2 = \begin{pmatrix} c_{11} & c_{12} \\ c_{21} & c_{22} \end{pmatrix} \tag{5.3}$$

If the values of Γ_1 and Γ_2 are v_1 and v_2 respectively, it follows that, in expectation, the prospect of having to play these games is equivalent to their values. Hence we may replace the matrix (5.2) with

$$\begin{pmatrix} a_{11} & v_1 \\ v_2 & a_{12} \end{pmatrix} \tag{5.4}$$

It is now possible to solve (5.4); this will give us the optimal strategies and value for (5.2).

For games of exhaustion, then, we start by obtaining solutions to all games in which both players start with only one unit of resources. As the game must terminate in one stage only, it can be solved directly. Using the value of this game, we can then solve all games in which the total resources of both players are three units. These solutions then help us to solve games with four units of resources, then five, and so on.

In general, for games with small amounts of resources (i.e., games which must terminate after a short number of stages) the values and optimal strategies can be obtained directly. For longer games, a recurrence relation can be obtained for the value of each stage. This recurrence relation is generally in the form of a difference equation which may or may not be easy to solve. If the difference equation can be solved, the values for each stage are used to give us the optimal strategies. If the difference equation cannot be solved, it may nevertheless be possible to obtain reasonable approximations of the values and optimal strategies.

5.2.2 Example. (The Inspection Game) Player I (the evader) wishes to carry out some forbidden action; there are N time periods in which this can be done. Player II (the inspector), who would like to prevent this, is allowed to carry out one inspection only, in any one of these time periods. The payoff is assumed to be $+1$ if the forbidden action is carried out undetected, -1 if the evader is caught (as he will be if he chooses to act at the same time that the inspector chooses to inspect) and 0 if the evader does not act at all.

In the first time period (stage) of the game, each player has two choices. Player I can act or not act; player II can inspect or not inspect. If I acts and II inspects, then the game is over and the payoff is -1. If I acts and II does not inspect, the game is over and the payoff is $+1$. If I does not act, and II inspects, then I will be free to act in the next time period (assuming $N > 1$) and the payoff is also $+1$. If I does not act, and II does not inspect, then we go on to the next stage of the game, which differs only in that there are fewer time periods left to play. Thus, the matrix for the first stage looks like this:

$$\begin{pmatrix} -1 & 1 \\ 1 & \Gamma_{N-1} \end{pmatrix} \tag{5.5}$$

This gives us the recursive definition

$$v_N = \text{value} \begin{pmatrix} -1 & 1 \\ 1 & v_{N-1} \end{pmatrix} \tag{5.6}$$

Now, it is clear that $v_{N-1} < 1$; hence the game matrix in (5.6) does not have a saddle point. We can, therefore, apply (2.5.24) to get the difference equation

$$v_N = (v_{N-1} + 1)/(-v_{N-1} + 3) \tag{5.7}$$

which, together with the initial condition

$$v_1 = 0 \tag{5.8}$$

defines v_N. We can solve this equation by the substitution

$$t_N = 1/(v_N - 1) \tag{5.9}$$

which gives us the new difference equation

$$t_N = t_{N-1} - \tfrac{1}{2} \tag{5.10}$$

$$t_1 = -1 \tag{5.11}$$

This equation has the obvious solution

$$t_N = -(N+1)/2$$

which gives us

$$v_N = (N-1)/(N+1) \tag{5.12}$$

Thus, (5.12) gives us the value of the game at each stage. We can then compute the optimal strategies for each player at each stage. In fact, the game matrix (5.4) now becomes

$$\begin{pmatrix} -1 & 1 \\ 1 & \frac{N-2}{N} \end{pmatrix} \tag{5.13}$$

and (2.46) and (2.47) give us the optimal strategies for $N \geq 2$

$$x^N = \left(\frac{1}{N+1}, \frac{N}{N+1} \right) \tag{5.14}$$

$$y^N = \left(\frac{1}{N+1}, \frac{N}{N+1} \right) \tag{5.15}$$

5.2.3 Example. (Women and Cats versus Mice and Men) In this fanciful game, team I consists of m_1 women and m_2 cats; team II consists of n_1 mice and n_2 men. At each stage, each team chooses a representative. One of the two representatives chosen is eliminated, according to the following rules: a women eliminates a man; a man eliminates a cat; a cat eliminates a mouse; and a mouse eliminates a woman. The game continues until one team is reduced to having only one type of player. When a team no longer has any choice, the other team is the obvious winner.

The game matrix, generally, will look like this

$$\begin{pmatrix} \Gamma(m_1-1, m_2; n_1, n_2) & \Gamma(m_1, m_2; n_1, n_2-1) \\ \Gamma(m_1, m_2; n_1-1, n_2) & \Gamma(m_1, m_2-1; n_1, n_2) \end{pmatrix} \tag{5.16}$$

and considerations similar to those in Example 5.2.3 lead to the recursive definition

$$v(m_1, m_2; n_1, n_2) = \frac{\begin{aligned}&v(m_1 - 1, m_2; n_1, n_2)v(m_1, m_2 - 1; n_1, n_2)\\&-v(m_1, m_2; n_1 - 1, n_2)v(m_1, m_2; n_1, n_2 - 1)\end{aligned}}{\begin{aligned}&v(m_1 - 1, m_2; n_1, n_2) + v(m_1, m_2 - 1; n_1, n_2)\\&-v(m_1, m_2; n_1 - 1, n_2) - v(m_1, m_2; n_1, n_2 - 1)\end{aligned}} \tag{5.17}$$

which, together with the boundary conditions

$$v(m_1, m_2; n_1, 0) = v(m_1, m_2; 0, n_2)$$
$$= +1 \quad \text{if} \quad m_1, m_2 > 0 \tag{5.18}$$

$$v(m_1, 0; n_1, n_2) = v(0, m_2; n_1, n_2)$$
$$= -1 \quad \text{if} \quad n_1, n_2 > 0 \tag{5.19}$$

defines the value inductively. Unfortunately, this is a nonlinear partial difference equation, which is not easy to solve. We can, however, obtain certain results; for instance, if $m_2 = n_1 = n_2 = 1$, we have

$$v(m, 1; 1, 1) = \frac{v(m - 1, 1; 1, 1) + 1}{-v(m - 1, 1; 1, 1) + 3} \tag{5.20}$$

and, by symmetry

$$v(1, 1; 1, 1) = 0 \tag{5.21}$$

But these equations are exactly those in Example 5.2.2. Hence

$$v(m, 1; 1, 1) = \frac{m - 1}{m + 1} \tag{5.22}$$

and the optimal strategies in such cases are also exactly as in 5.2.2.

5.3. Stochastic Games

These games are similar to the games treated in Section 5.2, but differ in several important respects. Although there are only a finite number of distinct positions, it is possible for the game to revert to a previous position, so that the games can theoretically continue indefinitely. Moreover, there is usually a payoff after each stage of the game, so that, once again, an infinite payoff is theoretically possible. There is, however, a randomization involved which assures that, for all choices of strategies, the probability of infinite play is zero, and the expected payoff is finite.

Specifically, a stochastic game is a set of p "*game elements*" or positions, Γ_k. Each game element is represented by an $m_k \times n_k$ matrix, whose entries are of the form

$$\alpha_{ij}^k = a_{ij}^k + \sum_{l=1}^{p} q_{ij}^{kl} \Gamma_l \tag{5.23}$$

with

$$q_{ij}^{kl} \geq 0 \tag{5.24}$$

$$\sum_{l=1}^{p} q_{ij}^{kl} < 1 \tag{5.25}$$

The entry α_{ij}^k, given by (5.23), means that if, in the kth game element, player I chooses his ith pure strategy and player II chooses his jth pure strategy, then there will be a payoff of a_{ij}^k units from II and I, plus a probability q_{ij}^{kl} for $l = 1, \ldots, p$ of having to play the lth game element next, and a probability

$$q_{ij}^{k0} = 1 - \sum_{l=1}^{p} q_{ij}^{kl} \tag{5.26}$$

that play will terminate. Condition (5.25) says that, at each stage, there is always a positive probability that play will terminate. Thus, the probability of infinite play will be zero, and all expected payoffs are finite.

5.3.1 Definition. A strategy for player I is a collection x^{kt} of m_k-vectors, for $k = 1, \ldots, p$ and all positive integers t satisfying

$$\sum_{i=1}^{m_k} x_i^{kt} = 1 \tag{5.27}$$

$$x_i^{kt} \geq 0 \tag{5.28}$$

A strategy will be said to be *stationary* if, for all k, the vectors x^{kt} are independent of t.

A strategy for player II is a similar collection of n_k-vectors y^{kt}.

Briefly, the number x_i^{kt} is the probability that player I will use his ith strategy, assuming that, at the tth stage of the game, he finds himself playing the game element Γ_k. As the element Γ_k may be played several times, player I need not use the same probabilities each time it is played. If, however, he does use the same randomized scheme each time the game element is played, the strategy is said to be stationary.

Naturally, a stationary strategy would be preferred from the point of view of simplicity.

Given a pair of strategies, an expected payoff can be calculated for any $k = 1, \ldots, p$ on the assumption that the first stage of the game will be the game element Γ_k. Thus, the expected payoff for a pair of strategies can be though of as a p-vector. As with ordinary matrix games, this will lead to the definition of optimal strategies and a value, the value being a p-vector $v = (v_1, v_2, \ldots, v_p)$.

It is clear that, if the value vector is to exist, one must be able to replace the game element Γ_k by the value component v_k. It follows that we must have

$$v_k = \mathrm{val}(B_k) \tag{5.29}$$

where B_k is the $m_k \times n_k$ matrix $\left(b_{ij}^k\right)$ defined by

$$b_{ij}^k = a_{ij}^k + \sum_{l=1}^{p} q_{ij}^{kl} v_l \tag{5.30}$$

This is, of course, an implicit definition only; we must show that there exists exactly one vector $(v_1, \ldots, v_p)$ satisfying (5.29) and (5.30).

5.3.2 Lemma. Let $A = (a_{ij})$ and $B = (b_{ij})$ be $m \times n$ matrices satisfying

$$a_{ij} < b_{ij} + k, \quad i = 1, \ldots, m, \quad j = 1, \ldots, n \tag{5.31}$$

for some k. Then $\mathrm{val}(A) < \mathrm{val}(B) + k$.

Proof. Let v be the value of B. Let y be an optimal strategy for II in B. Then, for all i

$$\sum a_{ij} y_j < \sum b_{ij} y_i + k \sum y_i \leq v + k$$

and so y gives a loss-ceiling smaller than $v + k$ in A.

5.3.3 Theorem. There exists exactly one vector $v = (v_1, \ldots, v_p)$ satisfying the relations (5.29) and (5.30).

Proof (Uniqueness). Suppose that there are two such vectors, v and w. Let k be the component for which $|v_k - w_k|$ is greatest, and, for definiteness, assume $v_k - w_k = c > 0$.

Define the two matrices B_k and $\overline{B}_k$ by

$$b_{ij}^k = a_{ij}^k + \sum q_{ij}^{kl} v_l$$

$$\overline{b}_{ij}^k = a_{ij}^k + \sum q_{ij}^{kl} w_l$$

It is seen that

$$\left| b_{ij}^k - \overline{b}_{ij}^k \right| \leq \sum q_{ij}^{kl} |v_l - w_l| < c$$

and so it follows from Lemma 5.3.2 that

$$\mathrm{val}(B_k) < \mathrm{val}(\overline{B}_k) + c$$

But since v and w are both assumed to satisfy (5.29) and (5.30), this means

$$v_k < w_k + c$$

But we had assumed $v_k - w_k = c$. This contradiction proves uniqueness.

Existence. We shall construct a sequence of vectors which converges to the desired value vector. We define, inductively

$$v^0 = (0, 0, \ldots, 0) \tag{5.32}$$

$$b_{ij}^{kr} = a_{ij} + \sum q_{ij}^{kl} v_l^r, \quad r = 1, 2, \ldots \tag{5.33}$$

$$v_k^{r+1} = \mathrm{val}(B_k^r) = \mathrm{val}(b_{ij}^{kr}). \tag{5.34}$$

We must prove, first, that the sequence of vectors $v^r = (v_1^r, \ldots, v_p^r)$ converges, and, second, that the limit has the desired properties (5.29) and (5.30). Let

$$s = \max_{k,i,j} \left\{ \sum_{l=1}^p q_{ij}^{kl} \right\} \tag{5.35}$$

By (5.25) (and by finiteness), $s < 1$. If we let

$$t_r = \max_k \left\{ \left| v_k^{r+1} - v_k^r \right| \right\}$$

it is easy to see (by Lemma 5.3.2) that $t_r \leq s t_{r-1}$ and so $t_r \leq s^r t_0$. It follows that the sequence of vectors v^r is Cauchy sequence and so must converge to a limit; call this limit v. Define now

$$w_k = \mathrm{val}(B_k) = \mathrm{val}\left(b_{ij}^k \right)$$

where

$$b_{ij}^k = a_{ij}^k + \sum q_{ij}^{kl} v_l$$

We shall see that $w_k = v_k$ for all k. In fact, for any $\varepsilon > 0$, we can choose r large enough so that, for all k

$$|v_k^r - v_k| < \varepsilon/2 \tag{5.36}$$

and

$$|v_k^{r+1} - v_k| < \varepsilon/2 \tag{5.37}$$

From (5.36) and Lemma 5.3.2, it is easy to see that, for all k, $|v_k^{r+1} - w_k| < \varepsilon/2$; this, together with (5.37), means

$$|w_k - v_k| < \varepsilon$$

for all k. But ε was arbitrary; hence $v_k = w_k$.

The second part of the proof of Theorem 5.3.3 is a constructive proof, in that it allows us to approximate the values of the game elements Γ_k in a reasonably efficient manner. If we assume that the game will continue as a stochastic game until we have played r times, and then necessarily terminate (if it has not already done so), we obtain a game of exhaustion (the *truncated* game) rather than a stochastic game. If we solve this game of exhaustion by the methods of Section 5.2, we obtain the values v^r, as well as optimal strategies for the matrix games B_k^r. The number s, defined by (5.35), is such that *the probability that the game continues beyond r stages*, whatever strategies may be employed, *will be at most s^r*. Thus, if r is large enough so that s^r is negligible, we can *approximate the stochastic game by truncating after r stages*. This is precisely what our sequence of vectors v^r does. Moreover, the optimal strategies x^{kr} and y^{kr} to the truncated games will converge, in the limit, to *optimal stationary strategies* to the stochastic game.

5.3.4 Example. Players I and II have, together, five units. At each stage of the game, player I chooses heads or tails, and II, in ignorance of I's move, makes a similar choice. If the choices match, II pays I either three units or 1 unit, depending on whether H or T was chosen. If the choices do not match, I pays II two units. After each stage, a coin is tossed to determine whether the game will continue or end; moreover, the game will terminate whenever either player is wiped out. We make the additional stipulation that neither player can pay more than he has.

This game can be represented by the four game elements Γ_k, where k is the amount held by I at the beginning of a given stage, as follows

$$\Gamma_1 = \begin{pmatrix} 3 + \frac{1}{2}\Gamma_4 & -1 \\ -1 & 1 + \frac{1}{2}\Gamma_2 \end{pmatrix} \tag{5.38}$$

$$\Gamma_2 = \begin{pmatrix} 3 & -2 \\ -2 & 1+\frac{1}{2}\Gamma_3 \end{pmatrix} \tag{5.39}$$

$$\Gamma_3 = \begin{pmatrix} 2 & -2+\frac{1}{2}\Gamma_1 \\ -2+\frac{1}{2}\Gamma_1 & 1+\frac{1}{2}\Gamma_4 \end{pmatrix} \tag{5.40}$$

$$\Gamma_4 = \begin{pmatrix} 1 & -2+\frac{1}{2}\Gamma_2 \\ -2+\frac{1}{2}\Gamma_2 & 1 \end{pmatrix} \tag{5.41}$$

If we use the inductive formulas (5.32)–(5.34), we obtain, as initial approximations to the values

$$v^0 = (0,0,0,0)$$

$$v^1 = (.33, -.13, -.29, -.5)$$

$$v^2 = (.26, -.19, -.29, -.53)$$

$$v^3 = (.26, -.19, -.31, -.55)$$

$$v^4 = (.26, -.19, -.32, -.55)$$

and v^4 is the value vector, correct to two decimal places. We can now use v^4 to determine the optimal strategies for each game element. In fact we have

$$B_1 = \begin{pmatrix} 2.72 & -1 \\ -1 & .91 \end{pmatrix}$$

$$B_2 = \begin{pmatrix} 3 & -2 \\ -2 & .84 \end{pmatrix}$$

$$B_3 = \begin{pmatrix} 2 & -1.87 \\ -1.87 & .72 \end{pmatrix}$$

$$B_4 = \begin{pmatrix} 1 & -2.10 \\ -2.10 & 1 \end{pmatrix}$$

and these matrix games have the optimal strategies

$$x^1 = (.34, .66), \quad y^1 = (.34, .66)$$
$$x^2 = (.38, .62), \quad y^2 = (.38, .62)$$
$$x^3 = (.40, .60), \quad y^3 = (.40, .60)$$
$$x^4 = (.50, .50), \quad y^4 = (.50, .50)$$

respectively. These vectors give the optimal (stationary) strategies to the stochastic game.

5.4.   Recursive Games

Recursive games, while similar to both games of exhaustion and stochastic games, present certain complications which are not found in the others. As with stochastic games, the game elements (positions) may be repeated; as with games of exhaustion, the termination probabilities at a stage may be zero. But these two facts conspire not only to make infinite play possible, but to have a positive probability. To keep values finite, it is, then, generally agreed that there is a payoff only when the game terminates. There may also be a payoff a_∞ stipulated in case of unending play.

The principal difficulty with these games, as opposed to stochastic games, is that approximation by truncation is not usually feasible. In fact, since infinite play may have positive probability, there is no number r such that the effects of truncation after r steps are negligible. Moreover, since the condition (5.25) is replaced by the weaker condition

$$\sum_{l=1}^{p} q_{ij}^{kl} \leq 1 \tag{5.42}$$

it follows that a system of equations similar to (5.29) and (5.30) need not have a unique solution. Similarly a sequence of vectors v^r, defined as in Theorem 5.3.3, need not converge to the true value of the game. We give as an example the recursive game with a single game element

$$\Gamma = \begin{pmatrix} \Gamma \\ -1 \end{pmatrix} \tag{5.43}$$

In this case it is clear that if we let $v^0 \geq -1$, then v^1, v^2, etc., will all equal v^0, while, if we let $v^0 \leq -1$, then v^1, v^2, etc., will all be equal to -1. Hence, while the sequence of values converges, it need not converge to the true value of the game, which is, of course, the larger of the two numbers a_∞ and -1.

To be entirely specific, a recursive game consists of a finite set of game elements Γ_k represented by matrices A_k whose entries are of the form

$$\alpha_{ij}^k = q_{ij}^{k0} a_{ij}^k + \sum_{l=1}^{p} q_{ij}^{kl} \Gamma_l \tag{5.44}$$

with the q's subject to the constraints

$$\sum_{l=0}^{p} q_{ij}^{kl} = 1 \tag{5.45}$$

$$q_{ij}^{kl} \geq 0 \tag{5.46}$$

If we now take the relations

$$v_k = \text{val}(B_k) \tag{5.47}$$

where B_k is the matrix $\left(b_{ij}^k \right)$, with

$$b_{ij}^k = q_{ij}^{k0} a_{ij}^k + \sum_{l=1}^{p} q_{ij}^{kl} v_l \tag{5.48}$$

we find that these relations will have solutions, even though these are not generally unique. In fact, it is not difficult to see that the function

$$f(v_1, v_2, \ldots, v_p) = \left(\text{val}(B_1), \ldots, \text{val}(B_p) \right)$$

is continuous (satisfying, indeed, a Lipschitz condition with constant 1). Moreover, if all the v_k are in the interval

$$\left[\min_{i,j,k} a_{ij}^k, \quad \max_{i,j,k} a_{ij}^k \right]$$

it is not difficult to see that all the entries b_{ij}^k, and hence the values of the matrices B_k, are also in this interval. By the Brouwer fixed point theorem, there must be a vector v such that $f(v) = v$. Unfortunately, there may be several such vectors, though in many interesting applications these are unique. Generally, the value vector will be that vector which is closest to the number a_∞. (See Everett [5.6] and Milnor and Shapley [5.10] for a more detailed treatment, as well as proofs.) Finally, it should be pointed out that optimal strategies need not exist, but only ε-optimal strategies. As strategies become more nearly optimal, the expected length of the game may increase without limit; thus there will be a discontinuity in the expectation, which will change from v to a_∞ as we approach a limiting strategy.

We give the following example, after Everett [5.6], to show some of these difficulties (this is one of a type of games known as Blotto games, after the protagonist, Colonel Blotto).

5.4.1 Example. Colonel Blotto, with three units, must capture an enemy outpost defended by two units. He must, however, be careful, while he attacks the enemy outpost, that the enemy does not capture his own camp. An attacker needs one more unit than the defending forces to be successful; if the attacking force is not large enough, it simply retreats to its own camp again, and the game recommences on the following day. The payoff is $+1$ if Blotto captures the enemy outpost without losing his own camp, and -1 if the enemy captures Blotto's camp. The payoff for infinite repetition is assumed to be zero.

This recursive game can be represented by a single game element Γ. The strategies in Γ simply correspond to a division into attacking and defending forces; thus Blotto will have four strategies, corresponding to 0, 1, 2, and 3 attacking units, respectively, while his opponent has three strategies. The matrix is

$$\begin{pmatrix} \Gamma & \Gamma & \Gamma \\ \Gamma & \Gamma & 1 \\ \Gamma & 1 & -1 \\ 1 & -1 & -1 \end{pmatrix} \tag{5.49}$$

The value of this game can be seen to be $+1$; Blotto's ε-optimal strategy will be of the form $(0, 1 - \delta - \delta^2, \delta, \delta^2)$. Now the smaller δ is, the larger will be the probability of victory for Blotto, and the larger the expected length of the game will be. Thus, it seems patience is most important here. If we let $\delta = 0$, however, then the game might be repeated indefinitely. Thus there is no optimal strategy for Blotto.

5.4.2 Generalizations. We can consider several generalizations of recursive and stochastic games. One possibility would be to permit payments in recursive games even when the games does not terminate. The difficulty then, of course, is that the expected payoff for a pair of strategies need not exist. It might be a divergent series, diverging to $+\infty$ or $-\infty$, or else oscillating. In the first two cases we could say the expectation is infinite, but, in the case of an oscillating series, there is, of course, no expectation.

A second possibility would be to let the game elements Γ_k have infinite numbers of pure strategies. Thus, the Γ_k may have the form of games on the square, or even more complicated forms. Everett [5.6] shows that these games will also have values and ε-optimal strategies under certain reasonable conditions; the ε-optimal strategies, however, might not be stationary.

5.5.   Differential Games

Another possible generalization of stochastic and recursive games is that of games in which the time interval between stages decreases, until, in the limit, a game is reached in which each player must make a move at each moment of time. Because the moves are continuous, it is natural to ask that the game elements change only slightly over small periods of time, i.e., that the game element change continuously. Thus, if a game element is represented as some point in euclidean space of some dimension, the strategies are generally though of as defining a differential motion for the point (game element).

More precisely, we can give a definition as follows: the game element is represented by an n-tuple $(x_1, \ldots, x_n)$ of real numbers, known as the *state variables*. At each moment t of time, player I chooses a p-tuple, $\varphi = (\varphi_1, \ldots, \varphi_p)$, of real numbers, subject generally to some constraints usually of the form $a_i \le \varphi_i \le b_i$ (where a_i and b_i are constant). Similarly, II chooses a q-tuple $\psi = (\psi_1, \ldots, \psi_q)$. The vectors φ and ψ are called the *control variables*. The control variables then influence the motion of the state variables, according to a system of differential equations (called the *kinematic equations*)

$$\dot{x}_i = f_i(x; \varphi; \psi), \quad i = 1, \ldots, n \tag{5.50}$$

where the $\dot{x}_i$ should be treated as forward derivatives of the x_i with respect to time.

The differential game continues, according to the kinematic equations, until termination, which is assumed to occur when the state variables reach some closed subset $\mathscr{C}$ of n-space, whose boundary is called the *terminal surface*. In practical applications, this may mean that player I is close enough to player II to catch him, or that a specified time period has terminated (if a game is to terminate after a specified time, then time is, of course, also a state variable). The payoff then may be of several forms, most common of which are the *terminal* and the *integral* payoff, as well as combinations of these two.

If the game begins at time $t = 0$, and ends at time $t = T$ at the point $(y_1, \ldots, y_n)$, then a terminal payoff is simply a function $G(y_1, \ldots, y_n)$, defined on the *terminal surface* of the game; an integral payoff will be of the form

$$\int_0^T K(x_1, \ldots, x_n) \, dt \tag{5.51}$$

The most general type of payoff which we will treat here will have the form of a function G plus an integral such as (5.51).

5.5.1 The Main Equation.

As with discrete multistage games, the technique for solving differential games consists in replacing game elements by their values, and then solving recurrence equations for these values. (These equations will now be differential equations.)

Suppose then that values do exist; the value for a game which starts at the point $x = (x_1, \ldots, x_n)$ will be denoted by $V(x_1, \ldots, x_n)$. Let us assume that I chooses the control variable $\overline{\varphi}$, while II chooses the control variable $\overline{\psi}$, at time $t = 0$. In that case, after a very small interval Δt of time, we find that the state variables are approximately at the position $x + \Delta x$, where

$$\Delta x_i = f_i(x; \overline{\varphi}; \overline{\psi}) \Delta t \tag{5.52}$$

and (if the game has an integral payoff) there will have a total payoff of approximately

$$K(x_1, \ldots, x_n)\Delta t \tag{5.53}$$

The game will start again from the point $x + \Delta x$ given by (5.52), with a payoff (5.53) already effected. If optimal strategies are used from the moment Δt on, the total payoff will be

$$K(x_1, \ldots, x_n)\Delta t + V(x + \Delta x)$$

We know, however, that

$$V(x + \Delta x) \cong V(x) + \sum V_i(x)\Delta x_i$$

(where V_i is the partial derivative of V with respect to x_i), or

$$V(x + \Delta x) \cong V(x) + \sum_{i=1}^{n} V_i(x)f_i(x; \overline{\varphi}; \overline{\psi}) \ \Delta t$$

Hence, assuming that $\overline{\varphi}$ and $\overline{\psi}$ are optimal choices of the control variables for the time $t = 0$, we have

$$V(x) \cong K(x)\Delta t + V(x) + \sum V_i(x)f_i(x; \overline{\varphi}; \overline{\psi}) \ \Delta t$$

or, letting $\Delta t \to 0$

$$K(x) + \sum V_i(x)f_i(x; \overline{\varphi}; \overline{\psi}) = 0 \tag{5.54}$$

or, equivalently

$$\max_{\varphi} \min_{\psi} \left\{ K(x) + \sum V_i(x)f_i(x; \varphi; \psi) \right\} = 0 \tag{5.55}$$

Equation (5.54), or the equivalent (5.55), is known as the *main equation*. It is generally possible, in Equation (5.55), to interchange the order of the two operators, *max* and *min*, although in practice there may be lower-dimensional sets (*singular surfaces*) on which this is not so. Thus, pure strategies are generally sufficient; there is need for randomization, usually, only at a finite number of points in any play of the game.

5.5.2 The Path Equations. Once the *main equation* (5.54) has been obtained, it is possible (as in games of exhaustion) to work backward from the terminal surface by differential equations (as, before, we worked backward by difference equations). We have, in fact

$$K + \sum V_i f_i = 0$$

When we differentiate the left-hand side of this with respect to x_j, we obtain the sum of the terms

$$K_j + \sum_i V_i f_{ij} \tag{5.56}$$

(where $K_j = \partial K / \partial x_j$ and $f_{ij} = \partial f_i / \partial x_j$)

$$\sum_i \frac{\partial V_i}{\partial x_j} f_i \tag{5.57}$$

$$\sum_{k=1}^{p} \frac{\partial}{\partial \varphi_k} \left(K + \sum_i V_i f_i \right) \frac{\partial \overline{\varphi}_k}{\partial x_j} \tag{5.58}$$

and

$$\sum_{l=1}^{q} \frac{\partial}{\partial \psi_l} \left(K + \sum_i V_i f_i \right) \frac{\partial \overline{\psi}_l}{\partial x_j} \tag{5.59}$$

Let us consider (5.58), supposing that the constraints on the control variables φ_k are constant. We know that $\overline{\varphi}_k$ will be either at an interior point, or at an end-point, of its constraint interval. If it is at an interior point, then

$$\frac{\partial}{\partial \varphi_k} \left(K + \sum V_i f_i \right) = 0$$

because $\overline{\varphi}$ was chosen to maximize the parenthesis. If on the other hand $\overline{\varphi}_k$ is at an end-point, then (except at points of singularity) it will remain at that endpoint, and so

$$\frac{\partial \overline{\varphi}_k}{\partial x_j} = 0$$

We see thus that the entire term (5.58) vanishes; similarly, (5.59) will vanish. Consider next the term (5.57). We have

$$\frac{\partial V_i}{\partial x_j} = \frac{\partial^2 V}{\partial x_i \partial x_j} = \frac{\partial V_j}{\partial x_i}$$

and so

$$\sum_i \frac{\partial V_i}{\partial x_j} f_i = \sum_i \frac{\partial V_j}{\partial x_i} \cdot \frac{dx_i}{dt} = \frac{dV_j}{dt} \tag{5.60}$$

Call the right side of this equation $\dot{V}_j$. If we add this to (5.54), and set equal to zero, we obtain the equations

$$\dot{V}_j = -\left\{ K_j(x; \overline{\varphi}; \overline{\psi}) + \sum_i V_i f_{ij}(x; \overline{\varphi}; \overline{\psi}) \right\} \tag{5.61}$$

which, together with the system

$$\dot{x}_j = f_j(x; \overline{\varphi}; \overline{\psi}) \tag{5.62}$$

are called the *path equations* of the differential game. These $2n$ equations, with the value of the function G as terminal conditions, are a formal solution of the game.

It is sometimes easier to use retrograde time $\tau = C - t$ rather than forward time t (since, in fact, we have terminal rather than initial values for the system (5.61) and (5.62)).

5.5.3 Example. Players I and II control the motion of a point in the euclidean plane, each imparting a velocity whose magnitude depends on the position of the point and whose direction is entirely under the player's control. The velocity of the point is equal to the vector sum of these velocities.

The game will terminate when the point reaches the x-axis; the payoff will be equal to the time needed to complete the game, plus the amount $x_0^2/8$, where x_0 is the abscissa of the point at which play terminates.

If we let $u = y$ and $w = x + y$ be the magnitudes of the velocities controlled by players I and II, respectively, we have the kinematic equations

$$\dot{x} = y \cos \varphi + (x + y) \cos \psi$$

$$\dot{y} = y \sin \varphi + (x + y) \sin \psi$$

and the payoff

$$\int_0^T dt + \frac{x_0^2}{8}$$

Thus we have $K = 1$ for all x, y.

It is clear that if we have $u > w$, player I can always extend the game indefinitely. Hence we are interested only in points in the positive quadrant.

The main equation for this game will be

$$y(V_1 \cos \overline{\varphi} + V_2 \sin \overline{\varphi}) + (x + y)(V_1 \cos \overline{\psi} + V_2 \sin \overline{\psi}) = -1 \tag{5.63}$$

Now to maximize the first term on the left of (5.63), we must have

$$\cos \overline{\varphi} = \frac{V_1}{\sqrt{V_1^2 + V_2^2}} \tag{5.64}$$

$$\sin \overline{\varphi} = \frac{V_2}{\sqrt{V_1^2 + V_2^2}} \tag{5.65}$$

and to minimize the second term we must have

$$\cos \overline{\psi} = -\cos \overline{\varphi} \tag{5.66}$$

$$\sin \overline{\psi} = -\sin \overline{\varphi} \tag{5.67}$$

If we introduce these values into (5.63), and simplify, we obtain

$$\sqrt{V_1^2 + V_2^2} = 1/x \tag{5.68}$$

We also have

$$f_{11} = \cos \psi$$

$$f_{21} = \sin \psi$$

$$f_{12} = \cos \varphi + \cos \psi$$

$$f_{22} = \sin \varphi + \sin \psi$$

and when we substitute (5.64)–(5.68) in these, we obtain the path equation

$$\dot{V}_1 = 1/x$$

$$\dot{V}_2 = 0$$

$$\dot{x} = -x^2 V_1$$

$$\dot{y} = -x^2 V_2$$

If, instead of forward time, we use retrograde time $\tau = T - t$, we have the retrograde path equations

$$\overset{\circ}{V}_1 = -1/x$$

$$\overset{\circ}{V}_2 = 0$$

$$\overset{\circ}{x} = x^2 V_1$$

$$\overset{\circ}{y} = x^2 V_2$$

where $\overset{\circ}{V}_1$ is the derivative of V_1 with respect to τ, etc.

We also have the initial conditions that, for $\tau = 0$

$$x = x_0, \qquad y = 0,$$

$$V_1 = \tfrac{1}{4} x_0, \qquad V_2 = \sqrt{\frac{1}{x_0^2} - \frac{x_0^w}{16}}$$

These conditions show that we must have $x_0 \leq 2$. This means that none of the paths terminates at a point $x_0 > 2$.

If we differentiate V_1 with respect to τ, we obtain

$$\overset{\circ\circ}{V}_1 = \frac{1}{x^2} \overset{\circ}{x} = V_1$$

and this equation has the solution

$$V_1 = C_1 e^\tau + C_2 e^{-\tau}$$

which give us, in turn

$$x = \frac{1}{C_2 e^{-\tau} - C_1 e^\tau}$$

If we set $x_0 = a$, we can then introduce the initial conditions that, to solve for C_1 and C_2, we obtain

$$x = \frac{8a}{(4 + a^2)e^{-\tau} + (4 - a^2)e^\tau}$$

(5.69)

$$V_1 = \frac{4 + a^2}{8a} e^{-\tau} - \frac{4 - a^2}{8a} e^\tau$$

To find y, we notice that we have

$$\mathring{y}/\mathring{x} = V_2/V_1$$

which (remembering V_2 is constant along any *optimal* path) gives us

$$dy/dx = V_2(0)/V_1$$

Now $V_2(0)$ is given above, and V_1 can be obtained as a function of x; this gives us

$$\frac{dy}{dx} = \frac{x}{\sqrt{[16a^2/(16 - a^4)] - x^2}}$$

which has a solution

$$y = C_3 - \sqrt{[16a^2/(16 - a^4)] - x^2}$$

We solve for C_3 by remembering that, for $y = 0$, $x = a$. We thus have

$$y = \frac{a^3 \pm \sqrt{16a^2 - 16x^2 + a^4 x^2}}{\sqrt{16 - a^4}}$$

or, equivalently

$$x^2 + \left(y - \frac{a^3}{\sqrt{16 - a^4}}\right)^2 = \frac{16a^2}{16 - a^4} \tag{5.70}$$

i.e., the *optimal path* is a circle on the y-axis (see Figure 5.2). The value $V(x, y)$ can be obtained by solving (5.70) for a; we then solve (5.69) for τ. Then $V(x, y) = \tau + a^2/8$. The optimal strategies are also computable: both players attempt to follow the tangent to the circle (5.70). Player I (the maximizer) pushes up (away from the x-axis); player II pushes down (toward the x-axis).

5.5.4 The Isotropic Rocket Escape Game. The following example is due to Isaacs. A more complete treatment can be found in (5.8). We give a very simple form which should, however, be sufficient to show how most of these games can be solved.

A pursuer II is trying to capture an evader I. It is assumed that the pursuer, having large mass, moves by a rocket thrust of constant magnitude, whose direction he can control; this causes an acceleration in his motion. The evader, on the other hand, has negligible inertia; he moves with a fixed velocity in any direction he desires.

It is assumed that the game continues until capture is effected, which in this case means the pursuer is within a certain distance l of the evader. The payoff is simply time to capture, which the evader naturally wishes to maximize.

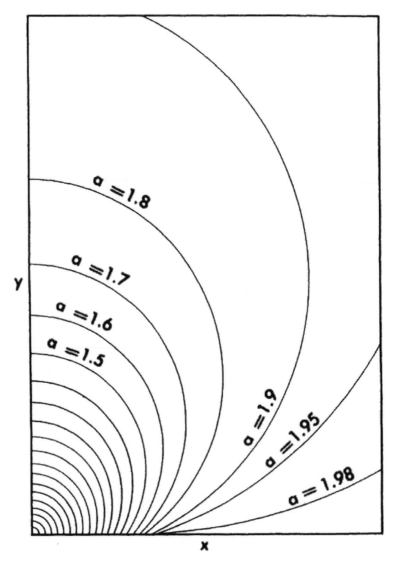

Figure 5.2.

We can represent a state of this game by six variables x_i. Four variables are needed to represent the positions of the two players – rectangular coordinates in the plane. Two additional state variables are needed to represent the pursuer's velocity (also a vector in the plane).

The kinematic equations here are, then

$$\dot{x}_1 = x_3, \qquad \dot{x}_2 = x_4$$
$$\dot{x}_3 = a \cos \psi, \qquad \dot{x}_4 = a \sin \psi$$
$$\dot{x}_5 = \cos \varphi, \qquad \dot{x}_6 = \sin \varphi$$

where a is the (constant) magnitude of the acceleration, and φ, ψ, are the two players' control variables.

The terminal surface $\mathscr{C}$ is given by

$$(x_5 - x_1)^2 + (x_6 - x_2)^2 = l^2$$

and we can parametrize it by five variables s_i related to the x_i by

$$x_1 = s_1, \qquad\qquad x_2 = x_2$$

$$x_3 = s_3, \qquad\qquad x_4 = s_4$$

$$x_5 = s_1 + l \cos s_5, \quad x_6 = s_1 + l \sin s_5$$

The payoff, time to capture, can be treated as a pure integral payoff, with $K = 1$. Applying our theory, we obtain the main equation (ME)

$$1 + V_1 x_3 + V_2 x_4 + a(V_3 \cos \psi + V_4 \sin \psi)$$

$$+ V_5 \cos \varphi + V_6 \sin \varphi = 0$$

Since φ must be chosen to maximize this expression, we have

$$\cos \hat{\varphi} = V_5/\rho, \quad \sin \hat{\varphi} = V_6/\rho, \quad \rho^2 = V_5^2 + V_6^2$$

and similarly, as ψ is chosen to minimize

$$\cos \hat{\psi} = -V_3/r, \quad \sin \hat{\psi} = -V_4/r, \quad r^2 = V_3^2 + V_4^2$$

so that the ME reduces to

$$1 + V_1 x_3 + V_2 x_4 - ar + \rho = 0$$

In turn, the retrograde path equations (RPEs) will take the form

$$\overset{\circ}{V}_1 = 0, \qquad \overset{\circ}{V}_2 = 0,$$

$$\overset{\circ}{V}_3 = V_1, \qquad \overset{\circ}{V}_4 = V_2,$$

$$\overset{\circ}{V}_5 = 0, \qquad \overset{\circ}{V}_6 = 0,$$

so that, on any optimal path, V_1, V_2, V_5, V_6 will be constant, as will also ρ. To obtain initial values for the V_j, we point out that the terminal payoff is zero; thus, on the terminal surface, $V = 0$ constantly, so that

$$\partial V / \partial s_j = 0, \quad j = 1, \ldots, 5$$

Using the chain rule for differentiation, this will give us

$$\sum_i \frac{\partial V}{\partial x_i} \frac{\partial x_i}{\partial s_j} = 0, \quad j = 1, \ldots, 5$$

or

$$V_1 + V_5 = 0, \quad V_2 + V_6 = 0$$

$$V_3 = 0, \qquad\qquad V_4 = 0$$

$$-l \sin s_5 V_5 + l \cos s_5 V_6 = 0$$

a system of five equations in six unknowns, with solution

$$-V_1 = V_5 = \alpha \cos s_5$$

$$-V_2 = V_6 = \alpha \sin s_5$$

$$V_3 = V_4 = 0$$

where α must yet be determined. We can substitute this in the ME to obtain two solutions for α; one of these will be positive and one negative. To determine which one we want, we note that when $s_5 = 0$ — which corresponds to termination with $x_5 = x_1 + l, x_6 = x_2$ — a small increase in x_5 will make V increase (it will, in effect, increase the distance between I and II). Thus, $v_5 > 0$ when $s_5 = 0$, and we conclude that $\alpha > 0$. Then $\alpha = \rho$, and substitution of the above in the ME gives us

$$1 - \rho s_3 \cos s_5 - \rho s_4 \sin s_5 + \rho = 0$$

or

$$\alpha = \rho = \frac{1}{s_3 \cos s_5 + s_4 \sin s_5 - 1}$$

Since this must be positive, it tells us that, at termination, we must have $s_3 \cos s_5 + s_4 \sin s_5 > 1$: the pursuer II must have its component of velocity, in the direction of the evader, at least equal to unity (which is the evader's own velocity).

We can now integrate directly. We have, for all $\tau \geq 0$

$$V_1 = -\rho \cos s_5 = -V_5$$

$$V_2 = -\rho \sin s_5 = -V_6$$

$$V_3 = -\rho \tau \cos s_5$$

$$V_4 = -\rho \tau \sin s_5$$

From these last two relations we obtain $r = \rho\tau$, and the optimal strategies $\hat\varphi, \hat\psi$ are

$$\cos\hat\varphi = \cos s_5, \quad \sin\hat\varphi = \sin s_5$$

$$\cos\hat\psi = \cos s_5, \quad \sin\hat\psi = \sin s_5$$

so that

$$\hat\varphi = s_5, \quad \hat\psi = s_5$$

Thus, under optimal play, we see that both I's velocity and II's acceleration are in the same direction; capture will be effected with II directly behind I.

The RPE now take the form

$$\overset{\circ}{x}_1 = -x_3, \quad \overset{\circ}{x}_3 = -a\,\cos s_5, \quad \overset{\circ}{x}_5 = -\cos s_5$$

$$\overset{\circ}{x}_2 = -x_4, \quad \overset{\circ}{x}_4 = -a\sin s_5, \quad \overset{\circ}{x}_6 = -\sin s_5$$

which have the solution

$$x_1 = s_1 - s_3\tau + \tfrac{1}{2}a\tau^2 \cos s_5$$

$$x_2 = s_2 - s_4\tau + \tfrac{1}{2}a\tau^2 \sin s_5$$

$$x_3 = s_3 - a\tau \cos s_5$$

$$x_4 = s_4 - a\tau \sin s_5$$

$$x_5 = s_1 + (l - \tau)\cos s_5$$

$$x_6 = s_2 + (l - \tau)\sin s_5$$

To see what this means, we can let

$$X = (x_5 - x_1, x_6 - x_2), \quad U = (x_3, x_4)$$

be the distance and velocity vectors. Some algebra shows that

$$\|X - \tau U\| = l - \tau + \tfrac{1}{2}a\tau^2$$

where $\|\cdot\|$ denotes the usual pythagorean (l_2) norm. Then the solution of the game depends on the behavior of the quadratic

$$Q(\tau) = l - \tau + \tfrac{1}{2}a\tau^2$$

It is obvious that $Q(0) = l > 0$, and $Q(\tau) > 0$ for large τ. On the other hand, $Q'(0) = -1$, so that Q reaches a minimum value for positive τ. This minimum value of Q may or may not be positive; the condition is $2al > 1$.

If this minimum value of Q is positive, then the pursuer can always effect capture. To get an idea of the situation, let us suppose $U = (0, 1)$. Then for any value of $\tau \geq 0$, the equation

$$\|X - \tau U\| = Q(\tau)$$

is represented by a circle with center $(0, \tau)$ and radius $Q(\tau)$.

If we draw these circles (see Figure 5.3), we can erase those parts of "later" circles which intersect "earlier" circles (i.e., those with smaller τ). For small values of τ, the radius function $Q(\tau)$ decreases, and the circles may be seen to be inside a narrowing envelope curve. After a while (for $\tau > 1/a$) the envelope widens again, and, eventually, the envelope disappears. The widening circles thus fill up the plane.

Each point, then, lies on one of these circles; when it lies on two circles, only the "earlier" one counts. Thus the figure shows us the time to capture, under optimal play, for given points. The optimal strategies call for the evader to move along the radius vector, toward the outside of the given circle. The envelope plays an important

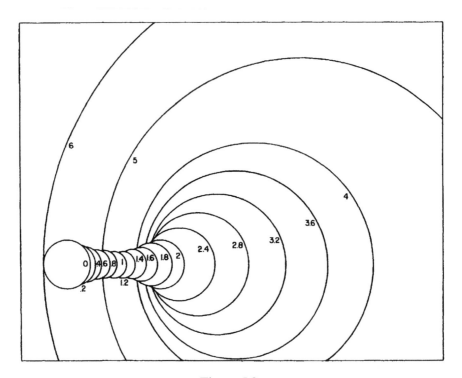

Figure 5.3.

role: for points on the outside, optimal play requires that the evader should actually move towards the side of the pursuer; this latter will "overrun," and must decelerate and return to effect capture.

In the second case, the minimum value of the function $Q(\tau)$ is negative or zero. In this case, the two branches of the envelope curve meet. It turns out that the evader can usually escape indefinitely; the pursuer is too clumsy to catch him. Nevertheless, there is a small area, directly in front of the pursuer, from which the evader cannot escape; this area is bounded by the envelope, and should be thought of as the "capture zone." The rest of the plane is the "escape zone."

Problems

1. Show that, for recursive games, the value vector is not the only vector which satisfies the functional equation (5.29) and, moreover, the method of successive approximations used for stochastic games need not converge to the value vector. Use the recursive game with two elements

$$\Gamma_1 = (-10), \quad \Gamma_2 = \begin{pmatrix} +20 & \Gamma_1 \\ \Gamma_1 & +20 \\ \Gamma_2 & \Gamma_2 \end{pmatrix}$$

2. Extend the results on recursive games to games in which there is a payoff after each play even if the game does not terminate.

(a) A *trap* is a game element, or set of game elements, such that once a play enters an element of the trap one player can force the play to remain in the trap in such a way as to accumulate payoffs and cause an unbounded expectation. Show that the game with three game elements

$$\Gamma_1 = \begin{pmatrix} \Gamma_2 & 10 & \Gamma_3 \\ -10 & 0 & -10 \\ \Gamma_3 & 10 & \Gamma_2 \end{pmatrix}$$

$$\Gamma_2 = (\Gamma_2 + 1), \quad \Gamma_3 = (\Gamma_3 - 2)$$

which has traps does not have a fixed point for the value mapping (5.29).

(b) If a game has all nonnegative payoffs, and has no traps, it will have a value.

(c) Even if a game has no traps, it might have no value if it has both positive and negative payoffs. Give an example. (The value should oscillate under the "best" strategies.)

3. Show that, if time is considered as continuous rather than discrete, a recursive game need not have a value, even if the functional equation (5.29) has a unique

solution. Consider the game with one game element

$$\Gamma = \begin{pmatrix} 0 & 1 & -1 \\ -1 & 0 & 1 \\ 1 & -1 & \Gamma \end{pmatrix}$$

where the game is played through an interval of time until it terminates. In other words, each player chooses a time at which to act; until that time, he has used his third pure strategy continually.

4. The study of differential games usually involves not only the path (5.61) but also a study of so-called singular surfaces. Consider the game played on the half-plane $y \geq 0$, with terminal surface $y = 0$ and terminal payoff

$$G(x^*) = \frac{1}{1 + x^{*2}}$$

Let the kinematic equations be

$$\dot{x} = \varphi \left(1 + 2\sqrt{|x|} \right) + \psi$$

$$\dot{y} = -1$$

where the control variables φ, ψ are restricted to the interval $[0, 1]$. Show that the y-axis is a singular surface in the sense that the optimal paths are a family of curves which *begin* on the y-axis. What should be done at a point on the y-axis (where two optimal paths meet)?

5. It seems quite reasonable to believe that any differential game with integral or continuous terminal payoff should have solutions in fixed strategies, except possibly on a lower-dimensional set (singular surface). Show that this is not so. Consider the game played on the half-plane $y \geq 0$, with terminal surface $y = 0$, and integral payoff $\int x\,dt$, whose kinematic equations are

$$\dot{x} = (\psi - \varphi)^2$$

$$\dot{y} = -1$$

and, as before, ψ, φ are restricted to the interval $[0, 1]$. Prove that, for any initial point (x_0, y_0), the values in pure strategies satisfy

$$v_1 \leq x_0 y_0$$

$$v_{11} \geq x_0 y_0 + \tfrac{1}{8} y_0^2$$

Hence the game has no solution in pure strategies for $y > 0$ (nor does it have any optimal paths).

6. A differential game of kind is one which can have only two possible outcomes, say, win and loss (for player I). Thus, we can think of the terminal surface $\mathscr{C}$ as divided into two sets $\mathscr{W}$ and $\mathscr{L}$ which are separated by an $(n-2)$-dimensional manifold $\mathscr{K}$. Player I tries to bring about termination of the game in $\mathscr{W}$ player II, in $\mathscr{L}$. In general, the playing space R^n will be divided into two parts, called the winning zone (WZ) and losing zone (LZ), such that, from a point in WZ, player I can force termination in $\mathscr{W}$; while from a point in LZ, II can force termination in $\mathscr{L}$. These parts are generally separated by a surface $\mathscr{N}$, which intersects $\mathscr{C}$ along $\mathscr{K}$.
   (a) On the assumption that $\mathscr{N}$ is smooth, show that the *normal vector* $(v_1, \ldots, v_n)$ to $\mathscr{N}$ satisfies the equation

$$\max_{\varphi} \min_{\psi} \sum v_i f_i(x, \varphi, \psi) = 0$$

(where v is oriented toward the WZ). Thus obtain a system of path equations (or retroactive path equations) for these games, similar to (5.61).
   (b) Consider the following game: Players I and II control the motion of points P and E, respectively, in the upper half-plane of R^2. The two points can move in any direction, at velocities 1 and $w < 1$, respectively. The game terminates with a win for I if ever the distance $\overline{PE}$ is less than d; it terminates with a loss for I if ever E reaches the line $y = 0$.
   (c) If we let (x_1, y_1) and (x_2, y_2) be the coordinates of P and E respectively, and let φ and ψ be the control variables, give the kinematic equations for this game; characterize the set $\mathscr{K}$.
   (d) Solve this game.

7. In an extensive game Γ, let σ_i be a (pure) strategy for player i, and let X be a vertex in i's personal move set S_i. Then X is said to be *possible for* σ_i if there exists an $(n-1)$-tuple of strategies (π_j) $(j \neq i)$ such that if i uses σ_i and the other players use π_j, then there is a positive probability that the play will pass through vertex X.
   Prove that vertex $X \in S_i$ is possible for strategy σ_i if and only if, at every $Y \in S_i$ on the path from the initial vertex A to X, σ_i chooses the alternative which leads to X.

8. (Kuhn's Theorem) In an extensive game Γ, player i is said to have *imperfect recall* if there are two vertices, X and Y, in the same information set S_{ij}, and a strategy σ_i of player i, such that X is possible for σ_i, but Y is not possible.
   A game Γ is said to have *perfect recall* if no player has imperfect recall in Γ.
   Prove that, in a game with perfect recall, *behavioral strategies are sufficient,* in the sense that any mixed strategy can be obtained by means of behavioral strategies (cf. Example 5.1.3, in which behavioral strategies are clearly insufficient).

9. Develop optimal strategies for the game of Example 5.2.3, with $m_1 = 4$, $m_2 = 3$, $n_1 = 2$, and $n_2 = 4$.

10. Obtain approximate values for the stochastic game

$$\Gamma_1 = \begin{pmatrix} 4 + 0.3\Gamma_4 & -1 \\ -2 & 1 + 0.2\Gamma_2 \end{pmatrix}$$

$$\Gamma_2 = \begin{pmatrix} 3 + 0.3\Gamma_4 & -1 \\ -2 & 2 + 0.2\Gamma_3 \end{pmatrix}$$

$$\Gamma_3 = \begin{pmatrix} 4 & -1 + 0.2\Gamma_1 \\ -2 & 1 + 0.2\Gamma_1 \end{pmatrix}$$

$$\Gamma_4 = \begin{pmatrix} 2 & -1 + 0.3\Gamma_3 \\ -2 + 0.2\Gamma_1 & 1 \end{pmatrix}$$

(Hint: carry out four iterations, starting with $v^0 = (0, 0, 0, 0)$.)

11. Modify Example 5.5.3 by letting the two velocity vectors be $u = 2y$ and $w = x + y$. Let the game terminate when the point reaches the line $x = y$, with payoff (to player I) equal to the time needed to complete the game, plus $x_0^2/10$, where x_o is the abscissa of the point at which play terminates.

12. Modify Example 5.5.4 by assuming that the pursuer is subject to a "downward" acceleration equal to $0.3a$: in other words, the derivative of x_4 will be $a(\sin \varphi - 0.3)$ rather than $a \sin \varphi$. Otherwise, let all the kinematic equations and payoff functions be as in the Example.

Chapter 6

Games with Incomplete Information

6.1. Introduction: The One-Stage Signaling Model

In our development to now, an assumption has been made throughout of complete information: both players (or all players, in case of more than two players) are aware of the game which is being played. In particular, they know what strategies are available, and what the payoffs will be for any particular choice of strategies. In real life, we can imagine that several things might go wrong with this scheme:

(a) a player may be unaware of one or more of the alternatives available at a given move;
(b) a player may be uncertain as to the payoffs associated with a given choice of strategies;
(c) a player may be uncertain as to the probabilities associated with the alternatives at a chance move.

Of course, it is difficult to deal with pure uncertainty, in which the player is unaware of some alternative, or in which he knows that something might, by chance, happen, but has no idea as to the probability of such an event. We therefore assume that each player has a (subjective) probability distribution as to those events for which he lacks complete information.

In three important papers, Harsanyi [6.4] points out that it is usually possible to reconcile the several players' information and subjective probability distributions by assuming that they can be obtained as conditional distributions — based on private information given the players — from an actual, objective distribution.

6.1.1 Example. Players I and II are to play a game. The actual payoff matrix of this game can be any one of eight 2×2 matrices, called $M_{ABC}, M_{ABC'}, \ldots, M_{A'B'C'}$, depending on whether three events, A, B, and C (or their complements, A', B', and C') have occurred. (It is possible for any combination of A, B, and C to occur.) In any case, we will let I's two choices be U or D, and II's two choices be L or R. Player I knows that A has occurred, and he assigns probabilities of 0.5 to (B, C), 0.25 to (B, C'), 0.25 to (B', C), and 0 to (B', C'). Player II knows that B has occurred, and he assigns probabilities of 0.4 to (A, C), 0.2 to (A, C'), 0.2 to (A', C), and 0.2 to (A', C').

In this case, it may be assumed that there is a basic distribution of the game, which must satisfy the conditions

$$P(B, C|A) = 0.5 \qquad P(B, C'|A) = 0.25$$
$$P(B', C|A) = 0.25 \quad P(B', C'|A) = 0$$
$$P(A, C|B) = 0.4 \qquad P(A, C'|B) = 0.2$$
$$P(A', C|B) = 0.2 \qquad P(A', C'|B) = 0.2$$

It may be checked that the distribution

$$P(A, B, C) = 0.2 \qquad P(A, B, C') = 0.1$$
$$P(A, B', C) = 0.1 \qquad P(A, B', C') = 0$$
$$P(A', B, C) = 0.1 \qquad P(A', B, C') = 0.1$$
$$P(A', B', C) = 0.25 \quad P(A', B', C') = 0.15$$

has the desired conditional probabilities.

Harsanyi now suggests that the game is actually part of the following game:

(a) Chance chooses A, B, C or their complements, according to the probability distribution P.
(b) I is told whether A occurred, but not whether B and C occurred.
(c) II is told whether B occurred, but not whether A or C occurred.
(d) I and II, with the given information, choose U or D, L or R, respectively.

As may be seen, in this somewhat complex game, each player has two information sets (A and A' for I, B and B' for II), two alternatives at each information set, and hence four pure strategies. The resulting 4×4 game can then be solved according to the standard techniques.

Continuing with the example, let the several 2×2 matrices be

$$M_{ABC} = \begin{pmatrix} 1 & 0 \\ 0 & 3 \end{pmatrix} \qquad M_{ABC'} = \begin{pmatrix} 2 & 1 \\ 1 & 4 \end{pmatrix}$$

$$M_{AB'C} = \begin{pmatrix} 1 & 3 \\ 2 & 1 \end{pmatrix} \qquad M_{AB'C'} = \begin{pmatrix} 1 & 0 \\ 4 & 0 \end{pmatrix}$$

$$M_{A'BC} = \begin{pmatrix} 3 & 2 \\ 0 & 3 \end{pmatrix} \qquad M_{A'BC'} = \begin{pmatrix} 2 & 4 \\ 1 & 1 \end{pmatrix}$$

$$M_{A'B'C} = \begin{pmatrix} 0 & 1 \\ 3 & 1 \end{pmatrix} \qquad M_{A'B'C'} = \begin{pmatrix} 2 & 0 \\ 3 & 4 \end{pmatrix}$$

In this case, some work shows that the eight 2×2 games reduce to the 4×4 matrix

$$
\begin{array}{c}
\begin{array}{cccc} LL & LR & RL & RR \end{array} \\
\begin{array}{c} UU \\ UD \\ DU \\ DD \end{array}
\begin{array}{c} = \\ \\ \\ \end{array}
\left(
\begin{array}{cccc}
1.3 & 1.45 & 1.1 & 1.25 \\
1.8 & 1.65 & 1.8 & 1.65 \\
1.1 & 0.7 & 2.0 & 1.95 \\
1.5 & 1.15 & 2.8 & 2.35
\end{array}
\right)
\end{array}
$$

where strategy UU for I means he chooses U in either information set, UD means he chooses U in A, D in A', etc., and similarly for II in cases B or B'.

It is now easy to see that this game has a saddle-point at (UD, LR). Thus, we understand that the optimal strategies are, for I (who is in A) to play U, for II (who is in B) to play L.

In this way, then, it is normally possible to reduce any game of incomplete information (i.e., one or both players do not know the actual game tree) to a game of imperfect information (both players know the game tree but they might not know where they are on the tree). We will assume that such a representation can be made. The following example is in such form.

6.1.2 Example. Each player has two alternatives, A and B, the choice to be made in ignorance of what the other player has chosen. One of these two alternatives is known as the *paying alternative*. If both chooses the paying alternative, then I wins (and II loses) \$1. If either, or both, choose the nonpaying alternative, then there is no payment. The difficulty is that neither player knows which of A and B is the paying alternative; *a priori*, it can be either A or B with probability 0.5.

We can model this game by introducing a chance move, at which either A or B is chosen (to be the paying alternative) with probability 0.5. The normal form is given by the matrix

$$
\begin{array}{c}
\begin{array}{cc} A & B \end{array} \\
\begin{array}{c} A \\ B \end{array}
\left(
\begin{array}{cc}
0.5 & 0 \\
0 & 0.5
\end{array}
\right)
\end{array}
$$

and it is clear that this game has the optimal strategies $x^* = y^* = (0.5, 0.5)$, and value $v = 0.25$.

Suppose, however, that player I knows the result of the chance move; player II does not know. In this case, the extensive form of the game changes in that player I's move are now in distinct information sets, so that he now has four pure strategies; the normal form becomes

$$
\begin{array}{c}
\begin{array}{cc} A & B \end{array} \\
\begin{array}{c} AA \\ AB \\ BA \\ BB \end{array}
\left(
\begin{array}{cc}
0.5 & 0 \\
0.5 & 0.5 \\
0 & 0 \\
0 & 0.5
\end{array}
\right)
\end{array}
$$

where I's strategy XY means that I chooses X if chance chose A, and Y if chance chose B. It is clear that the second row, AB, dominates all others. Hence the game has value 0.5, with optimal pure strategy AB for I. It does not matter what II does.

Let us suppose, next, that the game is to be played many times. There is now an important distinction to be made, depending on whether the chance move must be made each time that the game is played (so that, from trial to trial, the paying alternative changes), or whether the chance move is made once for all time. In the first case, the several trials of the game are quite independent, and so our above analysis holds: at each trial, player I mimics chance's move, and the expected payoff is 0.5 at each trial.

In the second case, the chance move is made once for all at the beginning of the first trial of the game. It may be easiest to imagine that the chance move determines the type of individual that player I is. The game will now be played several times, between the same players. After each trial, each player is told what the other player has done; player II is not, however, told what the payoff has been until after the final trial (since this would tell him what chance had chosen).

It is clear, now, that, in the first trial, player I has the same advantage as before. The difficulty, however, from I's point of view, is that, if I always plays the same way (i.e., mimicking chance), then II, who knows what I has chosen, will understand the situation; i.e., if I repeatedly plays A, it will become obvious that this is what chance has chosen. Thus, it should be expected that, for a game played n times, the value should be less than $n/2$, but more than $n/4$.

We will return to this game in Section 6.2, below. For the moment, we note that player I's problem here is essentially one of determining the extent to which his private information should be used. Clearly this information brings some advantage, but use of the information tends to reveal it. The problem, for player I, is to determine whether this information is more valuable to I or to II. Player II's problem, a more complicated one, is that of eliciting information from I's actions and then using this information.

To see this as simply as possible, we may consider the following *signaling* model: s $m \times n$ matrices, $A^1, \ldots, A^s$, are given, as is a probability vector $(p_1, \ldots, p_s)$. This information is given to both players. Then:

1. Chance chooses one of these matrices, A^h, according to the probability distribution $(p_1, \ldots, p_s)$.
2. Player I, knowing the value of h, chooses a row, $i(1 \leq i \leq m)$. (This is the *signal*.)
3. Player II, knowing the value of i but in ignorance of h, chooses a column, $j(1 \leq j \leq n)$.
4. There is then a payoff a_{ij}^h from II to I.

This model reduces the problem to its simplest form. I acts with private information; his action may (or may not) reveal this information to II, who can then base his own action on whatever information I's action has revealed. Of course I is free to try to deceive II, who uses the signal at his own risk.

6.1.3 Example. Consider the two matrices

$$A^1 = \begin{pmatrix} 4 & 1 \\ 3 & 0 \end{pmatrix} \quad A^2 = \begin{pmatrix} 1 & 3 \\ 2 & 5 \end{pmatrix}$$

which chance will choose with probability distribution (0.5, 0.5).

The situation is extremely clear here. In case $h = 1$, row 1 dominates row 2, while column 2 dominates column 1; if $h = 2$, then row 2 dominates row 1, while column 1 dominates column 2. Thus, player I, knowing the value of h, would seem to gain an advantage by playing $i = h$. This analysis, however, fails to take into account the fact that player II will act after receiving I's signal. Then II, by using $j = 3 - i$, would gain a definite advantage.

To see this, we note that, if I uses $i = h$, while II uses $j = 3 - i$, then with probability 0.5 the payoff will be $a_{12}^1 = 1$, and with probability 0.5 it will be $a_{21}^2 = 2$. Thus the expected payoff will be 1.5. On the other hand, if I always uses $i = 2$, then the expected payoff will be 2.5 against whatever strategy II may use. Thus the strategy $i = h$ is not optimal for I. The point is that the private information may be valuable to I, but it is even more valuable to II, and I is unable to take advantage of the information without revealing it in time for II to use it.

To compute optimal strategies for a game such as this, we note that player I has m^s pure strategies, since he can choose any of m alternatives in each of s information sets. Similarly, II has n^m pure strategies. Thus, even for as little as $m = n = s = 3$, each player would have 27 pure strategies, so that solution by means of mixed strategies, while possible, becomes prohibitively long even for small games.

Instead of using mixed strategies, however, we might use behavioral strategies. These will be vectors (x_i^h) and (y_j^i), with sm and mn components respectively, where x_i^h is the probability that I will choose row i, given that chance has chosen h, while y_j^i is the probability that II will choose column j, given that I has chosen i.

The expected payoff, in terms of behavioral strategies, is now

$$E(x, y) = \sum_h \sum_i \sum_j p_h x_i^h y_j^i a_{ij}^h$$

Now, I wishes to choose x so as to maximize this expression, while II wishes to choose y so as to minimize it, subject to the constraints that, for each $h = 1, \ldots, s$, the m-tuple $(x_1^h, \ldots, x_m^h)$ is a probability vector, while, for each $i = 1, \ldots, m$, the n-tuple $(y_1^i, \ldots, y_n^i)$ is a probability vector.

Let us now consider player I's problem. To express this as a linear program, we note that, whatever x player I chooses, he must fear that, for each i, II will choose that j which minimizes the sum

$$\sum_h p_h a_{ij}^h x_i^h$$

If we let u_i be the minimum (with respect to j) of this sum, then a gain-floor for I will be the sum of these u_i:

$$\sum_i u_i = \min_y \sum_j \sum_h \sum_i p_h x_i^h a_{ij}^h y_j^i$$

We can now express this as the program

$$\text{Maximize} \quad \lambda = \sum_i u_i \tag{6.1}$$

subject to the constraints

$$-\sum_h p_h a_{ij}^h x_i^h + u_i \leq 0 \quad \text{for each } i \text{ and each } j \tag{6.2}$$

$$\sum_i x_i^h = 1 \text{ for each } h \tag{6.3}$$

$$x_i^h \geq 0 \quad \text{for each } h \text{ and each } i \tag{6.4}$$

We consider, next, player II's problem. His fear is that, whatever y he chooses, player I will have chosen, for each h, that i which maximizes the expression

$$\sum_j p_h a_{ij}^h y_j^i$$

Let w_h be the maximum (with respect to i) of this expression. A loss-ceiling for II will now be the sum of these w_h:

$$\sum_h w_h = \max_x \sum_i \sum_j \sum_h p_h a_{ij}^h y_j^i x_i^h$$

We can therefore express the second player's problem as the linear program

$$\text{Minimize} \quad \mu = \sum_h w_h \tag{6.5}$$

subject to the constraints

$$-\sum_j p_h a_{ij}^h y_j^i + w_h \geq 0 \text{ for each } h \text{ and each } i \tag{6.6}$$

$$\sum_j y_j^i = 1 \text{ for each } i \tag{6.7}$$

$$y_j^i \geq 0 \quad \text{for each } i \text{ and each } j \tag{6.8}$$

It can now be seen that the two linear programs, (6.1–6.4) and (6.5–6.8), are dual to each other. Since they are, clearly, both feasible, it follows that the maximum of λ is equal to the minimum of μ, and this is clearly the value of the game.

6.1.4 Example. We can now return to the solution of example 6.1.3. The maximizing problem here is

$$\text{Maximize} \quad \lambda = u_1 + u_2$$

subject to

$$-2x_1^1 - 0.5x_1^2 + u_1 \le 0$$
$$-0.5x_1^1 - 1.5x_1^2 + u_1 \le 0$$
$$-1.5x_2^1 - x_2^2 + u_2 \le 0$$
$$-2.5x_2^2 + u_2 \le 0$$
$$x_1^1 + x_2^1 = 1$$
$$x_1^2 + x_2^2 = 1$$
$$x_i^h \ge 0$$

The minimizing problem is to

$$\text{Minimize} \quad \mu = w_1 + w_2$$

subject to

$$-2y_1^1 - 0.5y_2^1 + w_1 \ge 0$$
$$-0.5y_1^1 - 1.5y_2^1 + w_2 \ge 0$$
$$-1.5y_1^2 + w_1 \ge 0$$
$$-y_1^2 - 2.5y_2^2 + w_2 \ge 0$$
$$y_1^1 + y_2^1 = 1$$
$$y_1^2 + y_2^2 = 1$$
$$y_j^i \ge 0$$

It is not too difficult to see that these programs have the common value 2.5, obtained by setting

$$x_1^1 = 0, \quad x_2^1 = 1, \quad x_1^2 = 0, \quad x_2^2 = 1$$

$$u_1 = 0, \quad u_2 = 2.5$$

and

$$0 \le y_1^1 \le 0.17, \quad y_2^1 = 1 - y_1^1, \quad y_1^2 = 0.5, \quad y_2^2 = 0.5$$

$$w_1 = 0.75, \quad w_2 = 1.75$$

respectively. We note that I's optimal strategy calls for him to use row 2 at all times; i.e., he effectively ignores his private information, as the use of this information would disclose it to II — who can make better use of it than I. The optimal strategy for II is somewhat more complicated. Essentially, it seems to say that (since I's use of row 2 means nothing) II should use both columns with equal probability against row 2. In response to row 1, however, II will use a defensive strategy which almost always leads him to respond with column 2. (Of course optimal play for I means that row 1 will never be used.)

6.1.5 Example. Once again, let us assume that chance chooses, with probabilities $\frac{1}{2}$ and $\frac{1}{2}$, one of the two matrices

$$A^1 = \begin{pmatrix} 2 & 3 \\ 1 & 5 \end{pmatrix} \quad A^2 = \begin{pmatrix} 3 & 1 \\ 2 & 5 \end{pmatrix}$$

We have here the two programs

$$\text{Maximize} \quad \lambda = u_1 + u_2$$

subject to

$$x_1^1 + x_2^1 = 1$$
$$x_1^2 + x_2^2 = 1$$
$$u_1 - x_1^1 - 1.5x_1^2 \leq 0$$
$$u_1 - 1.5x_1^1 - 0.5x_1^2 \leq 0$$
$$u_2 - 0.5x_2^1 - x_2^2 \leq 0$$
$$u_2 - 2.5x_2^1 - 2.5x_2^2 \leq 0$$
$$x_i^h \geq 0$$

and

$$\text{Minimize} \quad \mu = w_1 + w_2$$

subject to

$$y_1^1 + y_2^1 = 1$$
$$y_1^2 + y_2^2 = 1$$
$$w_1 - y_1^1 - 1.5y_2^1 \geq 0$$
$$w_1 - 0.5y_1^2 - 2.5y_2^2 \geq 0$$
$$w_2 - 1.5y_1^1 - 0.5y_2^1 \geq 0$$
$$w_2 - y_1^2 - 2.5y_2^2 \geq 0$$
$$y_j^i \geq 0$$

We can, of course, proceed to solve this program directly. We might profitably, however, attempt to simplify it first.

In the first place, we note that $a_{21}^1 < a_{22}^1$ and $a_{21}^2 < a_{22}^2$. Thus, if I plays $i = 2$, optimal play requires that II play $j = 2$, and we have $y_1^2 = 0$, $y_2^2 = 1$.

Next, knowing this much, we see that for $h = 1$, $i = 1$ would lead to a payoff of at least 2, whereas $h = 1$, $i = 2$ would cause $j = 2$ and a payoff of 1. Thus it can never be good for I to choose $i = 2$ when $h = 1$, and so $x_1^1 = 1$, $x_2^1 = 0$.

Knowing this much, we have in effect (through row and column domination) reduced this to a game in which each player has only 2 (rather than 4) pure strategies. This game will have the matrix

$$\begin{pmatrix} 2.5 & 2 \\ 2 & 2.5 \end{pmatrix}$$

which clearly has value 2.25, obtained by assigning equal probabilities to both rows as well as to both columns. This gives the solution to the original game, which is

$$x_1^1 = 1, \quad x_2^1 = 0, \quad x_1^2 = 0.5, \quad x_2^2 = 0.5$$

$$y_1^1 = 0.5, \quad y_2^1 = 0.5, \quad y_1^2 = 1, \quad y_2^2 = 0$$

and it may be seen that these, together with

$$u_1 = 1.75, \quad u_2 = 0.5$$

$$w_1 = 1.25, \quad w_2 = 1$$

give the solution to the two linear programs.

6.2. The Multistage Model

We return now to the more complicated, multistage problem. A game is to be repeated many times. Player I knows the true state of nature; his actions will be made known to II. At each stage, player II, whose information is incomplete, tries to learn from I's previous actions.

6.2.1 Example. Let us consider once again the game of Example 6.1.2. Chance chooses one of the two matrices, A^1 and A^2, with equal probabilities. The game is now to be repeated n times.

For $n = 2$, it is reasonably easy to solve the game directly. To simplify the solution, we note that, at the second trial, player I should mimic chance — there is no danger in giving away information at that time. Thus I has effectively only one choice to make, at the first trial. Since this can depend on what chance has chosen, I has 4 pure strategies, labeled AA, AB, BA, and BB as before.

As far as player II is concerned, we note that he can gain by basing his choice at trial 2 on I's choice at trial 1. On the other hand, there is no point in making his choice at trial 2 depend on his own choice at the first trial: we have already seen that I's choice at trial 2 is independent of II's choice at trial 1. Thus II has, in effect,

8 strategies, called AAA, AAB, etc., where xyz means II chooses x at trial 1, and then chooses y or z at trial 2 depending on whether I chose A or B, respectively, at trial 1.

We can now form the payoff matrix, which turns out to be

$$
\begin{array}{c}
 \\
AA \\
AB \\
BA \\
BB
\end{array}
\begin{array}{cccccccc}
AAA & AAB & ABA & ABB & BAA & BAB & BBA & BBB \\
\left(\begin{array}{c}1\end{array}\right. & 1 & 1 & 1 & 0.5 & 0.5 & 0.5 & \left.\begin{array}{c}0.5\end{array}\right) \\
1 & 1.5 & 0.5 & 1 & 1 & 1.5 & 0.5 & 1 \\
0.5 & 0 & 1 & 0.5 & 0.5 & 0 & 1 & 0.5 \\
0.5 & 0.5 & 0.5 & 0.5 & 1 & 1 & 1 & 1
\end{array}
$$

It is not too difficult to solve this matrix game, which has value 0.75. Each player has many optimal strategies, among them $(0,0.5,0.5,0)$ for player I, and $(0,0,0.5,0,0,0,0,0.5)$ for II. It should be noted that any optimal strategy for II has to use either ABA or BBA (meaning that he plays, at stage 2, opposite to what I played at stage 1) with positive probability, as these are the only two columns which have entries smaller than 1 in the second row.

In theory, it is possible to compute the values $v_3, v_4, \ldots$ of the game with $3, 4, \ldots$ repetitions. In practice, the fact that it is to player II's advantage to base his strategy on what player I has done at previous stages means that this computation will be extremely complicated even for as few as four or five repetitions. Instead, we will try to look for the long-run average value

$$
w^* = \lim_{n \to \infty} \frac{1}{n} v_n
$$

which tells us what will "eventually" happen.

In the first place, we note that, if player I tries to make use of his information by mimicking chance at each trial of the game (or at least with substantially more than 0.5 probability), then player II will eventually notice this and, by using the opposite choice at each trial, decrease I's winnings to 0 in the long run. On the other hand, we recall that, if player I is not given information as to the outcome of the chance move, his optimal strategy is simply to play $(0.5, 0.5)$ at each trial, obtaining the value 0.25. Clearly, the information which player I has cannot hurt him (so long as II does not, also, have this information), and we conclude that I should be able to obtain at least a long-term average of 0.25 by choosing a long-term strategy which does not depend on the information obtained. This will keep II from using I's action to deduce something about the information I has. Thus, we must have

$$
w^* \geq 0.25
$$

Now consider player II. The point here is that II must make use of any information which I gives by using one of the two alternatives more frequently than the other. At the same time, II cannot be too predictable: except in extreme cases, II must still use some randomization. The idea is to modify the randomization periodically, on the basis of what I has done up to that point. More specifically, we can let x_1^k and x_2^k be the number of times I has used alternatives A and B, respectively, in the first k trials of the game. (Note $x_1^k + x_2^k$ must then be equal to k.) Then, on the $(k + 1)$st trial, II should choose A or B with probability x_2^k/k and x_1^k/k, respectively. Then, in the long run, the average payoff will be smaller than $0.25 + \delta$, for any positive number δ. Thus, it will follow that

$$w^* \leq 0.25$$

Since we had already proved the opposite inequality, it follows that this game has the long-run average value $w^* = 0.25$. We conclude that I's additional information, while it will definitely give him a short-run advantage, will lose its importance in the long run, since the value v^* is the same as that in the original (one-stage) game. (Of course, the short run may be quite lucrative; I may win large sums of money before II manages to obtain the necessary information.)

6.2.2 Example. Let us consider the same situation but with payoffs which are the negatives of those in Example 6.2.1. Thus, there is probability 0.5 that the paying alternative is either A or B. If both players choose the paying alternative, then the payoff is -1; otherwise the payoff is 0. The game matrix now becomes

$$
\begin{array}{cc}
 & \begin{array}{cc} A & B \end{array} \\
\begin{array}{c} A \\ B \end{array} & \begin{pmatrix} -0.5 & 0 \\ 0 & -0.5 \end{pmatrix}
\end{array}
$$

and it is clear that this game has optimal strategies $x^* - y^* = (0.5, 0.5)$ and value $v = -0.25$.

Once again, we shall suppose that player I knows the result of the chance move; player II does not know. The normal form now becomes

$$
\begin{array}{cc}
 & \begin{array}{cc} A & B \end{array} \\
\begin{array}{c} AA \\ AB \\ BA \\ BB \end{array} & \begin{pmatrix} -0.5 & 0 \\ -0.5 & -0.5 \\ 0 & 0 \\ 0 & -0.5 \end{pmatrix}
\end{array}
$$

In this case we see that the third row, BA, dominates all others. Hence the game has value $v = 0$, with optimal pure strategy BA for I. It does not matter what II does.

So far, the analysis does not differ much from that for the previous example. Suppose, however, that, as before, the game is to be repeated many times, with the chance move made, once for all times, at the beginning of the first trial. It turns out that, in this case, player I can maintain the initial advantage for all time: even if I plays, say, alternative B every time, thus making it clear that chance had chosen A (or vice versa), there is nothing that II can do to take advantage of the information thus obtained. We conclude that, in this case, the long-run value is $v^* = 0$.

The reader will note that there is a definite difference between the two examples. In 6.2.1, the additional information available to player I gives him an initial advantage but, in the long run, use of the information will redound to the opposing player's advantage, so that the long-run value is no greater than the value of the game without this information. In 6.2.2, the additional information gives I a permanent advantage, so that the long-run value is definitely greater than the value without information. It is this difference in behavior that interests us here.

Let us once again consider the game of Example 6.2.1, with the sole modification that chance has probabilities p and $1 - p$ (rather than $\frac{1}{2}$ and $\frac{1}{2}$) of choosing A and B respectively. If so, the game with no information will have the normal form

$$\Gamma(p) = \begin{pmatrix} p & 0 \\ 0 & 1 - p \end{pmatrix}$$

We see that, for $0 \le p \le 1$, this game has value

$$v_1(p) = p(1 - p)$$

with optimal strategies $x^* = y^* = (1 - p, \ p)$.

In similar fashion, the game of Example 6.2.2, with chance probabilities p and $1 - p$, will have the normal form

$$\Gamma(p) = \begin{pmatrix} -p & 0 \\ 0 & p - 1 \end{pmatrix}$$

and this game has value

$$v_2(p) = -p(1 - p)$$

with the same optimal strategies $(1 - p, \ p)$.

The reader will note that the function v_1 is concave whereas v_2 is convex. It turns out that this is the principal cause for the difference in behavior in the two games.

To see why this is so, we note that, in either of these games, player I has (at least) two alternative approaches. On the one hand, he can ignore whatever information he is given about chance's move; he does this by using the strategy $(1 - p, \ p)$, which is optimal in $\Gamma(p)$. Since he is acting in this way, it is impossible for II to deduce what chance has chosen, and I will, in expectation, receive the amount $v(p)$. Alternatively, he can make full use of the information he receives about chance's move. This will, in the long run, allow II to discover what the chance move was, and thus the game will become either $\Gamma(1)$ (if chance chose A) or $\Gamma(0)$ (if chance chose B). The expected payoff is then $pv(1) + (1 - p)v(0)$. Thus player I can, in effect, choose between the two quantities

$$v(p) \quad \text{and} \quad pv(1) + (1 - p)v(0)$$

The first of these quantities is larger if *v* is concave; the second is larger if *v* is convex. Thus, *it is to I's advantage to ignore the information if v(p) is a concave function of p; he should use the information if v is convex.*

More generally, we begin by defining the concavification of a function.

6.2.3 Definition. Let *f* be a continuous function defined over a compact convex set *A*. The *concavification* of *f*, notation cav *f*, is the minimal concave function, *g*, such that $g \geq f$ over the set *A*.

It is not obvious that such a function exists. However, we know (see appendix) that a function is concave if and only if the area under the graph of the function is convex. If *f* is not concave, we look for the convex hull of the area under the graph of *f*. This convex hull is itself the area under the graph of a new function, and this is the desired function, cav *f*. As an example, in the two games discussed above, we obtained the two payoff functions $v_1(p) = p(1 - p)$ and $v_2(p) = -p(1 - p)$, defined over the closed interval [0,1]. We noted above that v_1 is concave; it follows that v_1 is its own concavification, i.e., cav $v_1 = v_1$.

On the other hand, v_2 is not concave (it is in fact convex). The concavification is, here, cav $v_2 = 1$, i.e., the constant function 1.

6.2.4 Example. On a less trivial note, consider the function $f(p) = p^3$ over the interval $[-1, 1]$. In Figure 6.1, we see that the function curves down (i.e., the second derivative is negative, so the function is concave) from -1 to 0; from 0 to 1, however,

Figure 6.1.

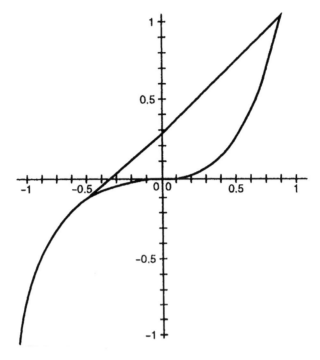

Figure 6.2.

the function curves up. The concavification is obtained by drawing a line from point (1,1) which will be tangent to the curve in the lower part of the graph, so as to "fill in" the convexity in the function. (See Figure 6.2.) Some elementary calculus show that this is the line from $(-0.5, -0.125)$ to $(1,1)$. This line has the equation $y = 0.75p + 0.25$, as is readily verified. The desired function is, then

$$\text{cav} f(p) = \begin{cases} p^3 & \text{if} \quad -1 \leq p \leq -0.5 \\ 0.75p + 0.25 & \text{if} \quad -0.5 \leq p \leq 1 \end{cases}$$

The reader should check that this is indeed a concave function; the diagram "shows" why it is minimal.

Let A and B, now, be two $m \times n$ matrices. We shall use $\Gamma(p)$ to represent the game with matrix $pA + (1 - p)B$. We use $\Omega(p)$ to represent the multistage game in which chance chooses either A or B with probability p and $1 - p$, respectively; in this latter case, player I is told of chance's decision but II is not told. As before, it is assumed that II is told of I's choice at each stage. Let $v(p)$ be the value of the matrix game $\Gamma(p)$, while $w^*(p)$ is the long-run average value of $\Omega(p)$.

We will prove that, in general

$$w^*(p) = \text{cav } v(p)$$

To do this, we must exhibit strategies, for each player, which will guarantee this value. More precisely, we shall exhibit an optimal strategy for player I, and δ-optimal strategies for player II. The development is due to Aumann and Maschler [6.2] and later to Kohlberg [6.5].

6.2.5 Lemma. Player I has a strategy which will obtain a long-term expectation equal to cav $v(p)$ or greater.

Proof. Let cav $v(p) = u$. We distinguish three cases:

1. $u = v(p)$
2. $u = pv(1) + (1 - p)v(0)$
3. $u = rv(q_1) + (1 - r)v(q_2)$, where $rq_1 + (1 - r)q_2 = p$, and q_1, q_2 are not 0 and 1 (in either order).

In case 1, $u = v(p)$. In this case, player I effectively ignores his information, playing whatever mixed strategy x is optimal in the game $\Gamma(p)$. This will, of course, give at least the expectation $v(p)$.

In case 2, player I uses his information, playing whatever mixed strategy is best for either A or B (whichever chance has chosen). The conditional expectation for the payoff is then either at least $v(1)$ (if chance chose A) or at least $v(0)$ (if chance chose B) and therefore the absolute expectation is at least $pv(1) + (1 - p)v(0)$.

In case 3, the situation is somewhat more complicated. Let us suppose, however, that the chance move was actually a two-step move:

(a) First, chance chooses C or D, with probability r or $1 - r$ respectively.
(b) Second, chance chooses A or B, with probabilities, conditional on C or D, given by

$$\text{Prob}(A|C) = q_1; \quad \text{Prob}(B|C) = 1 - q_1$$
$$\text{Prob}(A|D) = q_2; \quad \text{Prob}(B|D) = 1 - q_2$$

This is clearly equivalent to the single choice by chance, giving A with probability $rq_1 + (1 - r)q_2 = p$. We will, however, modify the situation by assuming that the two players are told of the outcome of the first step (C or D) but not that of the second step (A or B). In this case, I can use the optimal strategies for game $\Gamma(q_1)$ if C was chosen, or for game $\Gamma(q_2)$ if D was chosen. The conditionally expected payoff would then be at least $v(q_1)$ for C, and $v(q_2)$ for D. Now, since C and D have probabilities r and $1 - r$, it follows that the expected payoff will be at least $rv(q_1) + (1 - r)v(q_2) = u$, as desired.

Of course, in actuality, chance normally does not act in this way, with an intermediate "signal" C or D prior to choosing A or B. However, player I, knowing A or B, can in effect "create" such a signal. In fact, Bayes' theorem allows us to obtain the conditional probabilities

$$\text{Prob}(C|A) = \frac{rq_1}{rq_1 + (1-r)q_2}; \qquad \text{Prob}(D|A) = 1 - \text{Prob}(C|A) \qquad (6.9)$$

$$\text{Prob}(C|B) = \frac{r(1-q_1)}{r(1-q_1) + (1-r)(1-q_2)}$$

$$\text{Prob}(D|B) = 1 - \text{Prob}(C|B) \qquad (6.10)$$

Now, player I, knowing A or B, can carry out an experiment which will give outcome C or D with these conditional probabilities. Then, if this experiment has outcome C, player I will act *as if* he only knew C, i.e., he will use whatever strategy is optimal in $\Gamma(q_1)$, and continue using this mixed strategy for as long as the game is repeated. If, on the other hand, the experiment has outcome D, player I will act as though he only knew D, and use whatever strategy is optimal in $\Gamma(q_2)$ for as long as the game is repeated. The point of this is that player II will be unable to learn, from I's play, whether A or B has been chosen: at most, an empirical observation of I's strategies will enable II to determine whether I as playing *as in* $\Gamma(q_1)$ or *as in* $\Gamma(q_2)$. This gives II some information as to whether A or B has been chosen, but no more than I wants II to have.

Conditional on case C, the expected payoff will be at least equal to $v(q_1)$; conditional on case D, the expected payoff will be at least $v(q_2)$. Thus the absolute expectation will be at least $rv(q_1) + (1-r)v(q_2)$, or cav $v(p)$.

In any one of the three cases, then, we find that player I can obtain an expectation of at least cav $v(p)$. Thus, in any case

$$w^*(p) \geq \text{cav } v(p) \qquad (6.11)$$

More generally, we might assume that chance determines, prior to play of the game, what *type* of individual player I is; player I's type determines the payoff matrix. There are s possible types, with probabilities $p_1, p_2, \ldots, p_s$ of being chosen by chance; and, corresponding to these, there are s $m \times n$ payoff matrices, $A^1 = (a_{ij}^1), \ldots, A^s = (a_{ij}^s)$.

Let now $\Gamma(p)$ be the game with matrix $p_1 A^1 + p_2 A^2 + \cdots + p_s A^s$, and, as before, $\Omega(p)$ be the game in which chance chooses one of the matrices A^h with probabilities p_h, and player I is informed about this. Let $v(p)$ be the value of $\Gamma(p)$, and let $w^*(p)$ the long-run average value of $\Omega(p)$. Once again, we claim that

$$w^*(p) \geq \text{cav } v(p) \qquad (6.12)$$

To see this, we note that, if $u = \text{cav } v(p)$, then the point $(p; u)$ is a convex linear combination of at most s points $(q^k; v(q^k))$; i.e., there exist nonnegative numbers r_k whose sum is 1, such that

$$p = \sum r_k q^k \qquad (6.13)$$

$$u = \sum r_k v(q^k) \tag{6.14}$$

Once again we interpret chance's choice as a two-step process which, in step one, chooses one of $Q^1,\ldots,Q^s$ with probabilities $r_1,\ldots,r_s$, respectively; then, having chosen Q^k, it chooses one of $A^1,\ldots,A^s$ with probability $q_1^k,\ldots,q_s^k$, respectively. Now, *if* both players were advised of the outcome of step one only, they would in effect be playing the game $\Gamma(q^k)$, and they could both play optimally in this game, thus obtaining the expectation $v(q^k)$. Since this has probability r_k, the absolute expectation is $u = \text{cav } v(p)$.

In fact, player I is advised of the outcome of the two-step process, the matrix A^h chosen by chance. Player I can then, by Bayes' formula, obtain a probability

$$\text{Prob}\{Q^k | A^h\} = \frac{r_k q_h^k}{\sum r_l q_h^l} \tag{6.15}$$

Player I now uses a randomization mechanism which will choose according to these probabilities. If the mechanism chooses Q^k, I will then act as though he had received only the signal Q^k, rather than the complete information A^h; i.e., he will use whatever mixed strategy would be optimal in game $\Gamma(q^k)$. Player II can gain no more information from I's actions than that I is acting as in Q^k, i.e., that the several matrices A^h have probability q_h^k of being the "true" game. The best that II can do, then, is to play optimally in game $\Gamma(q^k)$. The expected payoff, as discussed above, will then be u.

6.2.6 Example. Consider the game with two matrices

$$A^1 = \begin{pmatrix} 3 & 1 \\ 4 & 2 \end{pmatrix} \quad A^2 = \begin{pmatrix} 2 & 6 \\ 1 & 4 \end{pmatrix}$$

and prior probabilities $p = (p, 1-p)$. The game $\Gamma(p)$ is then given by

$$pA^1 + (1-p)A^2 = \begin{pmatrix} 2+p & 6-5p \\ 1+3p & 4-2p \end{pmatrix}$$

It is not too difficult to see that this game $\Gamma(p)$:

(a) has a saddle-point at (row 1, column 1) for $0 \le p \le 0.5$
(b) has a saddle-point at (row 2, column 1) for $0.5 \le p \le 0.6$
(c) has a pair of mixed optimal strategies for $.6 \le p \le 0.667$
(d) has a saddle-point at (row 2, column 2) for $0.667 \le p \le 1$

Thus the function $v(p) = \text{val } \Gamma(p)$ is given by

$$v(p) = \begin{cases} p+2 & \text{for} \quad 0 \le p \le 0.5 \\ 3p+1 & \text{for} \quad 0.5 \le p \le 0.6 \\ \dfrac{13p^2 - 13p + 2}{p-1} & \text{for} \quad 0.6 \le p \le 0.667 \\ 4-2p & \text{for} \quad 0.667 \le p \le 1 \end{cases}$$

The graph of this function consists, then, of two line segments with slopes 1 and 3 respectively, followed by a concave curve (it may be checked that the second derivative of the curved part is always negative), followed by another line segment with slope -2. It reaches a maximum of 2.802 at $p = 0.6078$; the curved part, moreover, has an initial slope of 0.5 at $p = 0.6$ and a terminal slope of -5 at $p = 0.667$. (See Figure 6.3.) It may be seen that this function is not concave (a) because the slope of the second line segment is greater than that of the first, (b) because the slope of the final line segment is greater than the terminal slope of the curved part.

To obtain the concavification of this function, we must "fill in" the area in the angle between the first two line segments, by taking the segment from (0,2) to (0.6,2.8), and also the area between the curve and the final line segment by taking the line segment passing through (1,2) which is tangent to the curve. (See Figure 6.4.) Some elementary calculus show that the tangency is at $p = 0.636$, $v = 2.773$. Thus the function $u(p) = \text{cav } v(p)$ is given by

$$u(p) = \begin{cases} 2 + 1.333 & \text{for} \quad 0 \le p \le 0.6 \\ \dfrac{13p^2 - 13p + 2}{p-1} & \text{for} \quad 0.6 \le p \le 0.636 \\ 4.125 - 2.125p & \text{for} \quad 0.636 \le p \le 1 \end{cases}$$

Figure 6.3.

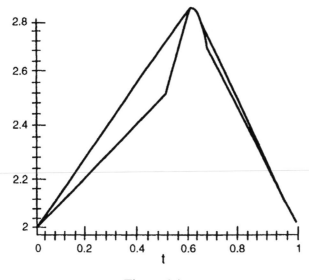

Figure 6.4.

Consider now the game $\Omega(p^\circ)$, where $p^\circ = (0.5, 0.5)$. We have $u(0.5) = 2.667$, which is not on the function $v(p)$. Rather, the point $(0.5, 2.667)$ can be expressed as

$$0.167(0, 2) + 0.833(0.6, 2.8) = (0.5, 2.667)$$

We note that both $(0, 2)$ and $(0.6, 2.8)$ are on the function $v(p)$.

Player I's strategy, now, consists in treating the chance move as though it were a two-step process. In step 1, chance chooses either Q^1 or Q^2 with probabilities 0.167 and 0.833 respectively. In step 2, chance chooses either A^1 or A^2 with conditional probabilities

$$\text{Prob}\{A^1|Q^1\} = 0, \qquad \text{Prob}\{A^2|Q^1\} = 1$$
$$\text{Prob}\{A^1|Q^2\} = 0.6, \quad \text{Prob}\{A^2|Q^2\} = 0.4$$

(It may be seen that this gives the desired probability 0.5 for each of the two types.) Now, I is actually told the outcome of the second step. He will act, however, as though he only knew the outcome of the first step. With this information, he will then choose optimal strategies for either game $\Gamma(0, 1)$ or $\Gamma(0.6, 0.4)$.

Of course, this two-step process is fictitious: Q^1 and Q^2 do not really exist, except in player I's mind. In reality, I is told the outcome of step 2 (either A^1 or A^2). Player I then applies Bayes' theorem to obtain the probabilities

$$\text{Prob}\{Q^1|A^1\} = 0, \qquad \text{Prob}\{Q^2|A^1\} = 1$$
$$\text{Prob}\{Q^1|A^2\} = 0.333, \quad \text{Prob}\{Q^2|A^2\} = 0.667$$

and, on the basis of these probabilities, chooses Q^1 or Q^2.

Player I's actual strategy, then, is as follows:

(a) If chance chooses A^1, then I will make believe he was told Q^2.
(b) If chance chooses A^2, then I will use a randomization mechanism which, with probability 0.333, will tell him to make believe he was told Q^1, and, with probability 0.667, will tell him to make believe that he was told Q^2.
(c) For Q^1, I will use the strategy which is optimal in the game $\Gamma(0,1)$, namely the first row. He will continue to use this through all trials of the game.
(d) For Q^2, I will use the strategy which is optimal in $\Gamma(0.6, 0.4)$, namely the second row. He will continue to use this through all trials of the game.

The reader should see what I's objective action will be. If chance chooses A^1, then I, playing as in $\Gamma(0.6, 0.4)$, will use row 2. If, instead, chance chooses A^2, then with probability 0.667, I will use row 2; with probability 0.333, he will play as in $\Gamma(0,1)$ and use row 1.

Against this strategy, we can also see what II's best reply would be. If II sees I use the first row, then II can deduce (assuming I is following this strategy) that chance has chosen A^2. Then II will choose column 1, with a consequent payoff of 2. If, on the other hand, II sees that I uses row 2, then II concludes that I is playing in $\Gamma(0.6, 0.4)$. In this game, no matter what II plays, the expected payoff is 2.8. Thus, with probability 0.167, the expected payoff will be 2; with probability 0.833, it will be 2.8. The marginal expectation is then 2.667 as per our theory.

We have therefore seen that player I can make use of his information so as to obtain an expected payoff which is at least equal to cav $v(p)$ i.e.

$$w^*(p) \geq \operatorname{cav} v(p)$$

In fact, however, we wish to prove that these two are equal. We must therefore exhibit a strategy for player II which will keep I from obtaining a higher long-run average than this.

We first note that we can, without loss of generality, assume that the *a priori* probabilities of the several types are all positive. In fact, if one of the types has zero probability, there is no point in including it in the analysis since it corresponds to something which both players know to be impossible. We next prove an auxiliary lemma.

6.2.7 Lemma. Let $(p_1^\circ, \ldots, p_s^\circ)$, the *a priori* probabilities of the several types, all be positive, and let $u = \operatorname{cav} v(p^\circ)$. Then there exist quantities $\theta_1, \ldots, \theta_s$ satisfying

$$p_1^\circ \theta_1 + \cdots + p_s^\circ \theta_s = u \tag{6.16}$$

and, for all probability vectors p

$$p_1 \theta_1 + \cdots + p_s \theta_s \geq v(p) \tag{6.17}$$

Proof. Since cav v is a concave function, we know that the set under the graph, i.e., the set

$$H = \{(p; z) | z \leq \text{cav } v(p)\}$$

is a convex set. By continuity, it is also closed. We will think of H as a closed convex subset of the s-dimensional space P, which consists of all $(p; z)$ satisfying $\Sigma p_h = 1$.

The point $(p^\circ; u)$ is on the boundary of set H, and so we can apply Theorem 2.4.2 in its stronger form (see problem 1 of Chapter 2): there exist numbers $\theta_1, \ldots, \theta_s$, Φ, not all 0, such that

$$p_1^\circ \theta_1 + \cdots + p_s^\circ \theta_s + u\Phi \leq p_1 \theta_1 + \cdots + p_s \theta_s + z\Phi \tag{6.18}$$

for all points $(p; z)$ in H.

Now, we see that Φ cannot be positive, since z can be made arbitrarily small and this would make the right-hand side of (6.18) smaller than the left. Hence $\Phi \leq 0$. On the other hand, if Φ were 0, then (6.18) would take the form

$$p_1^\circ \theta_1 + \cdots + p_s^\circ \theta_s \leq p_1 \theta_1 + \cdots + p_s \theta_s$$

where the θ_h are not all 0. In fact, however, the θ_h cannot even all be equal among themselves. The reason for this is that, if they were all equal to some common value τ, then the equation $\Sigma \theta_h p_h = \tau$ would be an identity in space p, and could not represent a separating hyperplane. We conclude that, in this case, some θ_h would be larger than their mean, and others would be smaller.

Now, since all $p_h^\circ > 0$, we can choose p so that $p_h > p_h^\circ$ whenever θ_h is smaller than average, and $p_h > p_h^\circ$ whenever θ_h is larger than average. This would also make the right-hand side of (6.18) smaller than the left.

We conclude, then, that $\Phi < 0$. The inequality (6.18) will be preserved if we divide by any positive constant (and in particular by $-\Phi$), so that there is no loss of generality in assuming that that $\Phi = -1$. Thus (6.18) now becomes

$$p_1^\circ \theta_1 + \cdots + p_s^\circ \theta_s - u \leq p_1 \theta_1 + \cdots + p_s \theta_s - z$$

We know that $v(p) \leq \text{cav } v(p)$, and so we can set $z = v(p)$. The inequality then becomes

$$p_1^\circ \theta_1 + \cdots + p_s^\circ \theta_s - u \leq p_1 \theta_1 + \cdots + p_s \theta_s - v(p) \tag{6.19}$$

Now, if the value of the left-hand side of (6.19) is 0, the lemma is proved. Suppose its value is different from zero, say δ. If so, we need merely replace each θ_h by

$$\theta_h' = \theta_h - \delta$$

Since both $p°$ and p are probability vectors, this decreases both sides of (6.19) by the amount δ. Thus the inequality is preserved, and the value of the left-hand side is now 0. This proves the lemma.

We note next that, if we subtract the same quantity, c, from each of the elements in any *one* of the matrices A^h, then the optimal strategies for the game are not changed, though the value of both $v(p)$ and cav $v(p)$ will be decreased by $p_h c$. In fact, it may be checked that whatever strategy guaranteed cav $v(p)$ in the original game will still guarantee the new value of this function in the new game; since player II's optimal strategy is based on obtaining information from I's actions, there will be no difference as far as this is concerned.

To simplify matters, now, we will normalize the game by, for each h, subtracting the amount θ_h (given by Lemma 6.2.7) from each entry d_{ij}^h in the matrix A^h. As mentioned above, this does not change the optimal strategies; the values $v(p)$ and cav $v(p)$ will decrease by the total amount $p_1\theta_1 + \cdots + p_s\theta_s$. Thus, for the newly modified games, we will have

$$\text{cav } v(p°) = 0 \tag{6.20}$$

and, for all p

$$v(p) \leq 0 \tag{6.21}$$

We will also assume that, for all h, i, and j, $|d_{ij}^h| \leq 1$. Again, there is no loss of generality in this, since it can be accomplished by dividing all entries d_{ij}^h by the largest of these. (This will not change the optimality of strategies in the game.)
The following lemma is also important:

6.2.8 Lemma. Let K be a nonempty closed subset of the s-dimensional Euclidean space R^s. For a given point x, let $C(x)$ be the point in K closest to x using the usual Euclidean metric. (In case of ties, any such point may be taken.)

Let g^t, $t = 1, 2, \ldots$, be a sequence of points in R^s, with norm not greater than 1, and let q^t be the mean of $g^1, \ldots g^t$:

$$q^t = \frac{1}{t}\sum_{k=1}^{t} g^k$$

Suppose that, for each t such that q^t is not in K, g^{t+1} is on the opposite side (from q^t) of the hyperplane which passes through $C(q^t)$ and is perpendicular to the line segment joining q^t to $C(q^t)$. Then d_t, the distance from q^t to K, is not greater than $c/\sqrt{t}$, where $c = \max(d_1, 2)$.

Proof. We prove this by induction on t; it is clearly true for $t = 1$. Suppose, now, that q^t is in K. In that case, we have

$$q^{t+1} = \frac{tq^t + g^{t+1}}{t+1}$$

We therefore have

$$q^{t+1} - q^t = \frac{g^{t+1} - q^t}{t+1}$$

Now, both g^{t+1} and q^t have norm less than or equal to 1, so the norm of their difference is at most 2. Since q^t belongs to K, the distance from q^{t+1} to K is at most equal to the norm of $q^{t+1} - q^t$, which is not greater than $2/(t+1)$, and this in turn is not more than $2/\sqrt{t+1}$, as desired.

Suppose, on the other hand, that q^t is not in K. By the induction hypothesis, the norm of $q^t - C(q^t)$ is not more than $c/\sqrt{t}$. By the hypothesis of the theorem, the angle $(q^t, C(q^t), g^{t+1})$ is either right or obtuse, which means that

$$(q^t - C(q^t)) \cdot (g^{t+1} - C(q^t)) \le 0$$

where $\cdot$ represents the usual dot product. Also, as in the previous case, the norm of $g^{t+1} - C(q^t)$ is not more than 2. Now

$$q^{t+1} - C(q^t) = \frac{t(q^t - C(q^t)) + g^{t+1} - C(q^t)}{t+1}$$

and so

$$|q^{t+1} - C(q^t)|^2$$
$$= \frac{t^2 |q^t - C(q^t)|^2 + |g^{t+1} - C(q^t)|^2 + 2t(q^t - C(q^t)) \cdot (g^{t+1} - C(q^t))}{(t+1)^2}$$
$$\le \frac{t^2(c^2/t) + 4}{(t+1)^2}$$
$$= \frac{c^2 t + 4}{(t+1)^2}$$
$$\le \frac{c^2}{(t+1)}$$

as desired.

6.2.9 Theorem. With the given normalization, II has a strategy to guarantee $w^* \le 0$.

Proof. We will develop a strategy for II which will keep I's expected winnings down to 0 (or, rather, smaller, as a long-run average, than any preassigned positive number).

For this purpose, II must keep track of what I has done up to a given moment. Suppose, then, that, at time t, player I uses row $i(t)$, while II uses column $j(t)$. Let $g^t = (g^t_1, g^t_2 \ldots, g^t_s)$ the vector with components

$$g^t_h = a^h_i(t)j(t) \tag{6.22}$$

which tell us what the payoff at time t would be if the actual type were h. Finally, let q^T be the mean of the g_t, i.e.

$$q^T = \frac{1}{T}\sum_{t=1}^{T}g^t \tag{6.23}$$

In general, II's problem consists in deducing, from I's play, what I's true type (the chosen matrix) is. The main idea is that, if I's previous play would have been good in type h, then this makes it more likely that the true type is indeed h. Of course I is free to attempt to deceive II by, e.g., bluffing as in poker, but in general this line of reasoning by II should be valid. Then II must choose a strategy which will be good (for II) in that state. Since, however, the state is never entirely known to II, his strategy must be good, given I's observed strategy, in all states. What this means is that — under the normalization given here — II's strategy must be such as to bring the expected payoff below 0, or at least arbitrarily close to 0, for all possible payoff matrices.

In line with the principle just stated, at time T II will assign, to the several possible types, probabilities proportional to the q_h^T. More precisely, let

$$r_h^T = \max\{q_h^T, 0\} \tag{6.24}$$

Then, assuming that, for at least one h, $q_h^T > 0$, define the probabilities p_h^T by

$$p_h^T = \frac{r_h^T}{\sum_1 r_1^T} \tag{6.25}$$

Now, at time $T + 1$, let II choose a one-shot strategy which is optimal in the game $\Gamma(p^T)$. (If, for all h, $q_h^T \leq 0$, it does not matter what II does at time $T + 1$.)

We can now apply Lemma 6.2.8. To do this, we let K be the nonpositive orthant of R^s. Then, for any point q^T which is not in K, we note that $C(q^T)$, the nearest point to q^T in K, is given by

$$C_h(q^T) = \begin{array}{ll} q_h^T & \text{if} \quad q_h^T \leq 0 \\ 0 & \text{if} \quad q_h^T \geq 0 \end{array}$$

It follows that

$$q^T - C(q^T) = r^T$$

and so the plane passing through $C(q^T)$, perpendicular to the line joining this point to q^T, is

$$\sum r_h^T z_h = 0$$

or, since p^t is a multiple of r^T

$$\sum r_h^T z_h = 0 \tag{6.26}$$

Now, y^{T+1} is an optimal strategy for the matrix game $\Gamma(p^T)$. By equation (6.21), we know this has value $v(p^T) \le 0$. Thus z^{T+1}, the expected value of g^{T+1}, the vector of payoffs for the several types, against any choice of a row by player I, will satisfy the inequality

$$\sum p_h^T z_h^{T+1} \le 0$$

and will therefore lie on the side of hyperplane (6.26) opposite to q^T.

Moreover, in the normalization of the game, all payoffs are smaller than 1 in absolute value, and hence z^{T+1}, being a weighted average of payoffs, is also smaller than 1 in absolute value. Thus Lemma 6.2.8 applies.

Now, if we could replace g^{T+1} by its expected value z^{T+1}, the theorem would be proved. In fact, it is now necessary to apply some results from martingale theory as well as the strong law of large numbers. All this will tell us that, with probability 1, q^T will eventually come arbitrarily close to the set K, i.e., player I's average winnings will be smaller than any preassigned $\varepsilon > 0$ for all sufficiently large T. The details are rather technical, and the reader is referred to Blackwell [1956] for these.

The above theorem was valid under the given normalization of the game matrices A^h. More generally, we have the following theorem:

6.2.10 Theorem. For the game $\Omega(p)$, where player I has complete information

$$w^*(p) = \text{cav } v(p)$$

We dispense with the proof, as it follows directly from the previous theorem, after reversing the normalization. Of course the normalization is not necessary in order to compute player II's strategy. If we dispense with the normalization, the thing to remember is that the quantities r_h^T mentioned above are not the maximum of q_h^T and 0; rather, we have

$$r_h^T = \max\{q_h^T - \theta_h, 0\} \tag{6.27}$$

where θ is as given in Lemma 6.2.7. The posterior probabilities p_h^T are then chosen proportional to r_h^T as before.

It may also be useful to notice that, if we let x_{ij}^T be the relative frequency with which the two players have (simultaneously) chosen row i and column j in the first T trials, then

$$q_h^T = \sum_i \sum_j d_{ij}^h x_{ij}^T \tag{6.28}$$

6.2.11 Example. Let us apply this technique to the game of Example 6.2.1.

As seen above, for a probability vector $p = (p_1, p_2)$, the game $\Gamma(p)$ is

$$\Gamma(p) = \begin{pmatrix} p_1 & 0 \\ 0 & p_2 \end{pmatrix}$$

This game has value $v(p) = p_1 p_2$, with optimal strategies $x^* = y^* = (p_2, p_1)$ for both players. As mentioned above, this value function is concave, hence it is its own concavification, and it follows that

$$w^*(p) = p_1 p_2$$

In this case, player I's optimal strategy in the game $\Omega(p)$ is the stationary strategy x^*: in effect, I ignores his private information (because this information is more valuable to II than to I).

Computation of II's optimal strategy depends on the *a priori* probabilities, p°. Let us suppose $p^\circ = (0.5, 0.5)$. In this case, we note that

$$v(p^\circ) = 0.25 = 0.25 p_1^\circ + 0.25 p_2^\circ$$

while, for all probability vectors p

$$v(p) = p_1 p_2 \leq 0.25 = 0.25 p_1 + 0.25 p_2$$

Thus we have here $\theta_1 = \theta_2 = 0.25$.

Now, let x_{ij}^T be the relative frequency with which players I and II have simultaneously chosen the ith row and the jth column in the first T trials of the game. Then, by (6.28)

$$q_1^T = x_{11}^T, \quad q_2^T = x_{22}^T$$

At trial $T + 1$, then, II assigns posterior probabilities p_h^T proportional to

$$r_1^T = \max\{x_{11}^T - 0.25, 0\}$$
$$r_2^T = \max\{x_{22}^T - 0.25, 0\}$$

assuming at least one of these is positive. Thus, at trial $T + 1$, II chooses a column according to an one-shot strategy which is optimal in $\Gamma(p^T)$, namely

$$y_1^{T+1} = p_2^T = r_2^T / (r_1^T + r_2^T)$$
$$y_2^{T+1} = p_1^T = r_1^T / (r_1^T + r_2^T)$$

Of course, if both x_{11}^T and x_{22}^T are smaller than 0.25, then II can choose his column at trial $T + 1$ arbitrarily.

This strategy of II will eventually bring I's average winnings below $0.25 + \varepsilon$ for any positive ε, in either of the two states (types).

6.2.12 Example. Let us consider the game with the same matrices as in the previous example, but with the *a priori* probabilities $p^\circ = (0.667, 0.333)$.

In this case we have $v(p^\circ) = 0.222$. To calculate the θ, we note that

$$v(p^\circ) = 0.111 p_1^\circ + 0.444 p_2^\circ$$

and

$$v(p) \le 0.111p_1 + 0.444p_2$$

for all p. Thus $\theta_1 = 0.111$ and $\theta_2 = 0.444$.

It follows from this that II's strategy will try to bring the payoff down to 0.111 (or below) for type 1, and to 0.444 for type 2. We will have

$$r_1^T = x_{11}^T - 0.111$$

$$r_2^T = x_{22}^T - 0.444$$

and otherwise the analysis continues as in the preceding example.

6.2.13 Example. We consider once again the game of Example 6.2.6. The two matrices (as given above) are

$$A^1 = \begin{pmatrix} 3 & 1 \\ 4 & 2 \end{pmatrix} \quad A^2 = \begin{pmatrix} 2 & 6 \\ 1 & 4 \end{pmatrix}$$

We have already constructed an optimal strategy for player I; we now look at player II's problem.

As seen above, the function $u(p, 1 - p) = \operatorname{cav} v(p, 1 - p)$, is given by

$$u(p, 1 - p) = \begin{cases} 2 + 1.333p & \text{for} \quad 0 \le p \le 0.6 \\ \dfrac{13p^2 - 13p + 2}{p - 1} & \text{for} \quad 0.6 \le p \le 0.636 \\ 4.125 - 2.125p & \text{for} \quad 0.636 \le p \le 1 \end{cases}$$

Near $p^\circ = (0.5, 0.5)$, the function u is given by $u = 2 + 1.333p$, or, setting $p_1 = p$, $p_2 = 1 - p$, $u = 3.333p_1 + 2p_2$. It is easily checked that we have

$$u(p^\circ) = 3.333p_1^\circ + 2p_2^\circ$$

and

$$u(p) \le 3.333p_1 + 2p_2 \quad \text{for all other } p.$$

Thus we have $\theta_1 = 3.333$, $\theta_2 = 2$.

Player II's technique, then, consists in trying to play so as to decrease the average payoff to 3.333 (in case chance has chosen type 1) or 2 (in case chance has chosen type 2). Specifically, II does this by keeping track of q_h^T, the average payoff per trial in either one of the two cases, or, equivalently, by keeping track of x_{ij}^T, the relative frequency with which the two have chosen (row i, column j) over the first T trials. Then, as discussed above, II computes

$$r_1^T = \max\{q_1^T - 3.333, 0\}, \quad r_2^T = \max\{q_2^T - 2, 0\}$$

Now, if both r_1^T and r_2^T are 0 or less, it does not matter what II does at trial $T + 1$. If, however, one or both of the r_h^T is positive, then at trial $T + 1$ II should choose an one-shot strategy optimal in the game $\Gamma(p^T)$, where p^T is a probability vector whose components are proportional to the r_h^T.

As an example, suppose that, over the first 100 trials, the two players have chosen (row 1, column 1) 23 times, (row 1, column 2) 28 times, (row 2, column 1) 31 times, and (row 2, column 2) 18 times. In that case

$$q_1^T = 3(.23) + 1(.28) + 4(.31) + 2(.18) = 2.57$$

$$q_2^T = 2(.23) + 6(.28) + 1(.31) + 4(.18) = 3.17$$

Thus $r_1^T = 0$, $r_2^T = 1.17$. The corresponding (proportional) probability vector is $p^T = (0, 1)$, and so, on the 101st trial, II should play *as in* $\Gamma(0, 1)$, i.e., as in game matrix A^2. In other words, he should choose column 1.

The above analysis is based on the premise that, at each repetition of the game, player II is told what row player I used. In general, of course, the information received may be considerably less (or more) than that. This can make the analysis much more complicated. Readers are referred to Kohlberg [6.5] for details.

Problems

1. Solve the following games according to the one-shot signaling model. Player I acts knowing the results of the chance move; player II acts in knowledge of I's move but in ignorance of the chance move.
   (a) $P(A) = 0.7$, $P(B) = 0.3$

$$A = \begin{pmatrix} 5 & 1 \\ 2 & 4 \end{pmatrix} \quad B = \begin{pmatrix} 2 & 3 \\ 4 & 3 \end{pmatrix}$$

   (b) $P(A) = 0.5$, $P(B) = 0.5$, A and B as in part (a).
   (c) $P(A) = 0.1$, $P(B) = 0.9$, A and B as in part (a).
   (d) $P(A) = 0.7$, $P(B) = 0.3$

$$A = \begin{pmatrix} 4 & 5 & 3 \\ 1 & 2 & 4 \end{pmatrix} \quad B = \begin{pmatrix} 2 & 6 & 2 \\ 4 & 5 & 1 \end{pmatrix}$$

   (e) $P(A) = 0.5$, $P(B) = 0.5$, A and B as in part (d).
   (f) $P(A) = 0.1$, $P(B) = 0.9$, A and B as in part (d).

2. Obtain the concavification of the following functions, over the given domains:
   (a) $f(x) = 2x^3 - 3x^2$ $-1 \leq x \leq 2$
   (b) $g(y) = \sin y$ $0 \leq y \leq 2\pi$
   (c) $h(t) = \sin t$ $0 \leq t \leq 3\pi$
   (d) $F(x, y) = xy$ $0 \leq x \leq 1, 0 \leq y \leq 1$

3. Let $u(x)$ be a continuous function defined over a compact subset, K, of n-dimensional Euclidean space. Show that, if $v(x)$ is defined over K, then a necessary and sufficient condition for v to be the concavification of u is that
   (a) v is concave
   (b) for every $x \in K$, there exist points $y^1, y^2, \ldots, y^n \in K$ (not necessarily all distinct) and non-negative numbers $r_1, \ldots, r_n$, such that
   (1) $\sum r_i = 1$
   (2) $\sum r_i y^i = x$
   (3) $\sum r_i u(y^i) = v(x)$

   (Hint: apply Theorems A.1.8–A.1.10 from the appendix to the "area under the curve.")

4. (a)–(f). Solve the several games in Exercise 1, above, using the repeated games model. (Assume that I knows the outcome of the chance move, and II is told at each repetition what action I took.) Obtain an optimal strategy for I, and describe the optimal strategy for II.

Chapter 7

Utility Theory

7.1. Ordinal Utility

In the preceding six chapters we have worked on the supposition that two players have directly opposite interests. Even when the two players form a closed system, with one losing whatever the other gains, it is not clear that their interests will always be directly opposed. If one player is a rich philanthropist, while the second is poor and deserving, it may be that the first player will prefer to let the second win (up to a certain amount). The fact is that it is not money, but rather the utility represented by the money, which should be considered in such a case. Similarly, the stakes in a game may be something that has no monetary value. Nonmonetary (say, sentimental) reasons may make it very valuable for one player, but practically worthless for another. It is this concept of personal value, or personal utility, which must be studied.

Let us consider, then, the fact that a person will prefer certain events to others, and that in other cases, he will be indifferent as to two events. This gives us the two relations p and i. The domain of these relations will be the set of events. Let us use A, B, C, etc., to denote events.

7.1.1 Definition. Given any two events A and B, we say A p B if A is preferable to B. We say A i B if A ɰ B and B ɰ A.

7.1.2 Utility Axioms. The relations p and i satisfy the following axioms:

1. Given any two events A and B, exactly one of the following must hold:
   (α). A p B;
   (β). B p A;
   (γ). A i B.
2. A i A for all A.
3. If A i B, then B i A.
4. If A i B and B i C, then A i C.
5. If A p B and B p C, then A p C.
6. If A p B and B i C, then A p C.
7. If A i B and B p C, then A p C.

Axiom 1 is also called the *trichotomy law*. Axioms 2, 3, and 4 mean that i is an equivalence relation. Axiom 5 (together with Axiom 1) means that p is an order relation. Finally, Axioms 6 and 7 are sometimes described by saying that p is transitive over i.

Essentially, the Axioms 7.1.2 represent a weak linear ordering of events, from the most desirable to the least desirable. It is usual to say that, if someone prefers the event A to the event B, then the event A has more utility than B, or the utility of A is greater than that of B.

Unfortunately, while this tells us that the utility of A is greater than that of B, it does not tell us how great this difference in utility is. For certain problems, this represents no difficulty (e.g., if it is merely a question of determining which event an individual will choose). On the other hand, if any risk is involved, it is clear that we must have some idea of the difference in utility between a given pair of events (and not merely which one is preferred by an individual). If, for instance, a person must choose between the event B and a lottery that will give either the event A or the event C with equal probabilities, and if, moreover, A p B p C, then the problem is to determine whether the possibility of gain (in case A happens) is enough to offset the risk of loss (in case C should occur).

If, now, there existed some commodity (called, perhaps, *utiles*) for which utility were linear (i.e., such that the utility of a certain quantity of utiles were directly proportional to that quantity) there would be no difficulty. We could simply decide the size of the side-payments (in utiles), which would induce our player to abandon A for B, and B for C, respectively, and then act accordingly. Unfortunately, we have seen that no commodity (not even money) behaves in this manner, and thus this idea is deficient. Yet we shall show that such a commodity can be dealt with and considered, just so long as we do not attempt to deal with it as with other commodities, which can be bought, sold, transferred, or destroyed.

7.2. Lotteries

We mentioned above that one of the principal cases in which we would like to be able to deal with utiles is that of risk-taking. Now, risk-taking depends essentially on the idea of a lottery.

7.2.1 Definition. Let A and B be any two events, and let $0 \le r \le 1$. Then by $rA + (1 - r)B$ we shall mean the *lottery* that has the two possible outcomes, A and B, with probabilities r and $1 - r$, respectively.

In a similar way, lotteries with three or more possible outcomes may be defined. It should be pointed out that a lottery is also an event. Thus there may be a lottery in which one of the outcomes is another lottery. Moreover, the combination of events by means of lotteries obeys all the usual laws of arithmetic (and linear algebra). We thus have the axioms

$$rA + (1 - r)B = (1 - r)B + rA \tag{7.1}$$

$$rA + (1 - r)\{sB + (1 - s)C\} = rA + (1 - r)sB + (1 - r)(1 - s)C \tag{7.2}$$

$$rA + (1 - r)A = A \tag{7.3}$$

Axiom (7.1), a commutative law, has an obvious meaning. Axioms (7.2) and (7.3), which resemble the distributive law, state that the manner in which a lottery or sequence of lotteries is held does not matter: only the final probabilities of the possible outcomes matter.

Now, in the lottery $rA + (1 - r)B$, it is clear that it should not make any difference if A is replaced by some C such that A i C. We have thus the axioms

If A i C, then, for any r, B

$$\{rA + (1 - r)B\} \text{ i } \{rC + (1 - r)B\} \tag{7.4}$$

If A p C, then, for any $r > 0$, B

$$\{rA + (1 - r)B\} \text{ p } \{rC + (1 - r)B\} \tag{7.5}$$

When we consider the lottery $rA + (1 - r)B$, we see that, for $r = 1$, the lottery is identical to the event A; while for $r = 0$, the lottery is identical to B. Now it seems reasonable that a very small change in r should cause only a very small change in the utility of the lottery (whatever is meant by this). This, together with the intermediate value theorem for continuous real-valued functions, gives us the axiom

7.2.2 Axiom. (Continuity Axiom) Let A, B, and C be events such that A p C p B. Then there exists some $r \in [0, 1]$ such that

$$\{rA + (1 - r)B\} \text{ i } C$$

It should be remarked that, in Axiom 7.2.2, we must actually have $r \in (0, 1)$. In fact, if $r = 0$ or 1, the lottery is equal to B or A, respectively, and by hypothesis A p C p B, so we cannot have a lottery equivalent (under i) to C in either of these two cases. This can be stated as a theorem:

7.2.3 Theorem. If A p C p B, and

$$\{rA + (1 - r)B\} \text{ i } C$$

then $0 < r < 1$. Moreover, r is unique.

Proof. We have seen that r cannot then equal either 0 or 1. To show uniqueness, suppose s is any other number in $(0, 1)$. We may assume $s < r$. We have $0 < r - s < 1 - s$, and so, since

$$B = \left\{ \frac{r - s}{1 - s} B + \frac{1 - r}{1 - s} B \right\}$$

and A p B, it follows that

$$\left\{ \frac{r - s}{1 - s} A + \frac{1 - r}{1 - s} B \right\} \text{ p } B$$

Now we have

$$rA + (1 - r)B = sA + (1 - s)\left\{\frac{r - s}{1 - s}A + \frac{1 - r}{1 - s}B\right\}$$

and thus by (7.5)

$$\{rA + (1 - r)B\} \ \text{p} \ \{sA + (1 - s)B\}$$

The axioms that we have given are now sufficient for the construction of a utility function. Let us consider two cases. The first possibility is that the individual we are dealing with may be indifferent to any choice, i.e., A i B for any two events A and B. Without asking whether such a person might exist, we point out that this is really a trivial case: all events have the same utility for this individual (who at any rate seems to be a vegetable rather than a person) and he hardly enters the framework of game theory. The second case is dealt with in the proof of the following theorem. (It should be pointed out that the theorem is true in both cases, not just the second.)

7.2.4 Theorem. There exists a function u mapping the set of all events into the real numbers, such that

$$u(A) > u(B) \quad \text{if and only if } A \ \text{p} \ B \tag{7.6}$$

$$u(rA + (1 - r)B) = ru(A) + (1 - r)u(B) \tag{7.7}$$

for any two events A and B, and any $r \in [0, 1]$. Moreover, the function u is unique up to a linear transformation; i.e., if there exists a second function, v, also satisfying (7.6) and (7.7), then there exist real numbers, $\alpha > 0$, and β, such that, for all A

$$v(A) = \alpha u(A) + \beta \tag{7.8}$$

Proof (Existence). If, for all A and B, A i B, we can simply let $u(A) = 0$ for any event. Suppose, then, that there exist two events, E_1 and E_0, such that E_1 p E_0. Now, by the Axioms 7.1.2, for any event A, there are five possibilities

(a) A p E_1;
(b) A i E_1;
(c) E_1 p A p E_0; $\hspace{2cm}$ (7.9)
(d) A i E_0;
(e) E_0 p A.

Let us define, first of all, $u(E_1) = 1$ and $u(E_0) = 0$. We can then define $u(A)$, for any event A, as will be explained. The five possibilities are treated in order.

In case (a), we have A p E_1 p E_0. By the continuity axiom, there exists $r \in (0, 1)$ such that

$$\{rA + (1 - r)E_0\} \ \text{i} \ E_1$$

We define

$$u(A) = 1/r$$

(7.10a)

In case (b) we must clearly define

$$u(A) = 1$$

(7.10b)

In case (c), there exists $S \in (0, 1)$ such that

$$\{sE_1 + (1 - s)E_0\} \text{ i } A$$

We then define

$$u(A) = s$$

(7.10c)

In case (d), we will clearly have

$$u(A) = 0$$

(7.10d)

Finally, in case (e), there exists $t \in (0, 1)$ such that

$$\{tA + (1 - t)E_1\} \text{ i } E_0$$

In this case, we define

$$u(A) = \frac{(t - 1)}{t}$$

(7.10e)

We have thus defined a function $u(A)$ for all events A, and must now show that it satisfies (7.6) and (7.7).

In fact, the proof that u does satisfy these conditions is quite lengthy, depending on whether each of the two events, A and B, lies in case (a), (b), (c), (d), or (e). We shall prove these assuming both are in case (c); the proofs in each of the other 14 cases are quite similar.

Assume, then, that both A and B are in case (c), and that $u(A) = s_1 u(B) = s_2$. If, now, $s_1 = s_2$, we see that both A and B are equivalent (under i) to the lottery $s_1 E_1 + (1 - s_1)E_0$ and hence A i B. If, on the other hand, $s_1 > s_2$, it is seen, as in the proof of 7.2.3, that

$$\{s_1 E_1 + (1 - s_1)E_0\} \text{ p } \{s_2 E_1 + (1 - s_2)E_0\}$$

and so A p B. Similarly, if $s_2 > s_1$, we see that B p A. This proves that u satisfies (7.2.6).

To prove (7.2.7), let $r \in (0, 1)$. We have

$$A \text{ i } \{s_1 E_1 + (1 - s_1)E_0\}$$

$$B \text{ i } \{s_2 E_1 + (1 - s_2)E_0\}$$

and so, by (6.2.4)

$$\{rA + (1 - r)B\} \text{ i } \{r[s_1 E_1 + (1 - s_1)E_0]$$

$$+ (1 - r)[s_2 E_1 + (1 - s_2)E_0]\}$$

Hence

$$\{rA + (1 - r)B\} \text{ i } \{[rs_1 + (1 - r)s_2]E_1$$

$$+ [r(1 - s_1) + (1 - r)(1 - s_2)]E_0\}$$

and so

$$u(rA + (1 - r)B) = rs_1 + (1 - r)s_2$$

or

$$u(rA + (1 - r)B) = ru(A) + (1 - r)u(B)$$

We must, finally, prove that u is unique up to a linear transformation. In fact, let v be any other function satisfying (7.6) and (7.7). Because E_1 p E_0, we must have $v(E_1) > v(E_0)$, and so we can define

$$\beta = v(E_0)$$
$$\alpha = v(E_1) - v(E_0) > 0$$

Now, suppose E_1 p A p E_0. If $u(A) = s$, we know

$$A \text{ i } \{sE_1 + (1 - s)E_0\}$$

and so

$$v(A) = v(sE_1 + (1 - s)E_0)$$
$$v(A) = sv(E_1) + (1 - s)v(E_0)$$
$$v(A) = s(a + \beta) + (1 - s)\beta = s\alpha + \beta$$
$$v(A) = \alpha u(A) + \beta$$

Similarly, for the other possibilities ((a), (b), (d), (e)), (7.8) may be shown to hold.

We have thus shown the existence of a utility function. The fact that this function is not unique does not detract much from it; in fact, the difference between any two utility functions is only as to the zero and unit of the scale — a difference quite similar to the difference between the Celsius and Fahrenheit temperature scales. By definition, the utility function, *u*, is linear with lotteries (7.7). The fact that, in a lottery with two outcomes, the first outcome may be given any probability between 0 and 1 means that the range of the utility function is a convex subset of the real line, i.e., an interval. There are 11 types of intervals; one of these, however, is the empty set, while another, consisting only of a single point, clearly applies only to the trivial case mentioned above (of an individual who is indifferent to any choice). This still leaves, however, nine types of intervals, and the question of which of these nine is the true utility space for a given individual remains open.

7.3. Commodity Bundles

From an abstract point of view, utilities are assigned to all types of events. This is as it should be, for players differentiate among all types of events. On the other hand, from the economist's point of view, the natural application of utility theory is toward the possession of commodities.

We assume a finite number, say *n*, of basic commodities. These commodities are held by different individuals in different quantities. It is natural to use an *n*-vector $q = (q_1, \ldots, q_n)$ to denote the amounts of these commodities held by an individual (or group of individuals). We shall call such an *n*-vector a *commodity bundle*. These commodity bundles are then treated as events: i.e., the event of holding a given bundle. Thus a utility function will be defined on the commodity bundles.

A *caveat* is in order here: we must distinguish between the lottery

$$rq' + (1 - r)q''$$

which gives the bundle q' with probability *r* and the bundle q'' with probability $1 - r$, and the bundle *q* defined by

$$q_j = rq'_j + (1 - r)q''_j, \quad j = 1, \ldots, n$$

which is a fixed bundle and not a lottery. For instance, q' could consist of two automobiles without tires, and q'' could consist of 10 tires. Letting $r = \frac{1}{2}$, it is clear that the bundle *q* (one car with five tires) is much preferable, at least for most people, to the lottery $rq' + (1 - r)q''$.

One obvious condition that the utility function must satisfy is monotonicity: if $q' \geq q$, we must have $u(q') \geq u(q)$. It is also sometimes assumed that the function *u* is continuous, and has continuous first- and second-order partial derivatives (since it is monotone, at least continuity and the first-order partials will exist almost everywhere).

The partial derivatives $\partial u/\partial q_j = u_j$ are called *shadow prices* (they represent, naturally, the price in utiles that would be "fair" for the particular commodity at the given level).

The jth commodity is said to be subject to the *law of diminishing returns* if, for all q

$$u_{jj} = \frac{\partial^2 u}{\partial q_j^2} \leq 0$$

Most commodities are assumed to be subject to this law.

The ith commodity is said to be *grossly* substitutable for the jth commodity if

$$u_{ij} = \frac{\partial u_i}{\partial q_j} = \frac{\partial^2 u}{\partial q_j \partial q_i} \leq 0$$

i.e., if a decrease in the amount of the jth commodity makes the shadow price for the ith commodity greater. Since we usually have $u_{ji} = u_{ij}$, it follows that *gross substitutability* is a symmetric relation.

The ith and jth commodities are said to be *complementary* if

$$u_{ij} = u_{ji} \geq 0$$

i.e., if an increase in the amount of one of them makes the value of the other greater.

A commodity, say the nth, is said to be *separable* if its utility does not depend on the other commodities, i.e., if there exist two functions, $w = (q_1, \ldots, q_{n-1})$ and $\varphi = \varphi(q_n)$, such that for all q

$$u(q_1, \ldots, q_n) = w(q_1, \ldots, q_{n-1}) + \varphi(q_n)$$

A separable commodity is a natural one to use as a medium of exchange. In some cases the function φ is linear; if this holds for two different persons, we say that the nth commodity is *linearly transferable* between the two players. The importance of such a case is that, if the units of utility for the two persons are suitably chosen, a transfer of this commodity from one person to the other will leave the total utility of the two constants. Thus the other $n - 1$ commodities can be divided so as to maximize this total utility; a transfer of the nth commodity (*side payment*) is then used to correct any inequity that this first process may have caused.

7.4.   Absolute Utility

In the field of welfare economics, it is sometimes desired to know whether a certain action "will help individual *a* more than it hurts individual *b*." Now this cannot be determined simply by measuring the increase and decrease of utility, which this action causes to the two individuals, for we have already pointed out that the units of utility scales are arbitrary and thus cannot be used for interpersonal comparisons.

The difficulty is that there is no absolute scale on which to measure utilities for this purpose.

Suppose, however, that the utility spaces of different people are bounded intervals. Then there will be an absolute unit of utility: the length of these intervals. We can normalize all the intervals so that they have 0 and 1 for their upper and lower end-points. It is then natural to compare utilities on these scales.

The problem, then, reduces to the question of whether the utility space of an individual is bounded or not. This is, of course, an open question. Isbell, who favors the bounded interval, suggests the following arguments in its favor:

1. Assume that my utility space is unbounded above. Let A be the event "nothing changes," and let B be the event "I am boiled in oil." As the space is not bounded above, there must be some event C such that

$$u(C) > 2u(A) - u(B)$$

But this means that

$$\{\tfrac{1}{2}C + \tfrac{1}{2}B\} \;\text{p}\; A$$

However, I can conceive of no event C, which would make me prefer the lottery $\{\tfrac{1}{2}C + \tfrac{1}{2}B\}$ to my present state in life. Hence, my utility space is bounded above.

2. Assume that my utility space is not bounded below. Let A be the event "nothing changes" and let B be the event "I win a million dollars." Now, if the space is not bounded below, there will be some event C such that

$$A \;\text{p}\{0.999999B + 0.000001C\}$$

But I can conceive of no such event C. Hence my utility space is bounded below.

3. Suppose, moreover, that the utility space is not bounded above. Then there exists a sequence of events, $A_1, A_2, \ldots$ such that

$$u(A_n) = 2^n$$

Then the lottery

$$\sum_{n=1}^{\infty} 2^{-n} A_n$$

will have infinite utility (the Petersburg paradox). Therefore, to avoid this paradox, the utility space must be bounded above. Similarly, it may be argued that the utility space is bounded below.

These are Isbell's arguments for bounded utility; it may be argued that the fact a person cannot conceive of an event C just described simply means his imagination is defective, and that the Petersburg paradox is precisely that — a paradox. The question remains open, naturally. It can only be said that, while things would be simpler with bounded utility scales, we can nevertheless obtain most of the desired results without any recourse to them.

Problems

1. Complete the proof of Theorem 7.2.2.

2. (Arrow's impossibility proof.) When only ordinal utilities are considered (and no lotteries) there is often no way to determine which action a group of people should take.

   Consider a group of m individuals, faced with n alternatives, $A_1, A_2, \ldots, A_n$. By a *profile of individual preferences* we mean a function that assigns to each individual i a weak ordering of the alternatives. By a *social welfare function* we mean a function F, which assigns an ordering (the "social" ordering) to each profile. The following axioms seem reasonable:

   $\mathscr{A}1$: F is defined for all profiles.
   $\mathscr{A}2$: Suppose that, for a given profile, A_j is preferred to A_k by the function F. Then A_j will continue to be preferred to A_k if the profile is changed in the following way:

   (a) The relative orders of alternatives other than A_j are not changed.
   (b) In each of the individual orderings, the position of A_j is improved.

   $\mathscr{A}3$: Let ϑ_1 be a subset of $\vartheta = \{A_1, \ldots, A_n\}$. Suppose that a profile is modified in such a way as to leave the relative orderings of the elements of ϑ_1 unchanged. Then the social ordering, F, will also leave the relative ordering of the elements of ϑ_1 unchanged.
   $\mathscr{A}4$: For each pair of alternatives, A_j and A_k, there exists a profile such that the social ordering prefers A_j to A_k.
   $\mathscr{A}5$: There is no individual, i, such that, for every A_j and A_k, the social ordering prefers A_j to A_k whenever i does so.

   Arrow's axioms $\mathscr{A}1 - \mathscr{A}5$ are inconsistent whenever $n \geq 3$ and $m \geq 2$:

   (a) With respect to a given function, F, a coalition S is said to be *decisive* for the ordered pair (A_j, A_k) if, whenever all members of S prefer A_j to A_k, then the social ordering does likewise. A set is decisive if it is decisive for at least one ordered pair (A_j, A_k).
   (b) Let V be a minimal decisive set (i.e., V is decisive but no proper subset of V is decisive). Then $V \neq \varnothing$.
   (c) Assume V is decisive for (A_j, A_k). Let $i \in V$ and let A_l be any other alternative. Then $\{i\}$ is decisive for (A_j, A_l), and hence, by the minimality of V, $V = \{i\}$.
   (d) The set $\{i\}$ is decisive for all pairs, contradicting $\mathscr{A}5$.

3. Assume $u(A) = 1$ and $u(B) = 0$. Find the $u(C)$ in each of the following cases:

(a) C ¡ $0.2A + 0.8B$
(b) D ¡ $(0.75C + 0.25B)$ and B ¡ $(0.5A + 0.5D)$
(c) D ¡ $(0.75C + 0.25B)$ and A ¡ $(0.4D + 0.6C)$

4. Assume $u(A) = 5$ and $u(B) = 2$. Find $u(C)$ in each of the cases of Problem 3.

5. In a two-person situation, with three commodities, assume

$$u_1(x, y, z) = 3x + 0.2xy - 0.6y^2 + 4y + z$$

$$u_2(x, y, z) = 4x - 0.5x^2 - 0.2xy + 2y + z$$

Let the two players have the endowments (initial bundles) (8, 1, 18) and (2, 7, 15). Find an efficient distribution of the total amount (10, 8, 33) available to the two players:

(a) Use Lagrange multipliers to maximize the sum of u_1 and u_2. Note that this maximum does not depend on how the third commodity (the *numéraire*) is divided between the two players.
(b) Use these Lagrange multipliers to determine "fair" prices for the first and second goods, in terms of the *numéraire*.
(c) Use these prices to determine a "fair" division of the third good.

Note: This problem can also be solved without the use of multipliers. Simply determine the prices in terms of derivatives: the prices should be proportional to the derivatives at the maximizing point; clearly the derivative with respect to z is equal to 1.

Chapter 8

Two-Person General-Sum Games

8.1. Bimatrix Games (Noncooperative)

We have so far studied, in general, only zero-sum games. In parlor games, or in situations in which the stakes are small monetary amounts (so that utility will be nearly linear with money), such models are accurate enough. On the other hand, when stakes are more complicated, as often happens in economic situations, it is not generally true that the interests of the two players are exactly opposed; very often, both can gain through cooperation. Such situations are called general-sum games; they include zero-sum games as a special case.

In general, a finite two-person general-sum game can be expressed as a pair of $m \times n$ matrices, $A = (a_{ij})$ and $B = (b_{ij})$, or equivalently, as an $m \times n$ matrix (A, B) each of whose entries is an ordered pair (a_{ij}, b_{ij}). The entries a_{ij} and b_{ij} are the payoffs (in utilities) to the players I and II, respectively, assuming they choose, respectively, their ith and jth pure strategies. A game in this form is called a *bimatrix game*.

It is clear that, in many cases, cooperation between the players can work out to both players' advantage. On the other hand, it is possible that, even where this cooperation is mutually beneficial, it is forbidden by the rules of the game (by, e.g., an antitrust law). We distinguish, therefore, two cases for bimatrix games:

1. The noncooperative case, in which any type of collusion, such as correlated strategies and side payments, is forbidden;
2. The cooperative case in which all such cooperation is permitted.

We shall treat, first, the noncooperative case. As in the case of zero-sum games, we can define mixed strategies for I and II as m- and n-vectors, respectively, with nonnegative components whose sum is 1. We define equilibrium pairs in the obvious manner:

8.1.1 Definition. A pair of mixed strategies (x^*, y^*) for the bimatrix game (A, B) is said to be in equilibrium if, for any other mixed strategies, x and y

$$xAy^{*T} \leq x^*Ay^{*T}$$

$$x^*By^T \leq x^*By^{*T}$$

We naturally inquire as to the existence of such pairs. The following theorem answers the question in the affirmative.

8.1.2 Theorem. Every bimatrix game has at least one equilibrium point.

Proof. Let x and y be any pair of mixed strategies for the bimatrix game (A, B). We define

$$c_i = \max\{A_i{\cdot}y^{\mathrm{T}} - xAy^{\mathrm{T}}, 0\}$$
$$d_j = \max\{xB_{\cdot j} - xBy^{\mathrm{T}}, 0\}$$

and

$$x'_i = \frac{x_i + c_i}{1 + \sum_k c_k}$$

$$y'_j = \frac{y_j + d_j}{1 + \sum_k d_k}$$

It is clear that the transformation $T(x, y) = (x', y')$ is continuous. Moreover, it is easily seen that both x' and y' are mixed strategies. We now show that $(x', y') = (x, y)$ if and only if (x, y) is an equilibrium pair. In fact, if (x, y) is an equilibrium pair, it is clear that for all i

$$A_i y^{\mathrm{T}} \leq xAy^{\mathrm{T}}$$

and so $c_i = 0$. Similarly, $d_j = 0$ for all j. Hence $x' = x$ and $y' = y$.

Suppose, on the other hand, (x, y) is not an equilibrium pair. This means either that there exists some $\bar{x}$ such that $\bar{x}Ay^{\mathrm{T}} > xAy^{\mathrm{T}}$ or that there exists some $\bar{y}$ such that $xB\bar{y}^{\mathrm{T}} > xBy^{\mathrm{T}}$. We shall assume the first case holds; the proof in the second case is identical. Since $\bar{x}Ay^{\mathrm{T}}$ is a weighted average of the terms $A_i{\cdot}y^{\mathrm{T}}$, it follows that there must be some i for which $A_i{\cdot}y^{\mathrm{T}} > xAy^{\mathrm{T}}$, and hence some $c_i > 0$. As the c_i are nonnegative, the sum $\sum_k c_k > 0$.

Now, xAy^{T} is a weighted average (with weights x_i) of the terms $A_i{\cdot}y^{\mathrm{T}}$. We must, therefore, have $A_i{\cdot}y^{\mathrm{T}} \leq xAy^{\mathrm{T}}$ for some i such that $x_i > 0$. Now, for this i, we have $c_i = 0$, and so

$$x'_i = \frac{x_i}{1 + \sum_k c_k} < x_i$$

and so $x' = x$.

In the second case we show that $y' \neq y$. Hence $(x', y') = (x, y)$ if and only if (x, y) is in equilibrium.

Now, the set of all strategy pairs is a closed, bounded, convex set, and hence the *Brouwer fixed point theorem* holds. As the transformation $T(x, y) = (x', y')$ is continuous, it must have a fixed point. This fixed point will be an equilibrium pair.

We may remark that the proof of Theorem 8.1.2 can easily be generalized to equilibrium *n*-tuples for finite *n*-person games. Let us note also that the proof is an existence proof and gives us no method of finding equilibrium pairs.

The problem of finding equilibrium pairs to bimatrix games has been approached from different sides, and some success has been obtained. The reader is referred to the bibliography for details and methods. Rather than give these methods, we prefer to discuss the difficulties with the notion of equilibrium pairs — difficulties which are most apparent when compared with the parallel notion of optimal strategies for zero-sum (matrix) games.

It has been pointed out (Theorem 2.1.2) that, for zero-sum games, equilibrium pairs were both interchangeable and equivalent in the sense that if (x, y) and (x', y') are equilibrium pairs for the matrix game A, then so are (x, y') and (x', y) and, moreover, $xAy^{\mathrm{T}} = x'Ay'^{\mathrm{T}}$. This effectively meant that the property of belonging to an equilibrium pair is intrinsic to the vector x (not depending on the vector y, which is in equilibrium with it), and so we consider optimal strategies rather than equilibrium pairs. We give now an example to show that this is not so for general-sum games.

8.1.3 Example. (The Battle of the Sexes) Consider the bimatrix game

$$\begin{pmatrix} (4, 1) & (0, 0) \\ (0, 0) & (1, 4) \end{pmatrix}$$

It is obvious that the pure strategies $x = (1, 0)$ and $y = (1, 0)$ are an equilibrium pair; so are $x' = (0, 1)$ and $y' = (0, 1)$. On the other hand, (x, y') and (x', y) are not equilibrium pairs. Moreover, the payoffs for the two equilibrium pairs (x, y) and (x', y') are distinct; (x, y) will be preferred by I, whereas II will prefer (x', y'). Thus it is not clear that these equilibrium pairs are entirely stable; even if I know that II will use the pure strategy y', he may insist on using x rather than x', hoping that this will induce II to switch to y. In general, it is difficult to tell what will happen. (It should be noticed that the mixed strategies $x'' = \left(\frac{4}{5}, \frac{1}{5}\right)$ and $y'' = \left(\frac{1}{5}, \frac{4}{5}\right)$ are also an equilibrium pair.)

Even when there is only one equilibrium pair (as is often the case), it is not clear that this equilibrium pair is exactly what we want:

8.1.4 Example. (The Prisoner's Dilemma) Consider the game

$$\begin{pmatrix} (5, 5) & (0, 10) \\ (10, 0) & (1, 1) \end{pmatrix}$$

In this game, it is easy to see that the second row dominates the first row, while the second column dominates the first column (if we adapt the definition of domination, given above for zero-sum games, in the obvious manner). Hence the only equilibrium pair is given by the second pure strategy of each player. This gives the payoff vector

(1,1). But, if both players play "wrong," i.e., choose the first pure strategy, the result (5,5) is much better for both. The trouble is that each can gain by double-crossing the other.

8.2.   Perfect Equilibrium Points

We saw above the difficulties that can arise with Nash equilibria. Because of this, some substantial work has been done toward *refining* the concept, with a view to eliminating many of the less attractive equilibria, and thus avoiding (some of) the pathologies. The first (and perhaps the most important) such refinement, the *perfect equilibrium*, is due to Selten [8.9].

8.2.1 Definition. Let (A, B) be an $m \times n$ bimatrix game. Let σ and τ be strictly positive vectors, with m and n components respectively, such that

$$\sum \sigma_i < 1 \quad \text{and} \quad \sum \tau_j < 1.$$

Then the perturbed game $(A, B; \sigma, \tau)$ is a constrained game, with payoff matrices A and B, where the two players are constrained to use mixed strategies x, y which satisfy

$$x_i \geq \sigma_i \quad \text{for all} \quad i = 1, \ldots, m$$

and

$$y_j \geq \tau_j \quad \text{for all} \quad j = 1, \ldots, n$$

respectively.

   The basic idea behind the perturbed game is, simply, that since players are not perfect, each player is liable to make mistakes, though perhaps with very small probabilities. Thus each pure strategy will be chosen with positive probability; the *perturbation vectors* σ and τ give lower bounds for these probabilities.

   Now, it is not too difficult to modify the proof of Theorem 8.1.2 so as to show that any perturbed game $(A, B; \sigma, \tau)$ has at least one equilibrium pair of mixed strategies. (This is Problem 7.) The following lemma, given without proof, is also useful:

8.2.2 Lemma. Let $(x; y)$ be an equilibrium pair for the perturbed game $(A, B; \sigma, \tau)$. Suppose row k is such that

$$\sum_j a_{kj} y_j < \max_i \sum_j a_{ij} y_j$$

Then $x_k = \sigma_k$.

(The lemma says, essentially, that if a row is less than optimal as a reply to the other player's strategy, then it must be used as little as possible. There is a similar statement concerning less-than-optimal columns.)

8.2.3 Definition. Let (A, B) be a bimatrix game. An equilibrium pair $(x^*; y^*)$ of mixed strategies to (A, B) is a *perfect equilibrium* if there exists (i) a sequence of perturbation vectors (σ^k, τ^k) converging to $(0, 0)$ together with (ii) a sequence of equilibrium pairs $(x^k; y^k)$ to the perturbed games $(A, B; \sigma^k, \tau^k)$, which converges to $(x^*; y^*)$.

Note that the definition requires only that *some* sequence of perturbed equilibria converge to $(x^*; y^*)$; it is not necessary that *all* such sequences converge. Thus perfectness is a relatively weak condition.

8.2.4 Theorem. Every bimatrix game has at least one perfect equilibrium in mixed strategies.

Proof. We have already seen that each perturbed game has an equilibrium pair of mixed strategies.

Let, now, a sequence of pairs of vectors $(\sigma^k; \tau^k)$ be given, converging to the pair $(0; 0)$. For each k, let $(x^k; y^k)$ be an equilibrium pair of (mixed) strategies for the perturbed game $(A, B; \sigma^k, \tau^k)$. Since the set of all pairs of mixed strategies is compact, this sequence must have a converging subsequence with limit $(x^*; y^*)$. By continuity, this $(x^*; y^*)$ will be an equilibrium pair for the original (unperturbed) game (A, B). This is the desired perfect equilibrium.

8.2.5 Example. Consider the bimatrix game

$$\begin{pmatrix} (0,0) & (0,0) \\ (0,0) & (1,1) \end{pmatrix}$$

It is easy to see that this game has the two pure-strategy equilibria (row 1, column 1) and (row 2, column 2). Of course, the first of these seems rather unsatisfactory as both players use dominated strategies; nevertheless, it satisfies the definition of equilibrium pair.

Consider, however, the perturbed game $(A, B; \sigma, \tau)$. It is clear that, so long as $y_2 > 0$, then row 1 cannot be an optimal reply to y, and so, by Lemma 8.2.2, equilibrium pair $(x; y)$ must have $x_1 = \sigma_1$. Similarly, we must have $y_1 = \tau_1$. Thus the perturbed equilibrium is of the form $(\sigma_1, 1 - \sigma_1; \tau_1, 1 - \tau_1)$, and, as the perturbation vectors converge to zero, the perturbed equilibria must converge to $(0,1; 0,1)$, that is, to (row 2, column 2). It follows that this is a perfect equilibrium; on the other hand, (row 1, column 1) is nonperfect.

For this last example, the perfectness concept works rather well: the equilibrium pair at (row 1, column 1) seems "unsatisfactory," giving a payoff of $(0,0)$ and thus it seems reasonable that it should be eliminated in favor of the other equilibrium with its payoff of (1,1). On the other hand, consider the following example:

8.2.6 Example. In the bimatrix game

$$\begin{pmatrix} (10,10) & (0,10) \\ (10,0) & (1,1) \end{pmatrix}$$

there are equilibria at (row 1, column 1) and at (row 2, column 2). Here, the same reasoning applies as in Example 8.2.5, and we conclude that the equilibrium at (row 2, column 2) is perfect, while that at (row 1, column 1) is nonperfect.

In this case, the "good" equilibrium payoff (10,10) seems to be discarded in favor of a much "poorer" (1,1). In fact, this example is somewhat analogous to the Prisoner's Dilemma, and is due mainly to the fact that the first row and column of the matrix are dominated (albeit weakly) by the second row and column. Indeed, it is easy to see that any equilibrium in which either player uses a dominated row or column with positive probability will be nonperfect.

More generally, a mixed strategy x of player I is said to be dominated if there is another strategy, x' which is at least as good as x against all columns of matrix A, and better than x against at least one column. There is a similar definition for dominated mixed strategy of player II. Then we have the following theorem:

8.2.7 Theorem. Let $(x; y)$ be an equilibrium pair of the bimatrix game (A, B). Then $(x; y)$ is perfect if and only if it is undominated (i.e., neither x nor y is dominated).

We will not prove this theorem. We will, however, make a pair of comments. The first is that an analogous theorem for games with three or more players turns out to be false. The second is that while it is relatively easy to check whether a pure strategy is dominated by another pure strategy, it is considerably more difficult to check whether a mixed strategy is dominated. One technique for this solves an auxiliary matrix (i.e., zero-sum) game. We refer the reader to van Damme [8.11] for a general discussion of this problem, including a proof of the theorem.

Instead of requiring merely that some sequence of perturbed equilibria converge to the given equilibrium, we might require a stronger condition:

8.2.8 Definition. An equilibrium pair $(x; y)$ of the bimatrix game is *strictly perfect* if for every sequence of strictly positive perturbation vectors $(\sigma^k; \tau^k)$, which converges to $(0;0)$, there exists a sequence $(x^k; y^k)$ of equilibrium pairs to the corresponding perturbed games $(A, B; \sigma^k, \tau^k)$, which converges to $(x; y)$.

Clearly, any strictly perfect equilibrium is a perfect equilibrium. Thus the concept is definitely stronger. Unfortunately, it may be too strong, because strictly perfect equilibria need not exist. As an example we note that the game

$$(A, B) = \begin{pmatrix} (1,1) & (1,0) & (0,0) \\ (1,1) & (0,0) & (1,0) \end{pmatrix}$$

has no strictly perfect equilibria. In fact, column 1 strictly dominates the two other columns; against this column, either row is optimal. Hence the equilibria are all those pairs (and only those pairs) in which II uses column 1 with probability 1. It will follow that, in the perturbed game $(A, B; \sigma, \tau)$, II must, in equilibrium, use $y = (1 - \tau_2 - \tau_3, \tau_2, \tau_3)$. Now, if II makes the "mistake" (column 2) more frequently than the other mistake (column 3), then I should use row 1 as much as possible, i.e., if $\tau_2 > \tau_3$, then the equilibrium $x = (1 - \sigma_2, \sigma_2)$. If, however, $\tau_3 > \tau_2$, then $x = (\sigma_1, 1 - \sigma_1)$. Thus, for some sequences of perturbation vectors, the equilibria must converge to (row 1, column 1); for others, they must converge to (row 2, column 1). We conclude that this game has no strictly perfect equilibria.

Subgame-Perfect Equilibria

A special type of equilibrium, the subgame-perfect equilibrium, refers specifically to games in extensive form. For an explanation of this, consider the game depicted in Figure 8.1.

It may be seen that *if* the game ever reaches position Z, then I should choose e over f, since that gives him a payoff of 3 rather than 1. This would give II a payoff of 1, so II, at y, should choose c rather than d. This in turn would give I a payoff of 3, so I is definitely better off, at X, choosing a. Thus, by the process of composition of subgame strategies, discussed in Theorem 1.4.5, the strategies (a, e) for I and c for II are in equilibrium. This is, in effect, the equilibrium guaranteed by Theorem 1.4.5.

Consider, however, the pair of strategies (a, f) for I and c for II. As mentioned above, e is always better than f, *if* position Z is ever reached. However, if II uses the pure strategy c, Z will never be reached, and so I's use of (a, f) is no worse than (a, e). Hence it is not difficult to see that the strategies (a, f) and c are also an equilibrium pair.

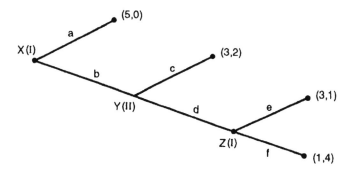

Figure 8.1.

The question is whether it is reasonable for these two strategies to be in equilibrium. Specifically, in what sense can it be considered "good" for player I to use alternative f? A logician might point out that the formal implication, "If play reaches position Z, then it is good for I to use f," is logically true because the antecedent ("play reaches position Z") is here false. On the other hand, there seems to be something wrong with a strategy that tells player I to use f. Thus, while it is true that these two strategies form an equilibrium pair (as may be checked directly), we will consider ourselves dissatisfied with such an equilibrium pair. We express this dissatisfaction by introducing a new concept, the *subgame-perfect equilibrium*.

For this purpose, we refer the reader to the definition of *subgame*, as in Section 1.4, above. Then:

8.2.9 Definition. Let Γ be a game in extensive form. A pair (σ, τ) of (pure) strategies is a subgame-perfect equilibrium if, wherever the game decomposes (at a node X), the corresponding restrictions of σ and τ are an equilibrium pair for the subgame Γ_X.

We note in this regard that since the game decomposes trivially at its initial node, then any subgame-perfect equilibrium is an equilibrium. We also note that, in the game of Figure 8.1, $((a,f),c)$ is not a subgame-perfect equilibrium since (f, φ) is not an equilibrium pair for the subgame Γ_z.

The obvious question is whether such equilibria always exist.

8.2.10 Theorem. Every finite game of perfect information has at least one subgame-perfect equilibrium in pure strategies.

Proof. It is not difficult to see that the proof of Theorem 1.4.5 is valid with the additional condition of subgame-perfectness. In fact, the equilibrium guaranteed by that theorem will be subgame-perfect.

When information is imperfect, randomization is, as we know, normally necessary. Unfortunately, it is not clear just how the idea of subgame-perfectness should be generalized to mixed strategies. On the other hand, generalization to *behavioral* strategies is quite straightforward. Now, we know that even a zero-sum game need not have equilibria (saddle-points) in behavioral strategies, let alone subgame-perfect equilibria. Kuhn's theorem (Problem 8 of chapter 5, page 115) tells us that all games with perfect recall have equilibria in behavioral strategies (in effect, with perfect recall, any probability distribution on the outcomes attainable via mixed strategies is also attainable via behavioral strategies). We then have the following theorem:

8.2.11 Theorem. Any finite two-person game with perfect recall has at least one subgame-perfect equilibrium pair of behavioral strategies.

Proof. The proof is by induction on the number of subgames that a game has. We omit the details, except to say that it proceeds along the same lines as that of Theorems 1.4.5 and 8.2.10.

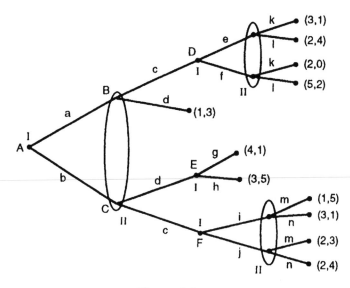

Figure 8.2.

8.2.12 Example. Consider the game of Figure 8.2.

The game begins at node A. As may be seen, it decomposes at nodes D, E, and F. We therefore look for equilibrium pairs for each of the subgames Γ_D, Γ_E, and Γ_F. We note that Γ_D may be reduced to normal form, obtaining

$$
\begin{array}{c}
 \\
e \\
f
\end{array}
\begin{array}{cc}
k & l \\
\begin{pmatrix} (3,1) & (2,4) \\ (2,0) & (5,2) \end{pmatrix}
\end{array}
$$

and it is easy to see that (f, l) is an equilibrium pair for this game, yielding of course the payoff (5,2).

Next, subgame Γ_E is trivial as it consists of only one move, by player I. Clearly he should choose g there, and (g, ϕ) is an equilibrium pair for this, with payoff (4,1).

We note next that Γ_F can be reduced to the normal form

$$
\begin{array}{c}
 \\
i \\
j
\end{array}
\begin{array}{cc}
m & n \\
\begin{pmatrix} (1,5) & (3,1) \\ (2,3) & (2,4) \end{pmatrix}
\end{array}
$$

For this game, it may be checked that $((0.2i, 0.8j), (0.5m, 0.5n))$ is an equilibrium pair, with payoff (2,3.4).

Replacing each of these subgames by the corresponding payoffs, we then obtain the quotient game $\Gamma/_{D,E,F}$, as seen in Figure 8.3.

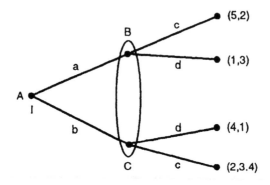

Figure 8.3.

This game reduces to its normal form

$$
\begin{array}{cc}
 & c \qquad\quad d \\
\begin{array}{c} a \\ b \end{array} &
\left(\begin{array}{cc}
(5,2) & (1,3) \\
(2,3.4) & (4,1)
\end{array} \right)
\end{array}
$$

It may be seen that this game has the equilibrium pair $((0.71a, 0.29b), (0.5c, 0.5d))$, with payoff $(3, 2.41)$.

Composition of these equilibrium pairs will give us the equilibrium pair of behavioral strategies:

$$(0.71a, 0.29b, f, g, 0.2i, 0.8j) \quad \text{for player I}$$

$$(0.5c, 0.5d, l, 0.5m, 0.5n) \quad \text{for player II}$$

It is easy to check that these form an equilibrium pair. What is more, they must be a subgame-perfect equilibrium since we specifically built them up from equilibrium pairs for each subgame.

A natural question to ask now deals with the relation between perfect equilibria (for games in normal form) and subgame-perfect equilibria (for games in extensive form). Obviously, the use of a common name for these two concepts suggests a close relation. In fact, however, while the relation seems strong enough, it turns out that there is no logical implication here. A subgame-perfect equilibrium need not be perfect; a perfect equilibrium need not be subgame-perfect.

To begin with, we note that any normal form ($m \times n$) game can be represented by an undecomposable game: player 1 moves first, choosing from among m alternatives; player 2, with no information as to player 1's move, then chooses among n alternatives. For such a game, clearly, any equilibrium point is trivially subgame-perfect. But of course many of these will fail to be perfect.

8.2.13 Example. To see that a perfect equilibrium need not be subgame-perfect, consider the game in Figure 8.4. The game decomposes at both nodes B and C. It is

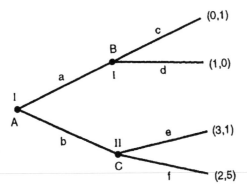

Figure 8.4.

clear that, at node *B*, player 1 should choose *d*, at node *C*, 2 should choose *f*, and at *A*, 1 should choose *b*. Thus $((b,d),f)$ is the only subgame-perfect equilibrium for this game.

Consider, however, the normal form of the game

$$
\begin{array}{c}
 \quad e \qquad\quad f \\
\begin{array}{c}
(a,c) \\
(a,d) \\
(b,c) \\
(b,d)
\end{array}
\left(
\begin{array}{cc}
(0,1) & (0,1) \\
(1,0) & (1,0) \\
(3,1) & (2,5) \\
(3,1) & (2,5)
\end{array}
\right)
\end{array}
$$

In this form, the pair $((b,d),f)$ is still a perfect equilibrium point, as may be readily verified. But the third and fourth rows of this game are identical. Hence $((b,c),f)$ is also a perfect equilibrium point, though clearly not subgame-perfect in the extensive form. The trouble seems to be that too much information (about the game) is lost in the passage to the normal form, and so the imperfectness is somehow lost in the process.

8.3. Evolutionary Stable Systems

An interesting application of game theory to biology has developed over the last few years, namely that of an *evolutionary stable system*.

Suppose the males of a certain species of birds can be divided into two types: a bold, pugnacious type, and a mild, timid type. We will call these, respectively, *hawks* and *doves*, though we repeat that they are members of a single species, rather than two different species. In fact, it is generally difficult if not impossible to distinguish between the two types, except for their behavior when two of these birds meet in a competitive situation.

Suppose, then, that two such males meet in competition — in the presence, say, of a prize, such as food, or a female, that they both want. If one is a dove, while the other is a hawk, the hawk will prepare for battle. The dove, as soon as he realizes what is happening, will retreat, allowing the hawk to capture whatever prize is at stake. We might assign a payoff of +30 to the hawk, and 0 to the dove in such case.

If two doves meet, both will ruffle their feathers, but will eventually come to a (more or less) amicable division of goods. We might assign a payoff of +10 to each.

If two hawks meet, they will fight for the prize at stake until one of them is near death. We can assign a payoff of +20 to the winner (this is not good as getting the prize without a fight, but better than having to share it) and −40 to the loser. Since both have (*a priori*) equal probabilities of winning, the expected payoff is −10 to each of the two.

We may represent this in the form of a bimatrix game (A, B):

$$
\begin{array}{c}
\begin{array}{ccc} & D & H \end{array} \\
\begin{array}{c} D \\ H \end{array}
\left(
\begin{array}{cc}
(10, 10) & (0, 30) \\
(30, 0) & (-10, -10)
\end{array}
\right)
\end{array}
$$

It is easy to see that this game has two equilibrium points in pure strategies, namely (D, H) and (H, D): The hawk gets everything without a fight, and the dove gets nothing but avoids a dangerous fight. There is also a third equilibrium point, in mixed strategies, in which both sides use the strategy (.333, .667). This gives an expected payoff of +3.33 to each player.

Now, what meaning can we give to an equilibrium point in such a setting? If these were humans in a dangerous setting (two teenagers, say, playing chicken, or two national leaders engaging in nuclear "brinkmanship"), it is easy to understand what the pure strategy pair (H, D) means: I has reason to believe that II is very timid, and hence I will play very aggressively; II has reason to believe that I is aggressive, and hence II will yield. In the case of animals, however, we feel that the participants are not capable of such reasoning. The pair (H, D) would suggest that *all* competitive encounters are between a hawk and a dove; since the animals cannot change their character to suit the adversary (especially since the adversary types are indistinguishable prior to play), such an interpretation is not viable.

Consider, however, the mixed strategy pair (.333, .667) for both players. This mixed strategy can be obtained if the species as a whole is divided into doves and hawks in a ratio of one dove for every two hawks. In such a case, it would follow that, in a random encounter between males of the species, each of the two adversaries has 1/3 probability of being a dove and 2/3 of being a hawk. The important point here is that the strategy is *symmetric*: since the probabilities are the same for each of players I and II, they can be assumed drawn from the same population.

Of course, a new question now arises. Granted that the (.333, .667) strategy is feasible, why should it ever come into being, or be preserved? We have just used animals' irrationality as a reason to discard the two (nonsymmetric) pure strategy equilibria; is this not also reason to discard *any* type of equilibrium in such settings?

The exact biological answers to such a question is of course much too complicated for a treatment here. Stripped to its essentials, however, it resides in the Darwinian idea of evolution and survival of the fittest, and says, in effect, that, in the long run, the gene pool of a species will evolve so as to optimize something — usually, so as to maximize the survival probabilities of that species in its environment. (Note, moreover, that the species is part of its environment, in the sense that each individual must live with the other members of the species.) We claim that the equilibrium ratio (.333, .667) mentioned above is optimal — a better word would be *stable* — in that sense.

In fact, let us assume that, over a period of many generations, a ratio $x =$ (.333, .667) of doves and hawks has been reached. This is the established gene pool. As mentioned above, the males of the species will occasionally meet, and battle for a prize, with an expected payoff

$$xAx^{\mathrm{T}} = (.333, .667)\begin{pmatrix} 10 & 0 \\ 30 & -10 \end{pmatrix}\begin{pmatrix} .333 \\ .667 \end{pmatrix} = 3.33$$

To explain the stability of this arrangement, we suppose a mutant gene tries to "invade" the species. We now distinguish two cases.

Case 1: Suppose now that a *mutant gene*, which gives only hawks, is somehow mixed into the population. We represent this by the vector $y = (0, 1)$. Against a population represented by x, this mutant gene can thrive. In fact, we find its expected payoff in competition, yAx^{T}, is itself equal to 3.33. Thus at first the mutant gene will be able to hold its own against the established gene pool: its survival seems as likely as that of the other.

After a while, however, the mutant gene runs into trouble. This is because we must think of it as competing against the established gene pool x in an environment represented not by x, but, rather, by the population as a whole. If we assume that the mutant gene has in fact thrived for a while, then the general population will be represented not by $x = (.333, .667)$, but, rather, by a mixture z of x and y, say by $z = (.32, .68)$. We then find that

$$xAz^{\mathrm{T}} = (.333, .667)\begin{pmatrix} 10 & 0 \\ 30 & -10 \end{pmatrix}\begin{pmatrix} .32 \\ .68 \end{pmatrix} = 2.93$$

whereas

$$yAz^{\mathrm{T}} = (0, 1)\begin{pmatrix} 10 & 0 \\ 30 & -10 \end{pmatrix}\begin{pmatrix} .32 \\ .68 \end{pmatrix} = 2.8$$

Thus the established gene does somewhat worse than before in the new environment. The mutant gene, however, does even worse than that. Consequently, the established gene will reproduce more rapidly than the mutant gene, which will either die out or become a negligible minority, i.e., *the mutant is selected against.*

Case 2: On the other hand, suppose a mutant gene appears, giving rise only to doves. In this case the mutant gene is represented by $y = (1,0)$, and it is easily seen that $yAx^T = 3.33$, so that, once again, the mutant gene can hold its own against the established population. After a while, however, the population mixture will change slightly due to the new genes, so that the fraction of doves increases, and the new population mixture might be something like $z = (.35, .65)$. We then have

$$xAz^T = (.333, .667)\begin{pmatrix} 10 & 0 \\ 30 & -10 \end{pmatrix}\begin{pmatrix} .35 \\ .65 \end{pmatrix} = 3.833$$

At the same time

$$yAz^T = (1,0)\begin{pmatrix} 10 & 0 \\ 30 & -10 \end{pmatrix}\begin{pmatrix} .35 \\ .65 \end{pmatrix} = 3.50$$

In this case, the mutant gene does well against the new environment, but the established gene does even better, and so this mutant is also selected against.

We see, then, that the gene type $x = (.333, .667)$ is stable in the sense that it can expel either of the two extreme mutants $(1,0)$ and $(0, 1)$. The question is whether it can expel all others, i.e., whether there might not be some alternative y which would prove impossible to expel.

In fact, x will be able to expel a mutant represented by any $y \neq x$. For, since x and y are both mixed strategies, we must either have $y_1 < x_1$ or $y_1 > x_1$.

Suppose $y_1 < x_1$. Then $y_2 < x_2$, and, in that case, the resulting environment, z, being a mixture of x and y, must also have $z_1 < x_1$ and $z_2 > x_2$. As in case 1 above, it will be easy to see that the first component of Az^T is larger than the second component of Az^T, i.e., doves do better than hawks in this environment. But the established gene has a greater share of doves than does the mutant gene, and it follows that the established gene will do better. Hence the mutant gene, so long as it has $y_1 < x_1$, will be selected against.

If, instead, we suppose that $y_1 < x_1$, the resulting environment will also have $z_1 < x_1$. In this case, it will be seen that hawks do better than doves in this environment. But the mutant gene has a greater share of doves than the established gene, and so will be selected against. In any case, we see that the established gene x can expel any mutant invader. We say that x is an *evolutionary stable strategy* (E.S.S.).

More generally, an *evolutionary system* is a bimatrix game (A, A^T), where A can be any *square* matrix. A *genotype* is any mixed strategy $x = (x_1, \ldots, x_n)$ for this game. If x is an established genotype, it may be that some different (mutant) genotype, $y \neq x$, will attempt to invade. If so, the environment will be changed, slightly, from x to $ry + (1 - r)x$, where r is a small positive number. Then x can expel y if, for all sufficiently small $r > 0$

$$xA(ry + (1 - r)x)^T > yA(ry + (1 - r)x)^T \tag{8.1}$$

This leads us to our definition:

8.3.1 Definition. Let (A, A^T) be an evolutionary system (symmetric matrix game). A mixed strategy x is an *evolutionary stable strategy* (E.S.S.) if, for any $y \neq x$, there exists $\delta > 0$ such that, for all $0 < r < \delta$, (8.1) holds.

To see how this works, we note first that, if (8.1) holds for all small positive r, then, by continuity

$$xAx^T \geq yAx^T \tag{8.2}$$

so that, as a strategy for player I, x is an optimal response to x (as a strategy for II). By symmetry of the game (i.e., since $B = A^T$), x is an optimal response for II against I's choice of x, i.e., the pair of strategies $(x; x)$ is in equilibrium. Thus *a necessary condition for x to be an evolutionary stable strategy is that $(x; x)$ be an equilibrium pair.*

This condition is not, however, sufficient. To obtain a sufficient condition, we distinguish two cases for the mutant genotype y, depending on whether (8.2) holds with equality or with strict inequality.

If (8.2) holds with strict inequality, then there will be no problem: the continuity of the bilinear form guarantees that (8.1) will hold for all sufficiently small values of r.

Suppose, however, that (8.2) holds with equality. In this case, we rewrite (8.1) as

$$rxAy^T + (1 - r)xAx^T > ryAy^T + (1 - r)yAx^T$$

Since, however, $xAx^T = yAx^T$, the second terms on both sides of this inequality cancel out, leaving

$$rxAy^T > ryAy^T$$

or, since $r > 0$

$$xAy^T > yAy^T$$

This condition will be sufficient. We thus have the following theorem:

8.3.2 Theorem. A necessary and sufficient condition for x to be an evolutionary stable strategy for the evolutionary system (A, A^T) is that

(a) the pair $(x; x)$ is in equilibrium
(b) if $y \neq x$ is a strategy such that $yAx^T = xAx^T$, then

$$xAy^T > yAy^T \tag{8.3}$$

A point to be made is that the definition of E.S.S. is a purely local definition, which does not tell us what will happen outside a neighborhood of the E.S.S. x. It may be that, once we leave that neighborhood, the mutant y turns out to be much more adaptable than the established type x. Once x has been reached, however, it is impossible to leave its neighborhood, and so x will be maintained.

The following example makes this clear.

8.3.3 Example. Consider the evolutionary system

$$\begin{pmatrix} (10, 10) & (0, 0) \\ (0, 0) & (5, 5) \end{pmatrix}$$

It is easy to see that both pure strategies are evolutionary stable. For example, $x = (0, 1)$, the second pure strategy, is stable because for all z close to x, the second component of Az is larger than its first component. Thus in any environment close to x, the second row is always superior to the first row. Outside a neighborhood of x, however, the alternative pure strategy, $(1, 0)$, will be better. In fact, the first row seems clearly a better strategy than the second.

Heuristically, we might feel that the second row represents an evolutionary "wrong turn": for this species, it would be better if all members belonged to the first type. However, some time in the past (when perhaps the external environment was different from the present one), the species developed to give rise only to the second type. Now any "first-type" member would come into an environment consisting only of second-type animals and would have difficulty competing against them. Thus the first type, if introduced at this time, will simply die out. But *if* it were possible to introduce a large number of first-type individuals now, they could thrive. Given this last example, we might ask as to the existence of pure E.S.S.s. The following is an easily verified sufficient condition:

8.3.4 Corollary. Let (A, A^T) be an evolutionary system. If there is any i such that $a_n > a_{ji}$ for all $j \neq i$, then the ith row is a pure E.S.S.

We omit the proof of 8.4 as it is a direct corollary of 8.3. The advantage of this condition is that it is easy to verify. Its drawback is that, while sufficient, it is not necessary, and there is no easily verifiable necessary condition.

In general, of course, we cannot expect pure E.S.S.s to exist — bimatrix games do not in general have equilibrium points in pure strategies. We turn instead to the question of existence of E.S.S.s in mixed strategies.

The obvious question is whether such an E.S.S. will exist for all evolutionary systems. This is actually two questions:

1. Will a symmetric bimatrix game (A, A^T) have a symmetric equilibrium pair $(x; x)$?
2. If so, will this x (or one such x, if there is more than one) satisfy condition (8.3)?

We answer question 1 first, in the affirmative.

8.3.5 Theorem. Any symmetric bimatrix game (A, A^T) has at least one symmetric equilibrium point $(x; x)$.

Proof. Consider the proof of Theorem 8.1.2. Since $B = A^T$, it will follow that if the iterative process starts with $y = x$, then all $c_i = d_i$, and so $y' = x'$. Thus we can think of the transformation T defined there as a mapping of pairs $(x; x)$ to pairs (x', x'). The Brouwer fixed-point theorem is applicable here also, and tells us that we will have some pair $(x; x)$ in equilibrium, i.e., there exists a symmetric equilibrium pair.

The second question cannot, unfortunately, be answered in the affirmative: there is no guarantee of existence.

We can best see this by considering the following system:

8.3.6 Example. Consider the evolutionary system

$$\begin{pmatrix} (2,2) & (1,2) \\ (2,1) & (1,1) \end{pmatrix}$$

It is easy to see that, for this bimatrix game, player I's payoff does not depend on his choice of strategy. Hence any strategy x will be an optimal response for I against any y of II. Similarly, any y of II will be an optimal response against any strategy x of I. It follows that any pair $(x; y)$ is in equilibrium, and so any $(x; x)$ is a symmetric equilibrium pair.

However, there is no equilibrium point which has the stability property (Condition 2 of Theorem 8.3.2). In fact, since the two rows of matrix A are equal, we will have

$$xAx^T = yAx^T$$

$$xAy^T = yAy^T$$

for any mixed strategies x and y. Thus there is no E.S.S. for this game.

The reader may well feel that the trouble with this last example is that it is very degenerate, i.e., it lies precisely in the fact that the rows of matrix A are equal. In effect, this means that the evolutionary system treats all genotypes equally. Under the circumstances, it will not distinguish between them, and so it can never select against any mutant. This suggests that if the entries in the matrix A were all different (something which will hold with high probability), then we might be able to guarantee an E.S.S.

We have, in fact, the following interesting result:

8.3.7 Theorem. If A is a 2×2 matrix such that $a_{11} \neq a_{21}$ and $a_{12} \neq a_{22}$, then the system has at least one E.S.S.

Proof. By Corollary 8.3.4, an E.S.S. in pure strategies will exist if $a_{11} > a_{21}$ or if $a_{22} > a_{12}$. Hence we need only consider the case when $a_{11} < a_{21}$ and $a_{22} > a_{12}$.

In this case, consider the mixed strategy x, given by (2.23):

$$x = \frac{A^*J^T}{JA^*J^T} \tag{8.4}$$

or, equivalently

$$x_1 = \frac{a_{12} - a_{22}}{a_{12} + a_{21} - a_{11} - a_{22}} \qquad x_2 = \frac{a_{21} - a_{11}}{a_{12} + a_{21} - a_{11} - a_{22}}$$

It is easily seen that this x is indeed a mixed strategy. Moreover, we note that the two components of the product Ax^T are both equal to the quantity v, given by (2.24):

$$A_1.x^T = A_2.x^T = v = \frac{|A|}{JA^*J^T}$$

It will follow that the product $yAx^T = v$ for all strategies y, i.e., all strategies of I are optimal responses to x as a strategy of II. In particular, x is an optimal response for I against x and, by symmetry, $(x; x)$ is an equilibrium pair.

It remains to prove the stability condition. Suppose y is any strategy other than x. We will have $y_1 = x_1 - \sigma$, $y_2 = x_2 + \sigma$, where σ is not zero. We know, moreover, that $xAx^T = yAx^T$, so that we have

$$\begin{aligned}
(x - y)Ay^T &= (x - y)A(x + (y - x))^T \\
&= (x - y)Ax^T + (x - y)A(y - x)^T \\
&= (x - y)A(y - x)^T \\
&= (\sigma, -\sigma)A(-\sigma, \sigma)^T \\
&= -\sigma^2(1, -1)A(1, -1)^T \\
&= -\sigma^2(a_{11} - a_{21} + a_{22} - a_{12})
\end{aligned}$$

In this last expression, we know that $\sigma \neq 0$ and that the expression inside the parentheses is negative. Hence $(x - y)Ay^T > 0$, and by Theorem 8.3.2, it follows that x is an E.S.S.

We have thus not only proved that "almost every" 2×2 evolutionary system has an E.S.S., but also obtained a rule for finding this E.S.S. In fact, the solution is exactly as for 2×2 matrix games (cf. 2.5.5, above): first we look for an E.S.S. in pure strategies; If no such E.S.S. exists, we apply formula (8.4) which is exactly the same as (2.23).

8.3.8 Example. Find an E.S.S. for the system (A, A^T), given

(a) $\begin{pmatrix} (3,3) & (5,4) \\ (4,5) & (2,2) \end{pmatrix}$

(b) $\begin{pmatrix} (4,4) & (5,3) \\ (3,5) & (2,2) \end{pmatrix}$

(c) $\begin{pmatrix} (8,8) & (0,10) \\ (10,0) & (3,3) \end{pmatrix}$

(d) $\begin{pmatrix} (4,4) & (5,2) \\ (2,5) & (3,3) \end{pmatrix}$

(e) $\begin{pmatrix} (4,4) & (2,5) \\ (5,2) & (3,3) \end{pmatrix}$

(f) $\begin{pmatrix} (4,4) & (3,2) \\ (2,3) & (5,5) \end{pmatrix}$

Solution:

(a) If this were an ordinary bimatrix game, it would have equilibrium points at (row 1, column 2) and (row 2, column 1). However, for an evolutionary system, such points cannot be considered. Rather, we must apply (8.4), giving us the E.S.S. (.75, .25), with payoff 3.5.

(b) We note here that the first row dominates the second. Thus (1,0) will be the only E.S.S. The payoff is 4.

(c) This is a version of the Prisoner's Dilemma. The species would do much better by choosing the first row (all doves), with a payoff of 8. Unfortunately, the second row dominates: even though doves do well, a hawk would thrive in their midst and eventually drive out the doves. It follows that the only E.S.S. is (0,1), with a payoff of 3.

(d) For this system, the first row dominates the second. This is the only equilibrium pair, and hence the only E.S.S.

(e) Here, the second row dominates the first row, and thus the second row is the only E.S.S.

(f) In this case, there are two E.S.S.s: each pure strategy is a stable equilibrium point. There is an additional equilibrium point $(x; x)$, where $x = (.5, .5)$, but this is not stable.

3×3 *and Larger Systems*

For 3×3 systems and larger, the situation can be considerably more complicated. As a relatively simple example, consider the system

$$\begin{pmatrix} (2,2) & (1,1) & (0,0) \\ (1,1) & (0,0) & (2,2) \\ (0,0) & (2,2) & (1,1) \end{pmatrix}$$

For this system, the first row and column clearly give a stable, symmetric equilibrium, which is therefore an E.S.S. There are equilibrium points at (row 2, column 3) and (row 3, column 2), but these are not symmetric and hence can be disregarded. Apart from this, however, there is a symmetric mixed strategy equilibrium at $x = (0, .333, .667)$, with payoff $xAx^T = 1.333$, and another one at $z = (.333, .333, .333)$, with payoff $zAz^T = 1$.

We check these last two for stability. For $x = (0, .333, .667)$, we note that $Ax^T = (.333, 1.333, 1.333)$. Thus the optimal responses to x are all y with $y_1 = 0$. If we restrict ourselves to such y (in effect discarding the first row, so that we are left with a 2×2 matrix), it is easy to see that the proof of Theorem 8.3.5 can be carried through directly, and it follows that x is an E.S.S.

For $z = (.333, .333, .333)$, we see that $Az^T = (1, 1, 1)$, and so all strategies are optimal replies to z. In particular, $y = (1, 0, 0)$ will be an optimal reply. Now $zAy^T = 1$ while $yAy^T = 2$, and we see that z is not stable.

In all then, the system has two E.S.S.s: the pure strategy (row 1), and the mixed strategy $(0, .333, .667)$. Of the other equilibrium points of the bimatrix game, two are not symmetric and hence inapplicable, while one is symmetric but unstable.

We note further that we cannot generalize Theorem 8.3.5 to 3×3 systems. As an example, consider the following system in which each column has all different entries:

$$\begin{pmatrix} (0,0) & (1,-1) & (-2,2) \\ (-1,1) & (0,0) & (3,-3) \\ (2,-2) & (-3,3) & (0,0) \end{pmatrix}$$

It is easy to see that this is a zero-sum game: it is in fact the game of Example 2.6.3, which, as we saw there, has a unique optimal strategy $x = (.5, .333, .167)$ for both players. It will follow that $(x; x)$ is the unique equilibrium pair. Since $A = -A^T$, the value of this game is 0, and, moreover, $xA = (0, 0, 0)$. Also, for all y, $yAx^T = 0$, and, by the skew-symmetry of A, $yAy^T = 0$. Thus x is not stable, and there is no E.S.S.

8.4. Correlated Equilibria

Let us consider the situation faced by two motorists as they approach an intersection: assuming they reach the intersection simultaneously, each one has the option of crossing without stopping, or of yielding and allowing the other to pass. Clearly, there is a loss of time (and utility) associated with yielding. On the other hand, insistence by both on crossing will lead to a large loss for both. The game can perhaps be represented by the matrix

$$\begin{pmatrix} (-10,-10) & (5,0) \\ (0,5) & (-1,-1) \end{pmatrix}$$

where, for both players, the first row or column represents crossing without stopping, and the second, yielding. It is not difficult to see that there are two equilibria in pure strategies, namely (row 1, column 2), and (row 2, column 1): one yields while the other passes without stopping. There is also an equilibrium in mixed strategies, in which both players use (1/16,15/16), with an expected payoff of $-5/8$ for both players.

Now, under some circumstances (e.g., if one of the two roads is clearly more important than the other, or if one of the vehicles is an emergency vehicle), one or the other of the two pure strategy equilibria seems very reasonable. In general, however, it is not clear which of these two equilibria should be chosen and even if the players (motorists) know which of the two is to be chosen, one of them might feel unhappy about this. ("Why should *I* be the one to yield?") The mixed strategy equilibrium seems fair enough, but is otherwise quite unsatisfactory. In the interest of fairness and efficiency, then, and to avoid misunderstandings, a traffic light might be installed.

In its simplest form, the traffic light is merely a randomization device which, with a certain probability, tells each of the two drivers whether to pass or stop. In a sense, this does not seem very different from a pair of mixed strategies, except for one thing: the two players' probability distributions can be correlated. Thus, the traffic light can, with probability .55, tell player 1 to cross (green light) while player 2 yields (red light); with probability .4, it tells 1 to yield while 2 crosses; with probability 0.05, it tells both players to stop. This gives an expected payoff of 2.7 to 1 and 1.95 to 2. Note that this would not be possible with mixed (uncorrelated) strategies.

Of course, the two motorists are not bound to obey the traffic light. (We assume there are no motorcycle policemen around to enforce it.) Nevertheless, we claim that this pair of strategies (obeying the light, i.e., stopping on red and crossing on green, for each player) is in equilibrium. We call this a *correlated equilibrium*.

To see that these two strategies are in equilibrium, we note that since the traffic light never is green for both players, then player 2 knows, on seeing a green light, that player 1 is seeing red. Thus, if 1 is indeed stopping on red, it follows that 2 can never gain by stopping on green.

Suppose, on the other hand, that 2 sees a red light. In this case he cannot be certain that 1 sees a green light, but there is a rather high (.917) probability that this is the case, and that (if 1 is crossing on green) there will be an accident if 2 does not stop. It is easy to calculate that 2 cannot gain by crossing on red. Thus, if 1 is obeying the light, 2's best response is also to obey the light.

In similar fashion, it can be seen that if 2 is obeying the light, then it is best for 1 also to obey the light. We conclude that these two strategies are in equilibrium.

It is important to notice, in passing, that if either one of the two players sees a red light, this player is not certain whether both players are seeing red. In other words, neither player has full information as to what the traffic light is doing: each player knows only the signal that the light has given him or her.

In general, let (A, B) be an $m \times n$ bimatrix game. A *correlated* mixed strategy is an $m \times n$ matrix $P = (p_{ij})$, such that

$$\sum_i \sum_j p_{ij} = 1 \tag{8.5}$$

$$p_{ij} \geq 0 \quad \text{for all} \quad i, j \tag{8.6}$$

This correlated strategy is to be thought of as a randomization device which, with probability p_{ij}, signals I to use the ith row and II to use the jth column. It gives no further information to either player.

The correlated strategy P is in equilibrium if, whenever player 1 is given the signal to use i, he or she cannot gain by disobeying, given the conditional probability that this signal induces on 2's action, and similarly, 2 cannot gain by disobeying a signal to use j.

Now, if 1 is given the signal to use row i, and uses i, then (assuming 2 is obeying the signal) 1's expected payoff will be

$$\frac{\sum_{j=1}^{n} a_{ij} p_{ij}}{\sum_j p_{ij}}$$

whereas, if he uses k, the expected payoff will be

$$\frac{\sum_{j=1}^{n} a_{kj} p_{ij}}{\sum_j p_{ij}}$$

Since I is not to gain by disobeying, it must be that for any pair of rows, i and k

$$\sum_{j=1}^{n} (a_{ij} - a_{kj}) p_{ij} \geq 0 \tag{8.7}$$

In a similar way, since j is not to gain by disobeying, it must follow that for any pair of columns, j and l

$$\sum_{i=1}^{m} (b_{ij} - b_{il}) p_{ij} \geq 0 \tag{8.8}$$

This gives us the following definition:

8.4.1 Definition. A *correlated equilibrium* for the bimatrix game (A, B) is a correlated mixed strategy, P, which satisfies conditions (8.7)–(8.8).

It may be noted that there are a total of $m(m-1)$ constraints of the form (8.7) and $n(n-1)$ of the form (8.8). Apart from this there is (8.5). All told, then, this is a system of $m(m-1)+n(n-1)+1$ constraints (not counting non-negativity constraints) in mn unknowns.

8.4.2 Example. Let us return to the example discussed above of the two motorists at an intersection. With $m=n=2$, the system will have five constraints and four variables

$$-10p_{11}+4p_{12} \geq 0$$
$$10p_{21}-4p_{22} \geq 0$$
$$-10p_{11}+4p_{21} \geq 0$$
$$10p_{12}-4p_{22} \geq 0$$
$$p_{11}+p_{12}+p_{21}+p_{22} = 11$$
$$b_{ij} \geq 0$$

It is not too difficult to obtain all solutions to this system: quite simply, both $2.5p_{11}$ and $0.4p_{22}$ must be no larger than the smaller of p_{12} and p_{21}. Apart from this, the four variables must be non-negative and add up to 1. Thus, the correlated strategy given above, $p_{11}=0$, $p_{12}=.55$, $p_{21}=.4$, $p_{22}=.05$, certainly is an equilibrium, but there are many others. Some seem very efficient — e.g., strategies such as

$$p_{11} = 0, p_{21} = .5, p_{12} = .5, p_{22} = 0$$

or

$$p_{11} = 0, p_{12} = .7, p_{12} = .3, p_{22} = 0$$

(this latter corresponding perhaps to an intersection between a busy route 1 and a small route 2) are both equilibria — but even a dangerous and inefficient one, such as

$$p_{11} = .1, p_{12} = .4, p_{21} = .3, p_{22} = .2$$

(where perhaps the traffic light malfunctions sporadically) is also an equilibrium. On the other hand, a strategy such as

$$p_{11} = .2, p_{12} = .5, p_{21} = .3, p_{22} = 0$$

is not an equilibrium, essentially because, when player 2 is given a green light, there is too large a probability that 1 also is receiving a green light. Thus player 2 would be better off not crossing on green, even if he knows that player 1 is obeying the light.

As can be seen from the example, bimatrix games can have very large sets of correlated equilibria. (Note that the game mentioned above had only three equilibria in (uncorrelated) mixed strategies.) The obvious question is whether all bimatrix games have at least one correlated equilibrium. It is not difficult to see that this must be so.

In fact, any pair (x, y) of mixed strategies for the two players in a bimatrix game can be thought of as representing a joint mixed strategy, $P = x^T y$, i.e., it is given by $p_{ij} = x_i y_j$ for each i and j. The reader can easily prove that if (x, y) is an equilibrium pair, then $P = x^T y$ is a correlated equilibrium. Since we already know (Theorem 8.1.3) that all bimatrix games have equilibrium pairs of mixed strategies, it follows that they also have correlated equilibria.

Apart from this, there is the question of computation. Let (A, B) be an $m \times n$ bimatrix game. To simplify our notation, we will let M and N be the sets of pure strategies for players 1 and 2, respectively, in (A, B). We consider, then, the auxiliary $mn \times (m(m-1) + n(n-1))$ matrix, C, which has one row for each pair $(i, j) \in M \times N$, and one column for each pair $(h, k) \in M \times M \cup N \times N$. The new matrix, C, has entries

$$c_{(i,j),(h,k)} = \begin{cases} a_{ij} - a_{kj} & \text{if } i = h \in M \text{ and } k \in M \\ b_{ij} - b_{ik} & \text{if } j = h \in N \text{ and } k \in N \\ 0 & \text{otherwise} \end{cases}$$

We will now consider C as a matrix game played between players I and II (not to be confused with players 1 and 2 of the bimatrix game). Let w be the value of this game, and let $P = (p_{ij})$ be an optimal mixed strategy for player I in this game.

Now, if $w \geq 0$, it is easy to see that P will satisfy conditions (8.7)–(8.8), and is hence a correlated equilibrium for (A, B). Suppose, on the other hand, that $w < 0$. In this case there is no P satisfying (8.7)–(8.8). But we have already seen that a correlated equilibrium always exists. It follows that $w \geq 0$, and hence the optimal P for game C will always be a correlated equilibrium for (A, B).

We note in passing that if $w = 0$, then any correlated equilibrium of (A, B) will indeed be an optimal strategy for C. If, however, $w > 0$ (which is certainly possible), then optimality of the strategy P is not necessary: it suffices that P guarantees the gain-floor 0 for player I.

8.4.3 Example. Consider the bimatrix game of Example 8.4.2. The auxiliary matrix game is here a 4×4 game

	(1,2)	(2,1)	(1′,2′)	(2′,1′)
(1,1′)	−10	0	0	−10
$C =$ (1,2′)	4	0	10	0
(2,1′)	0	10	0	4
(2,2′)	0	−4	−4	0

(where we have used unprimed symbols to denote rows of the original bimatrix game, and primed symbols to denote columns).

It is not difficult to see that this game has value $w = 2$ obtained by using the optimal strategies $(0,\frac{1}{2},\frac{1}{2},0)$ for I and $(\frac{1}{2},0,0,\frac{1}{2})$ for II. Then $(0,\frac{1}{2},\frac{1}{2},0)$, which is the unique optimal strategy for I, is certainly a correlated equilibrium in (A, B), yielding payoffs of $\frac{5}{2}$ to each player. However, since the value of C is positive, there are many other correlated equilibria (none of which are optimal strategies for I in C), as was noted above.

8.4.4 Example. Consider the bimatrix game

$$\begin{pmatrix} (3,1) & (2,5) & (6,0) \\ (1,4) & (3,3) & (2,6) \end{pmatrix}$$

In this case, we obtain the auxiliary matrix game

1	(1,2)	(2,1)	(1',2')	(1',3')	(2',1')	(2',3')	(3',1')	(3',2')
(1,1')	2	0	−4	1	0	0	0	0
(1,2')	−1	0	0	0	4	5	0	0
(1,3')	4	0	0	0	0	0	−1	−5
(2,1')	0	−2	1	−2	0	0	0	0
(2,2')	0	1	0	0	−1	−3	0	0
(2,3')	0	−4	0	0	0	0	2	3

It may be seen that this game has value 0, obtained by using the optimal strategies

$$P = \begin{pmatrix} 0 & .3 & .075 \\ 0 & .5 & .125 \end{pmatrix}$$

for player I, and $z_{1'2'} = z_{1'3'} = 0.5$, all other $z_{hk} = 0$, for player II. The optimal strategy for I is unique, and so this is the only strategy that guarantees 0 for I, i.e., it is the only correlated equilibrium. In fact, of course, it is really not correlated, as it corresponds to the equilibrium pair of mixed strategies (in the bimatrix game) (0.375,0.625) for player 1, and (0,0.8,0.2) for player 2. The expected payoffs will be 2.8 for 1 and 3.75 for 2. In this case, then, the correlation of strategies yields nothing new. (On the other hand, this allowed us to compute the equilibrium pair by means of linear techniques.)

Problems

1. (a) Consider a "supergame" consisting of 100 repetitions of the "Prisoner's Dilemma." In the supergame, each player has a very large number of pure strategies, each one of which decides what to do at the nth play of the game on the

basis of the supergame's "history," (i.e., what has happened on the first $n-1$ plays). Show that if two strategies are in equilibrium, use of these two strategies will cause both players to use the "double-cross" (second) strategy at each play of the game.

(b) Show, however, that, for $n>2$, the supergame strategy σ_n ("Play the first [cooperative] strategy for the first n trials, so long as my opponent does likewise, and double-cross the last $100-n$ trials") is better in the sense that, against the opponent's optimal response, it gives a better payoff.

(c) Show that if the supergame consists of X trials, where X is a random variable with an exponential distribution $P(n) = ke^{-\alpha n}$, then there exist equilibrium pairs which involve using the cooperative first strategy at each play of the game.

2. It was mentioned, in Problem 2.9, that two-person constant-sum games could be solved by the method of "fictitious play." Show that this method cannot be used to obtain equilibrium points of bimatrix games by considering the game (A, B), where

$$A = \begin{pmatrix} 2 & 0 & 1 \\ 1 & 2 & 0 \\ 0 & 1 & 2 \end{pmatrix}, \quad B = \begin{pmatrix} 1 & 0 & 2 \\ 2 & 1 & 0 \\ 0 & 2 & 1 \end{pmatrix}$$

(a) Assume the players start by choosing (1,1). There will be a run of (1,1), followed by a run of (1,3), then runs of (3,3), (3,2), (2,2), (1,1), etc., in a cycle.

(b) Each run is at least twice as long as the previous run.

(c) The empirical strategies do not converge at all, but cycle. Neither of the two cycles passes through $(\frac{1}{3}, \frac{1}{3}, \frac{1}{3})$, which is the unique equilibrium point.

3. Find equilibrium pairs for the following bimatrix games:

(a) $A = \begin{pmatrix} 5 & 1 \\ 2 & 4 \end{pmatrix}$ $B = \begin{pmatrix} 2 & 3 \\ 3 & 1 \end{pmatrix}$

(b) $A = \begin{pmatrix} 3 & 4 & 1 \\ 2 & 1 & 3 \end{pmatrix}$ $B = \begin{pmatrix} 2 & 5 & 3 \\ 4 & 2 & 5 \end{pmatrix}$

(c) $A = \begin{pmatrix} 4 & 2 \\ 1 & 3 \end{pmatrix}$ $B = \begin{pmatrix} 2 & 0 \\ 1 & 4 \end{pmatrix}$

4. Of the equilibrium pairs obtained in Problem 3, above, determine which are perfect equilibria. Which are strictly perfect?

5. For the following (symmetric) bimatrix games, find symmetric equilibrium pairs. Which of these are evolutionary stable systems?

(a) $A = \begin{pmatrix} 4 & 2 \\ 1 & 3 \end{pmatrix}$ $B = A^t$

(b) $A = \begin{pmatrix} 1 & 3 \\ 4 & 2 \end{pmatrix}$ $B = A^t$

(c) $A = \begin{pmatrix} 1 & 2 & 3 \\ 2 & 1 & 4 \\ 2 & 3 & 1 \end{pmatrix}$ $B = A^t$

6. Find all equilibrium pairs of correlated strategies for the games in Problem 1 above.

7. Prove that any perturbed game $(A, B; \sigma, \tau)$ has at least one equilibrium pair of mixed strategies. (*Hint*: modify the proof of Theorem 8.1.2.)

Chapter 9

Two-Person Cooperative Games

9.1. The Bargaining Problem

We consider now the case of games in which cooperation between the two players is allowed. This means that binding contracts can be made, that correlated mixed strategies are allowed, and that utility can be transferred from one player to the other (though not always linearly).

There is, generally, a set of outcomes which can be obtained by the two players by acting together. If we choose a pair of utility functions for the two players, this set can be mapped into a subset of two-dimensional Euclidean space R_2. Its image under this mapping will be closed and bounded above (at least, the sum of the coordinates will be bounded above). Moreover, since lotteries are permitted (by means of correlated mixed strategies) and utility is linear with lotteries, this image will be convex. The problem, then, is to choose a point from this set which will satisfy both players.

Given a two-person general-sum game, then, there is a certain subset S of R_2, called the *feasible set*. It is feasible in the sense that, given any $(u, v) \in S$, it is possible for the two players, acting together, to obtain the utilities u and v, respectively. Generally speaking, however, we find that the more one player gets, the less the other player will be able to get (though this need not always be so). Now, how much will one player be willing to give the other? How little will he be willing to accept, as the price of his cooperation?

Though it is of course impossible to determine how a person will act (in general, there are many personality differences to be taken into account), we can nevertheless set a minimum to the amount that a player will accept for himself. This is the amount that he can obtain by unilateral action, whatever the other player does. This is, of course, the maximin value of the game for that player. Let us call these two values u^* and v^*. Thus

$$u^* = \max_x \min_y xAy^{\mathrm{T}} \tag{9.1}$$

$$v^* = \max_y \min_x xBy^{\mathrm{T}} \tag{9.2}$$

(where x and y are taken over the sets of all mixed strategies) if the game is a bimatrix game with matrices (A, B).

We assume, then, that we are given a set S, together with maximin values (u^*, v^*). We want a rule which will assign, to such a triple (S, u^*, v^*), a "bargaining solution":

$$\varphi(S, u^*, v^*) = (\bar{u}, \bar{v})$$

We ask, naturally, how such a function φ is to be defined. While it is important to repeat that the outcome in any particular case will depend on the personalities and bargaining abilities of the two players, the following axioms seem reasonable conditions to lay on any such function. (These axioms are those of John Nash.)

9.1.1

N1   (Individual Rationality). $(\bar{u}, \bar{v}) \geq (u^*, v^*)$.

N2   (Feasibility). $(\bar{u}, \bar{v}) \in S$.

N3   (Pareto-Optimality).   If $(u, v) \in S$, and $(u, v) \geq (\bar{u}, \bar{v})$, then $(u, v) = (\bar{u}, \bar{v})$.

N4   (Independence of Irrelevant Alternatives). If $(\bar{u}, \bar{v}) \in T \subset S$, and $(\bar{u}, \bar{v}) = \varphi(S, u^*, v^*)$, then $(\bar{u}, \bar{v}) = \varphi(T, u^*, v^*)$.

N5   (Independence of Linear Transformations). Let T be obtained from S by the linear transformation

$$u' = \alpha_1 u + \beta_1$$

$$v' = \alpha_2 v + \beta_2$$

Then, if $\varphi(S, u^*, v^*) = (\bar{u}, \bar{v})$, we must have

$$\varphi(T, \alpha_1 u^* + \beta_1, \alpha_2 v^* + \beta_2) = (a_1 \bar{u} + \beta_1, \alpha_2 \bar{v} + \beta_2)$$

N6   (Symmetry). Suppose S is such that

$$(u, v) \in S \Leftrightarrow (u, u) \in S$$

Suppose also $u^* = v^*$, and $\varphi(S, u^*, v^*) = (\bar{u}, \bar{v})$. Then $\bar{u} = \bar{v}$.

Of these six axioms, N1, N2, and N3 are obvious and need no justification. N4 states that, if a point $(\bar{u}, \bar{v})$ is the solution to a bargaining problem, and then the feasible set is enlarged, the solution to the new problem will either be $(\bar{u}, \bar{v})$ itself or one of the new points in the set, not a point in the old, smaller set. As such (and, also, in other contexts) it has been open to criticism. N5 is natural enough if we assume that any utility function is as good as any other. It is open to criticism if absolute utility (7.4) is introduced. Finally, N6 is quite acceptable if the bargaining is between two equal entities; it may not be so acceptable if the bargaining is between unequal entities, e.g., between one person and an entire community.

Several other axioms might be suggested for a bargaining function; an example would be

N7　　(Monotonicity).　　If $T \subset S$, then

$$\varphi(T, x^*, y^*) \leq \varphi(S, x^*, y^*)$$

Unfortunately, N7 is not compatible with N2 and N3. Because it is difficult to quarrel with either of these, we conclude that N7 should be discarded (though someone might claim N7 is not very different from N4). In general, we need consider no further axioms because of the following remarkable theorem:

9.1.2 Theorem. There is a unique function φ defined on all bargaining problems (S, x^*, y^*) satisfying Axioms N1–N6.

In preparing the proof of this theorem, we give the following lemmas:

9.1.3 Lemma. If there are any points $(u, v) \in S$ such that $u > u^*, v > v^*$ then there is a unique point $(\bar{u}, \bar{v})$ which maximizes the function

$$g(u, v) = (u - u^*)(v - v^*)$$

over the subset of S for which $u \geq u^*$.

Proof. By hypothesis, this subset of S is compact. As g is continuous, it must have a maximum over this set. By hypothesis also, this maximum, M, will be positive. Suppose, then, that there are two points (u', v') and (u'', v'') which maximize $g(u, v)$. As $M > 0$, we cannot have $u' = u''$ as this would imply $v' = v''$. Suppose that $u' < u''$; this will mean $v' > v''$.

As S is convex, $(\hat{u}, \hat{v}) \in S$, where $\hat{u} = (u' + u'')/2$, $\hat{v} = (v' + v'')/2$. Now

$$
\begin{aligned}
g(\hat{u}, \hat{v}) &= \frac{(u' - u^*) + (u'' - u^*)}{2} \cdot \frac{(v' - v^*) + (v'' - v^*)}{2} \\
&= \frac{(u' - u^*)(v' - v^*)}{2} + \frac{(u'' - u^*)(v'' - v^*)}{2} \\
&\quad + \frac{(u' - u'')(v'' - v')}{4}
\end{aligned}
$$

Each of the first two terms in this last expression is equal to $M/2$. The third term, however, is positive. But this means $g(\hat{u}, \hat{v}) > M$, which contradicts the maximality of M. Hence the point $(\bar{u}, \bar{v})$ which maximizes g is unique.

9.1.4 Lemma. Let $S, (u^*, v^*)$ and $(\bar{u}, \bar{v})$ be as in Lemma 9.1.3. Let

$$h(u, v) = (\bar{v} - v^*)u + (\bar{u} - u^*)v$$

Then, if $(u, v) \in S$, the inequality $h(u, v) \leq h(\bar{u}, \bar{v})$ will hold.

Proof. Suppose $(u, v) \in S$, with $h(u, v) > h(\bar{u}, \bar{v})$. Let $0 < \varepsilon < 1$. By the convexity of S, $(u', v') \in S$, where $u' = \bar{u} + \varepsilon(u - \bar{u})$ and $v' = \bar{v} + \varepsilon(v - \bar{v})$. By the linearity of h, $h(u - \bar{u}, v - \bar{v}) > 0$. Now,

$$g(u', v') = g(\bar{u}, \bar{v}) + \varepsilon h(u - \bar{u}, v - \bar{v}) + \varepsilon^2(u - \bar{u})(v - \bar{v})$$

Letting ε approach zero, the last term becomes negligible. This means that $g(u', v') > g(\bar{u}, \bar{v})$. But this contradicts the maximality of $g(\bar{u}, \bar{v})$.

Lemma 9.1.4 says in effect that the line though $(\bar{u}, \bar{v})$ whose slope is the negative of the line joining $(\bar{u}, \bar{v})$ to (u^*, v^*) is a support for S; i.e., S lies entirely on or below this line.

We can now prove Theorem 9.1.2. We shall in fact show that, under the hypothesis of Lemma 9.1.3, the point $(\bar{u}, \bar{v})$ which maximizes $g(u, v)$ must be the solution to the bargaining problem. If this hypothesis does not hold, the problem is much simpler.

Proof of 9.1.2: Assume the hypothesis of Lemma 9.1.3 holds. Then the point $(\bar{u}, \bar{v})$ which maximizes $g(u, v)$ is well defined. It clearly satisfies Axioms N1 and N2 by construction. It satisfies N3 for, if $(u, v) \geq (\bar{u}, \bar{v})$, but $(u, v) \neq (\bar{u}, \bar{v})$, then $g(u, v) > g(\bar{u}, \bar{v})$. It satisfies N4 since, if it maximizes $g(u, v)$ over S, it will certainly maximize it also over the smaller set T. It must satisfy N5, for, if we set $u' = \alpha_1 u + \beta_1$ and $v' = \alpha_2 v + \beta_2$, then

$$g'(u', v') = [u' - (\alpha_1 u^* + \beta_1)][v' - (\alpha_2 v^* + \beta_2)] = \alpha_1 \alpha_2 g(u, v)$$

and so if $(\bar{u}, \bar{v})$ maximizes $g(u, v)$, it follows that $(\bar{u}', \bar{v}')$ maximizes $g'(u', v')$. Finally, $(\bar{u}, \bar{v})$ satisfies N6. In fact, suppose that S is symmetric in the sense of N6, and $u^* = v^*$. Then $(\bar{v}, \bar{u}) \in S$. But clearly $g(\bar{u}, \bar{v}) = g(\bar{v}, \bar{u})$. Since $(\bar{u}, \bar{v})$ is the unique point which maximizes $g(u, v)$, it follows that $(\bar{u}, \bar{v}) = (\bar{v}, \bar{u})$; i.e., $\bar{u} = \bar{v}$.

We know, thus, that $(\bar{u}, \bar{v})$ satisfies Axioms N1–N6. We must now show that it is the unique rule which satisfies these axioms. Let $(\bar{u}, \bar{v})$ be defined, then, as above. Consider the set

$$U = \{(u, v) | h(u, v) \leq h(\bar{u}, \bar{v})\}$$

(see Figure 9.1). By Lemma 9.1.3, $S \subset U$.

Let T be obtained from U by the linear transformation

$$u' = \frac{u - u^*}{\bar{u} - u^*}$$
$$v' = \frac{v - v^*}{\bar{v} - v^*} \tag{9.3}$$

It is easy to see that T is simply the set $\{(u', v') | u' + v' \leq 2\}$; moreover, $u^{*'} = v^{*'} = 0$. Because T is symmetric, it follows by N6 that the solution must lie on the line $u' = v'$; by N3, it must be the point $(1,1)$. Inverting the transformation 9.3, it follows by N5 that $(\bar{u}, \bar{v})$ must be the solution of (U, u^*, v^*). But since $(\bar{u}, \bar{v}) \in S$, it follows that $(\bar{u}, \bar{v})$ must be the solution of (S, u^*, v^*).

Let us suppose, now, that the hypothesis of Lemma 9.1.3 does not hold; i.e., there are no points $(u, v) \in S$ with $u > u^*$, $v > v^*$. By the convexity of S, we know that if there is any $(u, v) \in S$ with $u > u^*$, $v = v^*$, there can be no $(u, v) \in S$ with $v > v^*$. Under

Figure 9.1.

these conditions, we simply let $(\bar{u}, \bar{v})$ be the point in S which maximizes u, subject to the constraint $v = v^*$. Similarly, if there is $(u, v) \in S$ with $u = u^*, v > v^*$, there can be no $(u, v) \in S$ with $u > u^*$, and we let $(\bar{u}, \bar{v})$ be the point in S which maximizes v, subject to the constraint $u = u^*$. It is easily checked that these solutions satisfy the axioms; it is also easy to see that Axioms N1, N2, N3 will allow no other solution.

Theorem 9.1.2 shows, then, that Nash's axioms can be satisfied by a unique scheme; we have also shown what Nash's solution is. According to Lemma 9.1.3, if the boundary of S is smooth (i.e., has a tangent) at the point $(\bar{u}, \bar{v})$, the line of constant $h(u, v)$ is tangent to S at that point. Now the slope of the boundary of S, at any point, represents the rate at which utility can be transferred from one player to the other. What Nash's scheme tells us, in brief, is that additional utility must be divided between the two players in a ratio equal to the rate at which this utility can be transferred. Naturally, since utility is not assumed to be linearly transferable, there may be only one point at which utility is transferable at this given rate. Figure 9.2 illustrates this.

In case the utility is linearly transferable, of course, the problem becomes much simpler. In fact, we may assume (by a change of utility scales if necessary) that the rate at which utility can be transferred is 1:1 (i.e., player I can transfer one unit of utility to II by giving up one unit himself). Thus S contains all the points on or below

Figure 9.2.

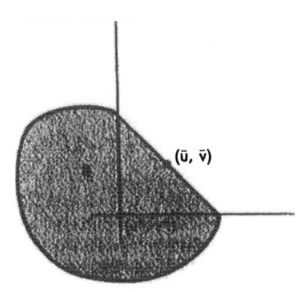

Figure 9.3.

a certain line $u + v = k$ (where k is the maximum possible utility the two players can obtain together) and above and to the right of (u^*, v^*). (See Figure 9.3.) The corresponding Nash solution will be $(\bar{u}, \bar{v})$, where $\bar{u} = (u^* - v^* + k)/2$, and $\bar{v} = (v^* - u^* + k)/2$.

It is easy to see that $\bar{u} + \bar{v} = k$, while $\bar{u} - \bar{v} = u^* - v^*$. Thus in this case, the relative positions of the two players are maintained, while their utilities are increased as much as possible (i.e., the excess utility is divided evenly between the two players).

9.1.5 Example. Two men are offered \$100 if they can decide how to divide the money. The first man is assumed to be very rich, while the second man has \$100 capital in all. It is assume that the utility of a sum of money is proportional to its logarithm. How should the money be divided?

As the first player is very rich, we can assume that the utility of \$$x$, with $x \leq 100$, is proportional to x. As the second player has only \$100, the utility he obtains from \$$x$ will be

$$\log(100 + x) - \log 100 = \log\frac{100 + x}{100}$$

(We are, of course, setting $u^* = v^* = 0$.) Thus the set S will consist of the convex hull of $(0,0)$ and the arc with equation

$$v = \log\frac{200 + u}{100}$$

As this function is convex, S consists simply of the area bounded by this curve (see Figure 9.4); of course, we are only interested in its intersection with the positive quadrant.

Now, we want the point in S which maximizes uv, i.e., the value of u which maximizes

$$g = u \log\frac{200 - u}{100}$$

When we differentiate and set equal to zero, we obtain the equation

$$\frac{u}{200 - u} = \log\frac{200 - u}{100}$$

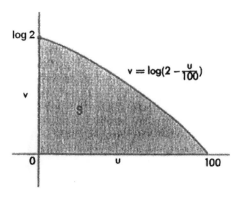

Figure 9.4.

This gives us, approximately, $u = 54.4$; i.e., player I should get approximately \$54.40, while II gets \$45.60.

In a sense, this result seems strange: it implies that the richer person should receive more than the poorer person, who, it might be argued, "needs the money more." In fact, however, this idea implies an interpersonal comparison of utility, which is not generally allowed. The fact is that player II's utility for money decreases rapidly, while I's does not. The result is that II will be very eager to get at least something, and can be "bargained down" by I.

9.2.   Threats

A serious objection can be raised to Nash's bargaining scheme, and it is that it does not take threats into account. The following example will best outline this objection.

9.2.1 Example. A worker has the choice of working, in which case he will be paid a subsistence salary while his boss has a profit of \$10, or not working, in which case he will starve, while his boss makes no profit. A reasonable assignment of utilities to these events would be $(0, 10)$ and $(-500, 0)$. The boss, however, may, if he desires, give part of the profit to the worker, so that (assuming utility to be linearly transferable) the set S will include all points below $u + v = 10$, and none other in the positive quadrant. As it is clear that $u^* = v^* = 0$, the Nash solution would seem to be $\bar{u} = \bar{v} = 5$. This disregards, however, the fact that the second player is in a much stronger position than his opponent. In fact, player I cannot prevent II from getting \$10 except by taking a very difficult step; a threat to stop work on his part would not be very believable, and as a result, he will probably continue to work for his subsistence salary.

It is difficult to deny that an example such as this tends to weaken Nash's bargaining solution. Some analysis of threats is in order if we are to correct these weaknesses. Quite generally, a threat is effective if it is believable, and if it tends to improve the position of the threatener vis-à-vis the person being threatened. Thus, a threat to kill someone is generally more effective than a threat to become angry, because the position of the killer is certainly improved vis-à-vis his victim (though in point of fact it may be worsened in absolute terms) while getting angry generally does no such thing. On the other hand, a threat to destroy the whole world, while it may possibly improve the threatened position in regard to others (reducing them all to equality in nothingness), is not very believable and hence not effective.

Nash suggests the following three-step bargaining scheme:

1. Player I announces a "threat" strategy x.
2. Player II, in ignorance of x, announces a threat strategy y.
3. Players I and II bargain. If they can come to an agreement, then that agreement take effect. If they do not come to an agreement, then they *must* use their threat

strategies *x* and *y*; the payoffs to the two players will be determined in this manner.

The question naturally arises as to the enforcement of these threats; if one of the players makes a wild threat, he may find himself averse to carrying out the threat later on. Some machinery may, however, be devised for this purpose. At any rate, we assume a player is bound to his threat in some manner or other.

In effect, what this means is that the maximum values u^* and v^* have been replaced by the threat values xAy^T and xBy^T. Axioms N1–N5 are now applied, and the result is the solution $(\overline{u}, \overline{v})$ where $(\overline{u}, \overline{v})$ is the point in S which maximizes

$$g(u, v) = (u - xAy^T)(v - xBy^T)$$

(subject, of course, to $u \geq xAy^T$).

A figure will probably explain the workings of this scheme best. In Figure 9.5, a typical set S (closed, bounded, and convex) is shown. The curve S^0 is the Pareto-optimal boundary of S (i.e., the subset of S which satisfies N3). From each point on S^0 at which a tangent to S^0 exists, a line is drawn whose slope is the negative of the tangent slope. If the tangent does not exist (as at point C) two lines are drawn, corresponding to the right and left tangents to S^0 at C. It is easy to see that, by the convexity of S, the lines in this family intersect, if at all, outside of S.

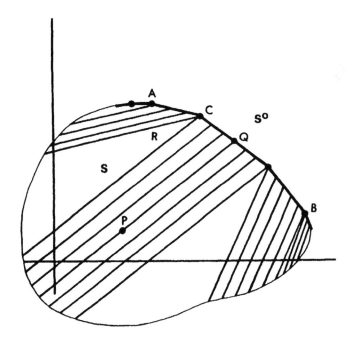

Figure 9.5.

Now, suppose the threat strategies x and y give, as threat value, a point such as P, lying on one of these lines. The arbitrated (Nash) value of the game is the point Q at which this line cuts S^0. If, on the other hand, the threat value is at a point such as R (lying inside the angle made by two lines from the same point in S^0) then the arbitrated value will be the point C at which these two lines meet S^0. Thus player I's aim, in choosing his threat strategy, will be to get a point lying on the lowest possible of this family of lines; player II's aim is to get a point on the highest possible of these lines.

To the question that arises as to the existence of equilibrium pairs of threat strategies, the answer can be given that they do exist. (It is, of course, necessary to define equilibrium pairs for such games; this is done in the obvious manner.)

9.2.2 Theorem. Any bimatrix game has at least one equilibrium pair of threat strategies (x, y).

The objective of the two players in choosing their threat strategies, however, are directly opposed. Hence it is not difficult to prove the following theorem as well.

9.2.3 Theorem. If (x', y') and (x'', y'') are equilibrium pairs of threat strategies, then so are (x', y'') and (x'', y'). Moreover, the Nash arbitrated payoff is the same for both (x', y') and (x'', y'').

We shall not give the proofs of Theorem 9.2.2 and 9.2.3 here; they are left as problems, pointing out only that 9.2.2 is a modification of 8.1.2, while 9.2.3 is a very slight modification of Theorem 2.1.2. By Theorem 9.2.3, we see that we can actually talk about *optimal* threat strategies (rather than merely equilibrium points).

An interesting problem is that of computing optimal threat strategies. In general, this can be complicated, since the arbitration value corresponding to a pair of threat strategies depends not only on the numbers xAY^{T} and xBy^{T}, but also on the form of the Pareto-optimal boundary of the set S. Since all that is required of S is that it be convex, there is no obvious method. In certain cases, however, it might be a simple problem.

If, in particular, utility is linearly transferable between the two players, the problem becomes quite simple. In fact, we can choose the utility scales so that utility is transferable at the rate of 1:1. The obvious adaptation of what we studied above tells us that, if x and y are the threat strategies, the arbitration value will be

$$\bar{u} = \frac{xAy^{\mathrm{T}} - xBy^{\mathrm{T}} + k}{2}$$

$$\bar{v} = \frac{xBy^{\mathrm{T}} - xAy^{\mathrm{T}} + k}{2}$$

where k is the maximum utility that can be obtained by the two working together. But this means player I will be trying to maximize $x(A - B)y^{\mathrm{T}}$, while II will be trying to minimize this same quantity. Thus, the optimal threat strategies for the bimatrix game (A, B) are the same as the optimal strategies for the matrix (zero-sum) game $A - B$, which we know how to solve.

9.2.4 Example. Consider the bimatrix game (A, B), given by

$$\begin{pmatrix} (1,4) & (-\frac{4}{3}, -4) \\ (-3, -1) & (4, 1) \end{pmatrix}$$

Assuming there is no transferable commodity, the set S is the convex hull of the four points (a_{ij}, b_{ij}) shown in Figure 9.6. The maximin security levels are seen to be 0 for both players, obtained by the mixed strategies $(\frac{3}{4}, \frac{1}{4})$ and $(\frac{1}{2}, \frac{1}{2})$ respectively. As S is nearly symmetric, it follows that the value, under the first Nash scheme, must be $(\frac{5}{2}, \frac{5}{2})$. On the other hand, this does not take player II's threat possibility into account; in fact, if II uses his first pure strategy, we see that I really can do very little against him.

We should, therefore, consider the threat possibility. We see that, on the Pareto-optimal boundary of S, utility is transferable (by randomization) at the rate 1:1. Thus we can consider the game $A - B$:

$$A - B = \begin{pmatrix} -3 & \frac{8}{3} \\ -2 & 3 \end{pmatrix}$$

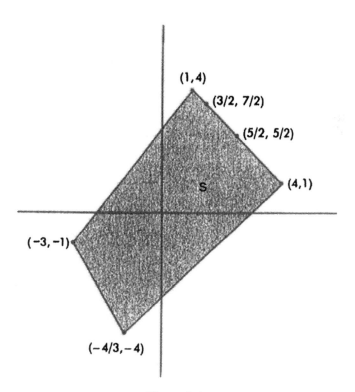

Figure 9.6.

This game has a saddle point at -2. As the maximum total utility for the two players is 5, we obtain the "arbitrated threat solution" $\left(\frac{3}{2}, \frac{7}{2}\right)$. This certainly takes II's stronger threat possibility into account.

9.3.   Time-Restricted Bargaining

An alternative approach to the bargaining problem, due to Rubinstein [1980], consists in treating the bargaining process as a game in extensive form. We will assume a set of feasible alternatives, S, closed, convex, and bounded above (as in 9.1). For simplicity, we shall assume that the conflict payoff is $(0,0)$. We will also make an assumption of nondegeneracy, so that S contains some positive points. We might then imagine a multistep process such as the following:

(1) Player I "makes an offer" of v, i.e., suggests a point $(u, v) \in S$.
(2) Player II can either accept v, in which case the bargain (u, v) will automatically come into effect, or counteroffer u' by suggesting the point $(u', v') \in S$. (Presumably, this would have $v' > v$.)
(3) Player I can either accept u', or make a new offer, v''.
(4) Player II can accept v'' or reject and counteroffer u'''.
(5) $-(\infty)$ The offers and counteroffers continue. If the two can never agree, the payoff will be $(0, 0)$.

The bargaining can continue in this fashion indefinitely, until finally one of the two accepts the last offer made by the other. Presumably the offers would increase — or at least not decrease — at each step, so that $v \le v'' \le v'''' \le \ldots$ and $u' \le u'''$ $\le u''''' \le \ldots$, but even this is not, strictly speaking, necessary.

It is not difficult to see that this game has many equilibrium pairs. In fact, let (u^*, v^*) be any Pareto-optimal point. Let σ be a strategy on the part of player I to offer v^* at some time, and never to offer any more, or accept any less than u^*. Let τ be a strategy on the part of II to offer u^* (or accept v^*) at some time, and never to offer any more than u^*, or to accept any less than v^*. Then (σ, τ) is an equilibrium pair. The problem is not in the lack of an equilibrium pair, but rather in the superabundance of them!

As in Section 8.3 above, such a multiplicity of equilibria suggests that we must look for a special type of equilibrium. Given that the game here is in extensive form, it seems reasonable to look for a subgame-perfect equilibrium. Now this is a game of perfect information; unfortunately, however, the game is not finite (the number of steps is unbounded, and there is possibly an infinite number of alternatives at each step). Hence Theorem 1.3.1 does not apply, and we cannot guarantee the existence of a perfect equilibrium pair.

Now, one of the problems with the game as described above lies in the fact that time does not matter: a bargain during the first stage is worth as much as the same bargain many stages later. Thus there is no reason for either player to yield in any way, since by refusing to yield now he or she can still hope to obtain more in the future, and in any case the same bargain will still be available later.

In real life, however, things are not so. In a strike, the outcome frequently hinges on the relative size of the union's strike fund and the firm's capital reserves, i.e., on which of the two parties can outwait the other. In other cases, even when the situation is not so critical as in a strike, still each bargainer realizes that a bargain today is worth considerably more (has greater utility) than the same bargain many time periods in the future. We consider therefore the same model as before, but with this modification:

Each step (offer or counteroffer) requires one period of time. (Acceptance can be immediate.) The two players have discount factors δ_1 and δ_2, per period, $0 < \delta_i < 1$. In the *primal* game, we will assume that Player I makes the first offer; in the *dual game*, player II makes the first offer.

It should be clear that this modification changes the character of the game considerably. In the extreme case, with δ_2 close to zero, player I can offer a very small (but positive) v. Player II would have to choose between accepting v immediately, or waiting one period of time — so that anything he can manage to obtain then must be discounted by δ_2. Clearly the discount factor, together with the advantage of the first move, has strengthened I's hand substantially.

Consider, now, the primal game. At stage 1, player I chooses a Pareto-optimal point (u^*, v^*) and offers v^* to II. Player II now chooses between accepting v^*, or refusing. If II refuses, then the game moves to stage 2, with II to make the next offer. It is not difficult to see that stage 2 of the primal game is strategically equivalent to stage 1 of the dual game—the alternatives are exactly the same, and the payoffs (except for the fact that they have all been multiplied by the corresponding discount factor) are the same.

Assuming, then, that II did not accept the offer at stage 1, it is now II's turn to make an offer, say u^{**}, corresponding to the Pareto-optimal point (u^{**}, v^{**}). Now I can accept u^{**}, or he can refuse—moving to stage 3, which is strategically equivalent to stage 1 of the same (primal) game.

Since the several stages of the game are, essentially, repetitions of either the primal or the dual game, it is possible to talk about stationary strategies: in a stationary strategy, a player's offer or acceptance is independent of previous stages played. Thus, a stationary strategy for I consists of a pair (v', u''): I will offer v' whenever it is his turn to act; he will accept any offer of u'' or more. Similarly, a stationary strategy for II is a pair (u', v''). II will offer u', and accept any offer of v'' or more. We look, now, for a *subgame-perfect equilibrium point, in stationary strategies*.

For an equilibrium, I's offer, v^*, should be large enough that II will accept it, but no larger. On the other hand, II should accept v^* if it is not smaller than the discounted amount $\delta_2 v^{**}$ which he would receive if I accepted II's counteroffer, u^{**}, after one stage of the game. Thus we obtain the relation

$$v^* = \delta_2 v^{**} \tag{9.4}$$

A similar consideration (for the dual game) gives us

$$u^{**} = \delta_1 u^* \tag{9.5}$$

Together with the fact that (u^*, v^*) and (u^{**}, v^{**}) must be Pareto-optimal points, this gives us a system of four equations in the four unknowns. Theorem 9.3.1 tells us that these values are what we are looking for.

9.3.1 Theorem. Any subgame-perfect equilibrium in stationary strategies is of the form $(\sigma, \tau) = ((v^*, u^{**}); (u^{**}, v^*))$, where (u^*, v^*) and (u^{**}, v^{**}) are nonnegative Pareto-optimal points satisfying (9.4) and (9.5).
Moreover, all such pairs are in equilibrium.

Proof. Let $((v_1, u_2); (u_3, v_4))$ be an equilibrium pair, and let (u_1, v_1), (u_2, v_2), (u_3, v_3), (u_4, v_4) be the corresponding Pareto-optimal points. It cannot be good for II to reject an offer which is larger than the (discounted) amount he will later demand, so $v_4 \leq \delta_2 v_3 < v_3$. By Pareto optimality, it follows that $u_3 < u_4$.

Given that II's strategy is (u_3, v_4), it is best for I to offer exactly v_4. For, if he offers $v_1 < v_4$, then II will refuse, and I will receive at most the (discounted) amount $\delta_1 u_3$, which is less than u_4. If he offers $v_1 > v_4$, then II will accept, and I will receive u_1, which by Pareto optimality is less than u_4. If I offers v_4, then II will accept and I will receive u_4.

For equilibrium, then, it is necessary that $v_1 = v_4$. Similarly, by looking at the dual game, equilibrium requires that $u_2 = u_3$. Now let u^{**} be the common value of u_2 and u_3, and let v^* be the common value of v_1 and v_4. Let (u^{**}, v^{**}) and (u^*, v^*) be Pareto-optimal points.

Assume, then, that at step 1, player I has offered v^*. II now can choose between getting v^* at stage 1 or v^{**} at stage 2. In fact he does accept v^*, and so $v^* > \delta_2 v^{**}$. But as mentioned above, $v^* \leq \delta_2 v^{**}$. Thus $v^* = \delta_2 v^{**}$. In similar fashion, $u^{**} = \delta_1 u^*$. We see then that, for an equilibrium pair, (9.4) and (9.5) must hold.

Conversely, suppose that (u^*, v^*) and (u^{**}, v^{**}) are Pareto-optimal points, with nonnegative components, satisfying (9.4) and (9.5). The discussion above shows that (v^*, u^{**}) is an optimal response to (v^{**}, u^*), and conversely. Thus this will be an equilibrium pair.

The question now is whether such as equilibrium pair exists. The following answers in the affirmative.

9.3.2 Theorem. There exists a unique pair (u^*, v^*) and (u^{**}, v^{**}) of positive Pareto-optimal points satisfying (9.4) and (9.5).

Proof. Let the nonnegative Pareto-optimal boundary of S be given by the equation

$$v = h(u), \quad 0 \leq u \leq a$$

Then $h(u)$ is a strictly decreasing, continuous, concave function of u, satisfying $h(a) = 0$. We note that (9.4) can be rewritten as

$$h(u^*) = \delta_2 h(u^{**})$$

and this can be combined with (9.5) into the single equation

$$g(u) = h(u) - \delta_2 h(\delta_1 u) = 0 \tag{9.6}$$

We wish to show that (9.6) has a unique solution.

Setting $u = 0$, we note that, by the nondegeneracy condition (that S contain at least one positive point), $h(0)$ is positive. Since $h(0) > 0$ and $\delta_2 < 1$, we see that $g(0) > 0$.

Set, now, $u = a$. Then $h(a) = 0$. Since $\delta_1 < 1$, we see that $\delta_1 a < a$, and, by the strict monotonicity of h, $h(\delta_1 a) > 0$. Since $\delta_2 > 0$, then $\delta_2 h(\delta_1 a)$ is positive, and hence $g(a) < 0$.

We see, then, that $g(0) > 0$, while $g(a) < 0$. Since h is continuous, so is g, and there must be some value of u which makes $g(u) = 0$. Call this value u^*.

To show uniqueness, assume h is differentiable. Then g is also differentiable, and we see

$$g'(u) = h'(u) - \delta_1 \delta_2 h'(\delta_1 u)$$

Now, h is concave and so h' is non-increasing; we also know h' is strictly negative. We note $\delta_1 < 1$, and so, for $u > 0$, $\delta_1 u < u$. Then, by the monotonicity of h'

$$h'(u) < h'(\delta_1 u)$$

and, since h' is negative

$$\delta_1 \delta_2 h'(\delta_1 u) > h'(\delta_1 u)$$

Thus $g'(u) < 0$ for all $u > 0$.

Of course, h need not be differentiable everywhere. However, h is known to be continuous, concave, and monotone. Hence g is continuous and it is the difference of two concave monotone functions. It follows that g is differentiable almost everywhere, and it is the integral of its derivative. Since this derivative is negative for almost all positive u, it will follow that g is also strictly monotone decreasing, and so (8.3.3) will have exactly one solution. We will call the points (u^*, v^*) and (u^{**}, v^{**}) so obtained the *Rubinstein primal and dual points*.

It remains to compute the points (u^*, v^*) and (u^{**}, v^{**}). These depend on the form of the Pareto-optimal boundary of S, but a numerical computation is, in general, quite straightforward. In the simplest case, the boundary is a straight line, and a computation is quite simple. In fact, suppose the boundary is the line

$$u + v = 1 \tag{9.7}$$

so that the problem reduces to dividing a sum of money, with utility (for both players) proportional to money. Then (9.6) reduces to the equation

$$1 - u^* = \delta_2(1 - \delta_1 u^*) \tag{9.8}$$

which has the solution

$$u^* = \frac{1 - \delta_2}{1 - \delta_1\delta_2} \tag{9.9}$$

9.3.3 Example. Let the Pareto-optimal boundary of S be the line $u + v = 1$.
Let $\delta_1 = 0.9$ and let $\delta_2 = 0.8$. Then (for the primal game)

$$u^* = 0.2/0.28 = 0.714$$

$$v^* = 1 - 0.714 = 0.286$$

$$v^{**} = 0.286/0.8 = 0.357$$

$$u^{**} = 1 - 0.357 = 0.643$$

Thus, in the primal game, I will offer 0.286, and keep 0.714 for himself. In the dual game, II will offer 0.643, and keep 0.357 for himself. As can be seen, there are two factors at work here:

(1) The first move is clearly an advantage. With the first move, I will get 0.714; with the second move, "only" 0.643.
(2) A higher discount factor (i.e., greater patience) is also an advantage. With the first move, I will get 0.714, while II, even with the first move, would only get 0.357.

9.3.4 Example. We consider once again the game of example 9.1.5, for which the Pareto-optimal boundary was given by

$$v = h(u) = \log(2 - 0.01u)$$

We will assume that both players have the same discount factor, $\delta = 0.9$. Equation (9.6) then takes the form

$$\log(2 - 0.01u^*) = 0.9 \log(2 - 0.009u^*)$$

which has the solution

$$u^* = 57.38$$

$$v^* = \log(1.4262) = 0.355$$

$$u^{**} = 51.64$$

$$v^{**} = \log(1.4836) = 0.394$$

Thus, in this game, player I expects to get \$57.38 if he makes the first offer, and \$51.64 if he makes the second offer. This contrasts with the Nash solution, which gives player I \$54.40. In this case, with equal discount factors, the first move gives an advantage which amounts to approximately \$3. (It would of course be smaller if the discount factors were close to 1.)

A good question now concerns what will happen if the discount factors are allowed to approach 1 (essentially, this is what will happen if the length of time periods approaches zero). In that case, the advantage of the first move would decrease, as is indicated by (9.4) and (9.5); i.e., we might expect that, in the limit (if the limit exists), $u^* = u^{**}$ and $v^* = v^{**}$. What is not clear is where the limiting point (u^*, v^*) would be. The following theorem answers this.

9.3.5 Theorem. Let the discount factors, δ_1 and δ_2, both approach 1, in such a way that the ratio $s = (\log\ \delta_1)/(\log\ \delta_2)$ is kept constant. Then the Rubinstein primal and dual points will both approach the point (u, v) which maximizes the product uv^s over the positive part of the set S. In particular, if $\delta_1 = \delta_2$, so that $s = 1$, then the Rubinstein points will approach the Nash solution (Theorem 9.1.2).

Proof. Let $s = (\log\ \delta_1)/(\log\ \delta_2)$. We now have, from (9.4)

$$v^{***s} = \delta_2^s v^{*s}$$

but

$$\delta_2^s = \delta_1$$

and so

$$v^{***s} = \delta_1 v^{*s}$$

and, multiplying by (9.5), we have

$$\delta_1 u^{**} v^{***s} = \delta_1 v^{*s} u^*$$

or, simply

$$u^{**} v^{***s} = u^* v^{*s} \tag{9.10}$$

We see, then, that the function uv^s has the same value at the two points (u^*, v^*) and (u^{**}, v^{**}). The point which maximizes uv^s will be a Pareto-optimal point, and must lie on the Pareto-optimal boundary of S, between these two points. As the discount factors approach 1, however, the distance between the two Rubinstein points will, by (9.4) and (9.5), approach 0, and so the points will both be very close to the maximizing point. In the limit, (u^*, v^*) and (u^{**}, v^{**}) will both converge to the point which maximizes uv^s.

In case $s = 1$, the limit point is the point which maximizes the product uv which is (of course) the Nash solution of Theorem 9.1.2.

A good question is as to why the two discount factors would approach 1 in exactly the form specified in Theorem 9.3.5. In fact, suppose that the reason the discount factors approach 1 is, not that the two players' time preferences for money change, but rather that the time periods become very short. In that case, the discount factors would be

$$\delta_1 = e^{-aT} \tag{9.11}$$

$$\delta_2 = e^{-bT} \tag{9.12}$$

where T is the length of the time period, and a, b, are factors which depend on the two players' time preferences. Then, as T approaches 0, $(\log \delta_1)/(\log \delta_2) = a/b$ would be constant.

A better answer to the same question is that the ratio $(\log \delta_1)/(\log \delta_2)$ need not be constant. What is needed is that this ratio approach the limit s. The proof of this, however, is somewhat complicated as it depends on the fact that, if s is close to r, then the two points which maximize uv^s and uv^r, respectively, will also be close. The proof of this is left to the reader.

9.3.6 Example. Let us consider once again the game of Example 9.1.5 and 9.3.3, but let us now set the discount factors

$$\delta_1 = 1 - 2\varepsilon$$
$$\delta_2 = 1 - \varepsilon$$

where ε is very small (i.e., infinitesimal). In this case, the logarithms of the discount factors are approximately given by $\log(\delta_1) = 2\varepsilon$, $\log(\delta_2) = \varepsilon$, and so the ratio is 2. It follows that, for very small ε, the solution point will be that point which maximizes the product uv^2, where, as mentioned above

$$v = \log(2 - 0.01u)$$

Some elementary calculations will then show that the maximum is obtained at the point $u = 38.62$, $v = 0.4786$. We see that, in this case, the difference in discount factors works seriously to player II's benefit: II now receives \$61.28 of the total \$100, and is much better off than before (when he received only \$45.60). It does not matter which of the two players moves first.

Problems

1. We think of a bargaining game as a noncooperative game, in which each player's "strategy" consists of making a demand. Then, if the two demands (u, v) are consistent (belong to S) the players receive their demand; otherwise they receive their minimax (or threat) values, (u^*, v^*).
   (a) Every point on the Pareto-optimal boundary of S is an equilibrium pair.
   (b) Assume $u^* = v^* = 0$, and let g be the characteristic function of S (i.e., $g = 1$ on S and $g = 0$ on the complement of S). Then, if the players demand (u, v) their payoffs will be $(ug(u, v), vg(u, v))$.
   (c) Let h be any non-negative function, and assume that the payoff, in case the demands (u, v) are made, is (uh, vh). Show that, if $(\bar{u}_h, \bar{v}_h)$ maximizes the function uvh, then $(\bar{u}_h, \bar{v}_h)$ is an equilibrium pair.
   (d) Let $h_1, h_2, \ldots$, be a sequence of nonnegative continuous functions which converge to g. Then, if $(\bar{u}, \bar{v})$ is the Nash solution (arbitrated value of the game, there exists a sequence of points $(\bar{u}_j, \bar{v}_j)$, each one of which is an equilibrium point of the demand game with payoff (uh_j, vh_j) which converges to $(\bar{u}, \bar{v})$).

2. Prove the existence of "optimal" threat strategies for the Nash modified bargaining model (Theorems 9.2.1 and 9.2.2).

3. Let S be the set of all (u, v) satisfying

$$(u - 1)^2 + 4v^2 \leq 4$$

   Find the Nash bargaining solution (a) assuming the conflict payoff is $(0, 0)$; (b) assuming that the conflict payoff is $(1, 0)$.

4. Let S be as in Exercise 3, above, and let the conflict payoff be $(0, 0)$. Find the time-constrained solution to this bargaining problem assuming the discount factors $\delta_1 = 0.9$, $\delta_2 = 0.88$, assuming (a) player I makes the first offer; (b) player II makes the first offer.

5. Let S be a nonempty compact convex subset of the first quadrant in the $u-v$ plane. For positive values of s, let $h(s)$ be the abscissa of the point (u, v) which maximizes the product $u^s v$. Prove that $h(s)$ is unique, and is a continuous function of s.

6. Let δ_1 and δ_2 both approach 1, in such a way that the ratio $(1 - \delta_2)/(1 - \delta_1)$ approaches the limit s. Assume the available alternatives form the compact convex set S, and let the conflict payoff be $(0, 0)$. Prove that, in that case, the two solutions of the time-constrained bargaining problem converge to the point (u, v) which maximizes the product $u^s v$, subject to being in the intersection of set S and the first quadrant of the plane. (Use the result from Problem 5 here.)

Chapter 10

n-Person Games

10.1. Noncooperative Games

In considering *n*-person games (where $n \geq 3$), the same distinction will arise as in considering two-person nonzero-sum games. That is, cooperation among some or all of the players may be forbidden or permitted by the rules of the game. Let us consider the first, i.e., the noncooperative case.

In the noncooperative case, the principal question is the existence of equilibrium *n*-tuples. This question is answered by the following theorem:

10.1.1 Theorem. Any finite *n*-person noncooperative game has at least one equilibrium *n*-tuple of mixed strategies.

We will not give a proof of 10.1.1 here. It may be seen, however, that there is no difficulty in extending the proof of Theorem 8.1.2 to this case.

Although Theorem 10.1.1 is certainly a valuable result, it may be pointed out that all the difficulties which were observed for equilibrium points of bimatrix games are also present here. Moreover, the computation of equilibrium *n*-tuples (where $n \geq 3$) is incomparably more difficult than the computation of equilibrium pairs.

In general, there is no great difference between the theory of noncooperative *n*-person games and noncooperative two-person general-sum games.

10.2. Cooperative Games

Let us consider the case in which cooperation is permitted. We have just seen that, in the noncooperative case, there is very little difference between two-person and *n*-person games. In fact, the *n*-person game is merely a generalization of the two-person case. In cooperative games, however, we find that a new idea appears: *coalitions*. In the two-person case, there is only one possible coalition. In the *n*-person case, there are many possible coalitions and this means that, if a coalition is to form and remain for some time, the different members of the coalition must reach some sort of equilibrium or stability. It is this idea of stability that must be analyzed in any meaningful theory.

We can see the meaning of this stability best by taking an example. Consider the following three-person game: Players 1, 2, and 3 are faced with the simple problem of forming a coalition. If any two of them succeed in forming a coalition, then the third player must pay each of them one unit. If no two-player coalition is formed, then there is no payoff at all.

In this simple example, there is really very little to be said. Assuming nothing about the personal relations among the three players, any one of the three possible two-player coalitions might be formed, and there is no way to distinguish among them. Hence the three payoffs $(-2, 1, 1)$, $(1, -2, 1)$, and $(1, 1, -2)$ corresponding to these three coalitions seem to be a "natural" result, and may be considered in some way a "solution" of the game. Alternatively, the payoff $(0, 0, 0)$ which is the average of these three payoffs may also be thought of as some sort of solution to the game. (We shall see later on in what sense these are solutions.)

Let us suppose now that the game is changed slightly, so that, if the coalition $\{2, 3\}$ is formed, player 1 must pay 1.1 units to 2, and 0.9 unit to 3. This would seem to improve 2's outlook, since he can now gain more from the same action. However, closer analysis will show that this is not so. In fact, the coalition $\{2, 3\}$ now becomes almost impossible to form (unless there is some external difficulty to complicate communications between 1 and 3) since both 1 and 3 stand to gain from forming $\{1, 3\}$. Hence 2 is in a worse position than before; in fact, one of the two possible coalitions he might join is very unstable. He might, however, remedy this by giving 3 a side payment of 0.1 unit (assuming such a side payment is possible). This will reduce the game to the previous game.

The foregoing example should serve to illustrate, first, the importance of side payments in this game and, second, the need for some sort of stability among the different payoffs in a solution. This latter question will be treated later on; at first we shall deal with side payments.

Once again, the existence of a linearly transferable commodity is in question. Let us assume that this commodity does exist; the theory developed on this assumption will be called the *side-payments theory*.

In the side-payments theory, the utility functions for the several players can be chosen so that the rate of transfer of utility among any two of them is 1:1. As pointed out in the two-person case this means that, if a subset S of the players (coalition) can obtain a total utility v, this utility can be divided among the members of S in any way possible. Thus we will be interested in the total utility which can be attained by any one of these coalitions.

10.2.1 Definition. For an *n*-person game, we shall let $N = \{1, 2, \ldots, n\}$ be the set of all *players*. Any nonempty subset of N (including N itself and all the one-element subsets) is called a *coalition*.

10.2.2 Definition. By the *characteristic* function of an *n*-person game we mean a real-valued function v defined on the subsets of N, which assigns to each $S \subset N$ the maximin value (to S) of the two-person game played between S and $N - S$, assuming that these two coalitions form.

Thus $v(S)$ is the amount of utility that the members of S can obtain from the game, whatever the remaining players may do. From this definition, it follows that

$$v(\varnothing) = 0 \tag{10.1}$$

for obvious reasons. Now if S and T are disjoint coalitions, it is clear that they can accomplish at least as much by joining forces as by remaining separate. Hence we have the *superadditivity* property

$$v(S \cup T) \geq v(S) + v(T) \quad \text{if} \quad S \cap T = \varnothing \tag{10.2}$$

In dealing with two-person games, we have seen that we generally dispense with the extensive form and deal only with the normal form. This is because the normal form allows us to study mixed strategies, which are the essence of such games, with greatest ease. Now, the essence of n-person games is not the randomization; rather, it is the formation of coalitions. Thus we shall study, not the normal form, but the characteristic function. In fact the characteristic function tells us the capacities of the different coalitions, assuming optimal (maximin) randomization, and thus is best suited to study coalitions. In general, we shall identify the game with its characteristic function.

10.2.3 Definition. By an *n-person game in characteristic function form* is meant a real-valued function v, defined on the subsets of N, satisfying conditions (10.1) and (10.2).

Some authors have dispensed with (10.2) and studied functions satisfying only the condition $v(\varnothing) = 0$. Very often such functions are called *improper* games, whereas those which also satisfy (10.2) are called *proper* games. Unless specifically stated, throughout this book we shall deal only with proper games.

As mentioned above, the amount $v(S)$ is the maximin value of the game played between S and $N - S$. Let us suppose that the normal form of the game is constant-sum (i.e., the sum of the utility payoffs to all the players is always the same). The game between S and $N - S$ is then strictly competitive, and it follows, if it is finite, that the minimax theorem holds. Thus we will have

$$v(S) + v(N - S) = v(N) \tag{10.3}$$

This leads to the following definition.

10.2.4 Definition. A game (in characteristic function form) is said to be constant-sum if, for all $S \subset N$

$$v(S) + v(N - S) = v(N)$$

It should be noted that a game might be constant-sum in its characteristic function form without being constant-sum in its normal form. (It is even possible for the game to be constant-sum in its normal form without being constant-sum in its characteristic function form, though this cannot happen with finite games.)

Let us suppose, then, that an n-person game is played. Without inquiring into the particular coalition structure obtained, we might like to know the possible payoff vectors. Now, assuming that some type of understanding is arrived at, the players will have the total utility $v(N)$ to divide. (In a constant-sum game, it is not necessary that an understanding be reached for this to happen.) This can be divided in any way at all, but it is of course clear that no player will accept less than the minimum which he can attain for himself.

10.2.5 Definition. An *imputation* (for the *n*-person game v) is a vector $x = (x_1, \ldots, x_n)$ satisfying

(i) $\sum\limits_{i \in N} x_i = v(N)$

(ii) $x_i \geq v(\{i\})$ for all $i \in N$

We shall use the notation $E(v)$ for the set of all imputations of this game.

It might be argued that, except in the constant-sum case, condition (i) in this definition is too strong, since it is quite likely that the players will fail to reach any understanding, and will wind up obtaining less than the total amount $v(N)$ which they can get. While this is a valid objection, it will be seen later on that many of our "solution concepts" eliminate all vectors which do not satisfy this condition.

The question now is this: Which of all the imputations will be obtained? This is, of course, a difficult (or even impossible) problem. There is one condition, however, in which this problem is trivial: that is when the set $E(v)$ has only one element. In this case, that unique imputation will be the obvious result. It does not matter what coalitions form. This gives rise to the distinction between *essential* and *inessential* games. It is clear (by superadditivity, repeated *n* times) that

$$v(N) \geq \sum_{i \in N} v(\{i\}) \tag{10.4}$$

Now, $E(v)$ will have only one point if equality holds in (10.4). Thus we have

10.2.6 Definition. A game v is *essential* if

$$v(N) > \sum_{i \in N} v(\{i\})$$

It is *inessential* otherwise.

It is the essential games which are of interest to us.

10.3.   Domination, Strategic Equivalence, and Normalization

Given a game v, let x and y be two imputations. Suppose the players of the game are confronted by a choice between x and y. Can we find any criterion to determine when one of these will be chosen rather than the other?

It is of course clear that, unless $x = y$, there will be some players who prefer x to y (those i such that $x_i > y_i$). Because the two vectors are imputations, however, there will also be some players who prefer y to x. Hence it is not enough merely to state that *some* players prefer x to y. On the other hand, it is not possible that *all* players will prefer x to y (since the sum of the components of either x or y is $v(N)$). What is necessary is that the players who prefer x to y be actually strong enough to enforce the choice of x.

10.3.1 Definition. Let x and y be two imputations, and let S be a coalition. We say x *dominates* y *through* S (notation: $x \vdash_s y$) if

(i) $x_i > y_i$ for all $i \in S$

(ii) $\sum\limits_{i \in S} x_i \leq v(S)$

We say x dominates y (notation: $x \vdash y$) if there is any coalition S such that $x \vdash_s y$.

Thus condition (i) states that the members of S all prefer x to y; condition (ii) states that they are capable of obtaining what x gives them.

It is easy to see that the relation $\vdash_s$ (for any given S) is a partial order relation. On the other hand, the relation $\vdash$, while it is irreflexive, is neither transitive nor antisymmetric (since the coalition S may be different in different cases). This is a serious difficulty; it will certainly complicate things for us later on.

Since we plan to analyze games by means of the domination relation, we are naturally interested in games whose imputation sets have the same domination structure.

10.3.2 Definition. Two n-person games u and v are said to be *isomorphic* if there exists a $1-1$ function f mapping $E(u)$ onto $E(v)$ in such a way that, for $x, y \in E(u)$, and $S \subset N$, $x \vdash_s y \leftrightarrow f(x) \vdash_s f(y)$.

It may be difficult to tell whether two games are isomorphic in this sense. We have, however, the following criterion:

10.3.3 Definition. Two n-person games u and v are said to be *S-equivalent* if there exist a positive number r and n real constants $\alpha_1, \ldots, \alpha_n$ such that, for all $S \subset N$

$$v(S) = ru(S) + \sum_{i \in S} \alpha_i$$

Essentially, if two games are *S*-equivalent, we can obtain one from the other simply by performing a linear transformation on the utility spaces of the several players. It is easy to prove that *S*-equivalent implies isomorphism:

10.3.4 Theorem. If u and v are *S*-equivalent, they are isomorphic.

Proof. Assume u and v are *S*-equivalent. Consider the function, defined on $E(u)$

$$f(x) = rx + \alpha$$

where r and $\alpha = (\alpha_1, \ldots, \alpha_n)$ are as in 10.3.3, It is easy to check that, if $x \in E(u)$, $f(x) \in E(v)$. Moreover, if $x \vdash_s y$, it is clear that $f(x) \vdash_s f(y)$. Hence f is the desired isomorphism.

Thus S-equivalence is sufficient for isomorphism. (The converse of this theorem is true in the restricted setting of constant-sum games. The proof is however rather lengthy and we will therefore not include it here. See References [10.2] and [10.6].)

It is obvious that S-equivalence is, indeed, an equivalence relation. It is interesting to choose one particular game from each equivalence class.

10.3.5 Definition. A game v is said to be in $(0,1)$ *normalization* if

(i) $v(\{i\}) = 0$ for all $i \in N$

(ii) $v(N) = 1$

10.3.6 Theorem. If u is an essential game, it is S-equivalent to exactly one game in $(0, 1)$ normalization.

Theorem 10.3.6 whose proof we omit (it is trivial), tells us that we can choose a game in $(0, 1)$ normalization to represent each equivalence class of games. The advantage of choosing such a game is that, in such a game, the value $v(S)$ of a coalition tells us directly its strength (i.e., the extra amounts which its members gain in forming it), and that all its imputations are probability vectors.

Other types of normalization have been used in the literature of n-person games. A common normalization is $(-1,0)$ normalization, in which $v(\{i\}) = -1$ and $v(N) = 0$. We shall, however, deal with games in $(0,1)$ normalization throughout. As our interest is in essential games, this entails no loss of generality.

The set of n-person games in $(0,1)$ normalization is the set of all real-valued functions, v, defined on the subsets of N, such that

$$v(\varnothing) = 0 \tag{10.5}$$

$$v(\{i\}) = 0 \quad \text{for all} \quad i \in N \tag{10.6}$$

$$v(N) = 1 \tag{10.7}$$

$$v(S \cup T) \geq v(S) + v(T) \quad \text{if} \quad s \cap T = \varnothing \tag{10.8}$$

These four conditions determine a $(2^n - n - 2)$-dimensional convex set. If the game is constant-sum, we have the additional condition

$$v(N - S) = v(N) - v(S) \quad \text{for all} \quad S \subset N \tag{10.9}$$

This gives $2^{n-1} - 1$ new equations, so that the set of constant-sum games has dimension $2^{n-1} - n - 1$. Thus it is clear that the dimension of the set of n-person constant-sum games is the same as that of the set of $(n - 1)$-person games. In fact, the two sets are congruent.

Indeed, if u is an $(n-1)$-person game in $(0,1)$ normalization, it may be expanded to an n-person constant-sum game v by the addition of a new player n, setting

$$v(S) = \begin{cases} u(S) & \text{if} \quad n \notin S \\ 1 - u(N-S) & \text{if} \quad n \in S \end{cases}$$

It is easily checked that this game is constant sum.

There are two special types of games which are of interest:

10.3.7 Definition. A game v is said to be *symmetric* if $v(S)$ depends only on the number of elements in S.

10.3.8 Definition. A game v in $(0,1)$ normalization is said to be *simple* if, for each $S \subset N$, we have either $v(S) = 0$ or $v(S) = 1$. A game is simple if its $(0,1)$ normalization is simple.

Essentially, a simple game is one in which every coalition is either winning (value 1) or losing (value 0), with nothing in between. As such, simple games are applicable to political science, as they include voting "games" in elections and legislatures.

Among simple games, we distinguish a special class: the weighted majority games:

10.3.9 Definition. Let $(p_1, p_2, \ldots, p_n)$ be a nonnegative vector, and let q satisfy

$$0 < q \le \sum_{i=1}^{n} p_i$$

Then the *weighted majority game* $[q; p_1, p_2, \ldots, p_n]$ is the simple game v defined by

$$v(S) = \begin{cases} 0 & \text{if} \quad \sum_{i \in S} p_i < q \\ 1 & \text{if} \quad \sum_{i \in S} p_i \ge q \end{cases}$$

10.4. The Core

Let us proceed, then, to the analysis of games in terms of the domination relation. An obvious first idea is to study the undominated imputations.

10.4.1 Definition. The set of all undominated imputations for a game v is called the *core*. The notation for the core is $C(v)$.

10.4.2 Theorem. The core of a game v is the set of all n-vectors x satisfying

(a) $\sum_{i \in S} x_i \ge v(S)$ for all $S \subset N$

(b) $\sum_{i \in N} x_i = v(N)$

Proof. If we let $S = \{i\}$, condition (a) reduces to $x_i \geq v(\{i\})$. This, together with condition (b), means that all such vectors are imputations.

Let us suppose that x satisfies (a) and (b), and that $y_i > x_i$ for all $i \in S$. But this, together with (a), means that

$$\sum_{i \in S} y_i > v(S)$$

and so it is not possible that $y \vdash_s x$. Hence $x \in C(v)$.

Conversely, suppose that y does not satisfy (a) and (b). If it fails to satisfy (b), it is not even an imputation and hence not in $C(v)$. Suppose, then, that there is some nonempty $S \subset N$ such that

$$\sum_{i \in S} y_i = v(S) - \varepsilon$$

where $\varepsilon > 0$. Let

$$\alpha = v(N) - v(S) - \sum_{i \in N - S} v(\{i\})$$

It is easy to see, by superadditivity, that $\alpha \geq 0$. Finally, let s be the number of elements in S. Now define z by

$$z_i = \begin{cases} y_i + \dfrac{\varepsilon}{s} & \text{if} \quad i \in S \\ v(\{i\}) + \dfrac{\alpha}{n - s} & \text{if} \quad i \notin S \end{cases}$$

It is easily seen that z is an imputation and, moreover, $x \vdash_s y$. Hence $y \notin C(v)$.

Theorem 10.4.2 shows that $C(v)$ is a closed convex set (as it is characterized by a set of loose linear inequalities). This is most interesting, since classical economic theory usually gives the core as "solution" to most game-theoretic problems. Any imputation in the core is stable, in that there is no coalition with both the desire and the power to change the outcome of the game.

Generally, of course, the core may have more than one point. This is not a great handicap; it simply means that more than one outcome is stable. A greater difficult with the core, however, is that it need not exist (i.e., it may be empty).

10.4.3 Theorem. If v is an *essential* constant-sum game, $C(v) = \varnothing$.

Proof. Suppose $x \in C(v)$. For any $i \in N$, we know

$$\sum_{i \in N - \{i\}} x_j \geq v(N - \{i\})$$

but, by the constant-sum property

$$v(N - \{i\}) = v(N) - v(\{i\})$$

and, since x is an imputation, we must have $x_i \leq v(\{i\})$. Because v is essential, then

$$\sum x_i \leq \sum v_i(\{i\}) < v(N)$$

but this means $x \notin E(v)$. The contradiction proves $C(v)$ is empty.

10.4.4 Example. Player 1 (a seller) has a horse which is worthless to him (unless he can sell it). Players 2 and 3 (buyers) value the horse at \$90 and \$100, respectively.

If 1 sells the horse to 2 at a price of x, he will effectively make a profit equal to x, while 2's profit is $90 - x$. The total profit of the coalition $\{1, 2\}$ is then \$90. Thus

$$v(\{1, 2\}) = 90$$

Similarly

$$v(\{1, 3\}) = 100$$

On the other hand, a single player, or the two buyers together, can obtain no profit. Thus

$$v(\{i\}) = v(\{2, 3\}) = 0$$

Finally, the coalition of all three players can do no better than to assign the horse to player 3 (with, possibly, some side payments which do not change the total amount of utility). Thus

$$v(\{1, 2, 3\}) = 100$$

We see from this that the core consists of all vectors (x_1, x_2, x_3) satisfying

$$x_1 + x_2 \geq 90$$
$$x_1 + x_3 \geq 100$$
$$x_1 + x_2 + x_3 = 100$$
$$x_i \geq 0$$

It may be seen that this requires $x_1 \geq 90$, $x_2 = 0$, $x_1 + x_3 = 100$, $x_3 \geq 0$. Thus

$$C(v) = \{(t, 0, 100 - t) | 90 \leq t \leq 100\}.$$

Heuristically, this says that player 3 will purchase the horse at a price of at least \$90. Player 2 (the "low bidder") is priced out of the market — but not before "bidding up" the price to at least \$90.

10.4.5 Example. (The Assignment Game) Let $N = M \cup M'$, where M and M' are disjoint sets with m elements each, say $M = \{1, 2, \ldots, m\}$ and $M' = \{m + 1, m + 2, \ldots, 2m\}$.

Player $i \in M$ (the ith seller), $i = 1, 2, \ldots, m$, owns a house which he values at a_i, dollars. Player $m + j$, $j = 1, 2, \ldots, m$ (the jth buyer) wishes to purchase a house, and values i's house at b_{ij} dollars. Thus, a coalition between players $i \in M$ and $m + j \in M'$ can obtain a profit

$$v(\{i, m + j\}) = c_{ij} = \begin{cases} b_{ij} - a_i & \text{if } b_{ij} \geq a_i \\ 0 & \text{if } b_{ij} \leq a_i \end{cases}$$

For any $S \subset M$ or $S \subset M'$, we will have $v(S) = 0$. For other S (which contain both sellers and buyers), $v(S)$ will be equal to the maximum total profit generated by sales of houses among the members of S, subject to the natural constraints that no seller can sell to more than one buyer, and no buyer can buy from more than one seller. Thus, if S has no more sellers than buyers, we *assign* buyer $m + j(i)$ to seller i, and the total profit is

$$\sum_{i \in S \cap M} c_{i, j(i)}$$

Then

$$v(S) = \max \sum_{i \in S \cap M} c_{i, j(i)}$$

where the maximum is taken over all such assignments. A slightly different treatment gives us

$$v(S) = \max \sum_{m + j \in S \cup M'} c_{i(j), j}$$

in case S has more sellers than buyers.

For the set N in particular, we have

$$v(N) = \max \sum_{i=1}^{m} c_{i, j(i)}$$

where the maximum is taken over all permutations $(j(1), j(2), \ldots, j(m))$ of the set M. This can also be written as

$$v(N) = \max \sum_{i=1}^{m} \sum_{j=1}^{m} p_{ij} c_{ij}$$

where $P = (p_{ij})$ is a permutation matrix, i.e., a matrix of 0's and 1's such that there is exactly one in each row and column. Essentially, we will have $p_{ij} = 1$ if the jth buyer buys from the ith seller, and $p_{ij} = 0$ otherwise.

Consider, now, the linear program

$$\text{Maximize} \quad \sum_{i=1}^{m} \sum_{j=1}^{m} q_{ij} c_{ij} \tag{10.10}$$

subject to

$$\sum_{i=1}^{m} q_{ij} = 1 \quad \text{for} \quad j = 1, \ldots, m \tag{10.11}$$

$$\sum_{j=1}^{m} q_{ij} = 1 \quad \text{for} \quad i = 1, \ldots, m \tag{10.12}$$

$$q_{ij} \geq 0 \tag{10.13}$$

Clearly, every permutation matrix $P = (p_{ij})$ satisfies the constraints (10.11)–(10.13). On the other hand, not every feasible $Q = (q_{ij})$ is a permutation matrix as some of the components might be fractional. It can nevertheless be shown, without too much trouble, that the extreme points of the program (10.10)–(10.13) are precisely the permutation matrices. Since the maximum will be found at an extreme point, we conclude that $v(N)$ is the maximum of the program (10.10)–(10.13).

Consider, next, the dual of the program (10.10)–(10.13), which is to

$$\text{Minimize} \quad \sum_{i=1}^{m} y_i + \sum_{j=1}^{m} z_j \tag{10.14}$$

subject to

$$y_i + z_j \geq c_{ij} \quad i, j = 1, \ldots, m \tag{10.15}$$

Generally, there is no nonnegativity restriction on the dual variables y_i, z_j. It is not difficult to see, however, that the minimum vector is not unique. In fact, if $(y^*; z^*) = (y_1^*, \ldots, y_m^*; z_1^*, \ldots, z_m^*)$ is a minimizing vector, then for any real number t, so is $(y'; z')$, where

$$y'_i = y_i^* - t, \quad z'_j = z_j^* + t$$

for all i and j.

In particular, we may choose

$$t = \min_i y_i^*$$

and then

$$\min_i y'_i = 0$$

Then all $y'_i \geq 0$, and, for some k, $y'_k = 0$. For every j, we have

$$z'_j = y'_k + z'_j \geq c_{kj} \geq 0$$

and so all $z'_j \geq 0$. Thus $(y'; z')$ is a minimizing vector for (10.14)–(10.15), with all components non-negative.

Now by duality

$$\sum_{i=1}^{m} y'_i + \sum_{j=1}^{m} z'_j = v(N)$$

and so $(y'; z')$ is an imputation for the game v. We shall show that it belongs to $C(v)$.

For any S, we have

$$v(S) = c_{i_1 j_1} + \cdots + c_{i_q j_q}$$

where $i_1, \ldots, i_q, j_1, \ldots, j_q$ are distinct members of S. Then

$$\sum_{i \in S} y'_i + \sum_{m+j \in S} z'_j \geq y'_{i_1} + \cdots + y'_{i_q} + z'_{i_1} + \cdots + z'_{i_q} \geq c_{i_1 j_1} + \cdots + c_{i_q j_q} = v(S)$$

and we conclude that $(y'; z') \in C(v)$.

In some cases, (y', z') will be the only imputation in the core. More usually, however, the imputation is not unique. In fact, any minimizing vector $(y^*; z^*)$ with nonnegative components will lie in the core. The usual rule is that the optimal assignment of sellers to buyers is unique; the market price, however, is not unique. There are, however, numerous special cases which we cannot treat here.

10.4.6 Example. (Simple Games) Let v be a simple game in $(0,1)$ normalization. We say player i is a veto player if

$$v(N - \{i\}) = 0$$

where, as usual, N is the set of all players.

Suppose, first, that v has no veto players. Then, for every $i \in N$, we have $v(N - \{i\}) = 1$. For an imputation x to lie in the core, we must have

$$\sum_{j \in N} x_j = v(N) = 1$$

$$\sum_{j \neq i} x_j \geq v(N - \{i\}) = 1$$

Thus $x_i = 0$ for all i, and so x cannot be an imputation. This contradiction proves $C(v) = \varnothing$.

Conversely, suppose v has one or more veto players; let S be the set of all these. Let x be such that

$$\sum_{i \in S} x_i = 1$$

$$x_i \geq 0 \quad \text{for all} \quad i \in S$$

$$x_i = 0 \quad \text{for} \quad i \notin S$$

Now, if T is a winning coalition, we must have $S \subset T$, so

$$\sum_{i \in T} x_i \geq \sum_{i \in S} x_i = 1 = v(T)$$

and it may be seen that $x \in C(v)$. Thus $C(v) \neq \varnothing$, and we conclude that, for a simple game v, the core is nonempty if and only if there is at least one veto player.

10.5. Balanced Collections

Theorem 10.4.3 gives one case in which the core is empty. It seems of interest to determine, more generally, those games which have nonempty cores. It is easily seen that, for any n, the set of n-person games with nonempty cores (thought of as a subset of Euclidean space of dimension $2^n - 1$) is a convex cone. In fact, suppose v and w have nonempty cores, with $x \in C(v)$, $y \in C(w)$. It is easily checked that, for nonnegative scalars r and s

$$rx + sy \in C(rv + sw)$$

In order to characterize this convex cone, we note that $C(v) \neq \varnothing$ if and only if the linear program

$$\text{Minimize} \quad \sum_{i=1}^{n} x_i = z \tag{10.16}$$

subject to

$$\sum_{i \in S} x_i \geq v(S) \quad \text{for all} \quad S \subset N \tag{10.17}$$

has a minimum $z^* \leq v(N)$. For, in such a case, any minimizing x^* lies in the core. Conversely, if $x \in C(v)$, then x satisfies (10.17) and, moreover, $\sum_{i \in N} x_i = v(N)$. Thus the minimum must be $z^* \leq v(N)$.

Consider, then, the dual program to (10.16–10.17):

$$\text{Maximize} \quad \sum_{S \subset N} y_S v(S) = q \tag{10.18}$$

subject to

$$\sum_{\substack{S \\ i \in S \subset N}} y_s = 1 \quad \text{for all} \quad i \in N \tag{10.19}$$

$$y_S \geq 0 \quad \text{for all} \quad S \subset N \tag{10.20}$$

Both of the programs (10.16)–(10.17) and (10.18)–(10.20) are feasible, and so the minimum, z^*, must equal the maximum, q^*. Thus, $C(v) \neq \varnothing$ if and only if the maximum $q^* \leq v(N)$. Or, in slightly different words

10.5.1 Theorem. A necessary and sufficient condition for the game v to have a nonempty core is that, for every nonnegative vector $(y_S)_{S \subset N}$ satisfying (10.19), we have

$$\sum_{S \subset N} y_S v(S) \le v(N) \qquad (10.21)$$

As it stands, Theorem 10.5.1 is not a very efficient tool. In fact, it requires us to solve a linear program which is no simpler than its dual (10.16)–(10.17). It turns out, however, that we can characterize the extreme points of the program (10.18)–(10.20) with a rather elegant concept: that of a balanced collection.

10.5.2 Definition. Let $\mathscr{C} = \{S_1, S_2, \ldots, S_m\}$ be a collection of nonempty subsets of $N = \{1, 2, \ldots, n\}$. We say that $\mathscr{C}$ is *N-balanced* (or, simply, *balanced* when there is no confusion as to the specific N) if there exist positive numbers $y_1, y_2, \ldots, y_m$ such that, for each $i \in N$

$$\sum_{\substack{j \\ i \in S_j}} y_j = 1 \qquad (10.22)$$

Then $y = (y_1, \ldots, y_m)$ is the *balancing vector* for $\mathscr{C}$; the y_j are *balancing coefficients*.

10.5.3 Examples

(a) The collection $\{N\}$ is clearly balanced, and in fact any partition of N (any collection of nonempty disjoint sets whose union is N) is balanced. The balancing coefficients here are all 1.

(b) Let $N = \{1, 2, 3\}$. Then the collection

$$\mathscr{C} = \{\{1, 2\}, \{1, 3\}, \{2, 3\}\}$$

is balanced, with balancing vector $\{\frac{1}{2}, \frac{1}{2}, \frac{1}{2}\}$. More generally, for any N, the collection of all $\binom{n}{s}$ sets with s elements is balanced, with all balancing coefficients equal to $\binom{n-1}{s-1}^{-1}$.

(c) Let $N = \{1, 2, 3, 4\}$. Then

$$\mathscr{C} = \{\{1, 2\}, \{1, 3\}, \{1, 4\}, \{2, 3, 4\}\}$$

is balanced, with the balancing vector $y = (\frac{1}{3}, \frac{1}{3}, \frac{1}{3}, \frac{2}{3})$.

Essentially, a balanced collection is a "generalized partition" — as seen from Example 10.5.3a. There are, of course, many more N-balanced collections than partitions of N. The following properties of balanced collections are of interest.

10.5.4 Theorem. The union of balanced collections is balanced.

Proof. Let

$$\mathscr{C} = \{S_1, \ldots, S_m\}$$
$$\mathscr{D} = \{T_1, \ldots, T_k\}$$

be balanced collections, with balancing vectors $(y_1, \ldots, y_m)$ and $(z_1, \ldots, z_k)$, respectively. Then

$$\mathscr{C} \cup \mathscr{D} = \{R_1, \ldots, R_q\}$$

where $q \leq m + k$. For any t, $0 < t < 1$, define now

$$w_j = \begin{cases} ty_l & \text{if} \quad R_j = S_l \in \mathscr{C} - \mathscr{D} \\ (1-t)z_p & \text{if} \quad R_j = T_p \in \mathscr{D} - \mathscr{C} \\ ty_l + (1-t)z_p & \text{if} \quad R_j = S_l = T_p \in \mathscr{C} \cap \mathscr{D} \end{cases}$$

It is easy to verify that $(w_1, \ldots, w_q)$ is a balancing vector for $\mathscr{C} \cup \mathscr{D}$. Thus the union of two balanced collection is balanced, and, by induction, the union of any number of balanced collections is balanced.

10.5.5 Lemma. Let $\mathscr{C}$ and $\mathscr{D}$ be balanced collections such that $\mathscr{C} \subset \mathscr{D}$ but $\mathscr{C} \neq \mathscr{D}$. Then there exists a balanced collection $\mathscr{B}$ such that

$$\mathscr{B} \cup \mathscr{C} = \mathscr{D}$$

but $\mathscr{B} \neq \mathscr{D}$. Moreover, the balancing vector for $\mathscr{D}$ is not unique.

Proof. Let

$$\mathscr{C} = \{S_1, S_2, \ldots, S_k\}$$
$$\mathscr{D} = \{S_1, S_2, \ldots, S_k, \ldots, S_m\}, \quad m > k$$

have the balancing vectors $(y_1, \ldots, y_k)$ and $(z_1, \ldots, z_m)$, respectively. For $t > 0$, define

$$w_j = (1+t)z_j - ty_j, \quad j = 1, \ldots, k$$
$$w_j = (1+t)z_j, \qquad\quad j = k+1, \ldots, m$$

For small $t > 0$, we see that all $w_j > 0$. Moreover, for $i \in N$

$$\sum_{\substack{j \\ i \in S_j \in \mathscr{D}}} w_j = (1+t) \sum_{\substack{j \\ i \in S_j \in \mathscr{D}}} z_j - t \sum_{\substack{j \\ i \in S_j \in \mathscr{C}}} y_j = 1$$

and so w is a balancing vector for $\mathscr{D}$. Since $w_j > z_j$ for $k+1 \leq j \leq m$, we see that z is not unique.

Now, there must be some j, $1 \leq j \leq k$, such that $y_j > z_j$, as otherwise, for any $i \in S_{k+1}$, we would have

$$1 = \sum_{i \in S_j \in \mathscr{C}} y_j \leq \sum_{i \in S_j \in \mathscr{C}} z_j < \sum_{i \in S_j \in \mathscr{D}} z_j = 1$$

which is a contradiction.

Set, then

$$t = \min \left\{ \frac{z_j}{y_j - z_j} \middle| y_j > z_j \right\}$$

Let

$$\mathscr{C}' = \{ S_j | S_j \in \mathscr{C}, (1 + t) z_j = t y_j \}$$

and set

$$\mathscr{B} = \mathscr{D} - \mathscr{C}'$$

Clearly, $\mathscr{C}'$ is a nonempty subcollection of $\mathscr{C}$, and so we have

$$\mathscr{B} \neq \mathscr{D}, \quad \mathscr{B} \cup \mathscr{C} = \mathscr{D}$$

Moreover, $w_j > 0$ for all $S_j \in \mathscr{B}$, and it is easily checked that (for this value of t) w is a balancing vector for the collection $\mathscr{B}$. Thus $\mathscr{B}$ is as desired.

We next define a *minimal balanced collection* as a balanced collection which is such that no proper subcollection is balanced. Then

10.5.6 Theorem. Any balanced collection is the union of minimal balanced collections.

Proof. By induction on m (the number of sets in the collection).

The theorem is clearly true when $m = 1$, as the only balanced collection with one set is $\{N\}$, which is clearly minimal. Suppose, then, that it is true for all collections with $m - 1$ or fewer elements. Let $\mathscr{D}$ be a balanced collection with m elements.

If $\mathscr{D}$ is itself minimal, then it is clearly the union of minimal balanced collections. If $\mathscr{D}$ is not minimal, then it has a proper balanced subcollection, $\mathscr{C}$, and, by Lemma 10.5.5, there exists another proper balanced subcollection, $\mathscr{B}$, such that $\mathscr{B} \cup \mathscr{C} = \mathscr{D}$. Since $\mathscr{B}$ and $\mathscr{C}$ are proper subcollections, they each have $m - 1$ or fewer elements, and thus each can be expressed as a union of minimal balanced collections. But then $\mathscr{D}$ is also the union of minimal balanced collections.

10.5.7 Theorem. A balanced collection has a unique balancing vector if and only if it is minimal.

Proof. Lemma 10.5.5 tells us that the balancing vector will be unique only for minimal balanced collections. Conversely, suppose that

$$\mathscr{C} = \{S_1, S_2, \ldots, S_m\}$$

has the two (distinct) balancing vectors y and z. Without loss of generality, we may assume that $y_j > z_j$ for at least one value of j. As before, choose

$$w = (1 + t)z - ty$$

where $t = \min\{z/(y_j - z_j)|y_j > z_j\}$, and we see that w is a balancing vector for

$$\mathscr{B} = \{S_j|(1 + t)z_j > ty_i\}.$$

As $\mathscr{B}$ is a proper subcollection of $\mathscr{C}$, we see that $\mathscr{C}$ is not minimal.

We see, then, that the balanced collections are precisely the unions of minimal balanced collections, and these in turn are those for which the balancing vector is unique. In fact, the minimal balanced collections correspond to the extreme points of the program (10.18)–(10.20), as our next theorem shows:

10.5.8 Theorem. The extreme points of (10.18)–(10.20) are the balancing vectors of the minimal balanced collections.

Proof. Let $y = (y_S)_{S \subset N}$ satisfy the constraints (10.19)–(10.20). Then it is a balancing vector for the collection

$$\mathscr{C} = \{S|y_s > 0\}.$$

If $\mathscr{C}$ is not minimal, let $\mathscr{B}$ be a balanced proper subcollection, with balancing vector $z = (z_S)_{S \subset N}$. We know $z_S > 0$ only if $y_S > 0$, and so, for small values of t, both

$$w = (1 - t)y + tz$$

and

$$w' = (1 + t)y - tz$$

satisfy the constraints (10.19)–(10.20). Now, $w \neq w'$, as $w_S < w'_S$ for any $S \in \mathscr{C} - \mathscr{B}$. But

$$y = \tfrac{1}{2}(w + w')$$

and so y is not extreme in (10.18)–(10.20).

Suppose, on the other hand, that $\mathscr{C}$ is a minimal balanced collection. If y is not extreme, we can write

$$y = \tfrac{1}{2}(w + w')$$

where $w \neq w'$, and both w and w' satisfy (10.19)–(10.20). Because of the nonnegativity constraints (10.20), we must have $w_S = w'_S = 0$ whenever $y_S = 0$. Thus w and w' will be balancing vectors for $\mathscr{B}$ and $\mathscr{B}'$, respectively, where

$$\mathscr{B} = \{S|w_S > 0\}$$

and

$$\mathscr{B}' = \{S | w'_S > 0\}$$

are both subcollections of $\mathscr{C}$. Since $\mathscr{C}$ is minimal, then $\mathscr{B}$ and $\mathscr{B}'$ must both coincide with $\mathscr{C}$. But then by Theorem 10.5.7, $w = w' = y$. This contradiction proves that y must be extreme.

10.5.9 Corollary. A minimal N-balanced collection has at most n sets.

Proof. Let $\mathscr{C}$ be a minimal balanced collection. Then its balancing vector y is a basic feasible point of (10.18)–(10.20). As we saw in Chapter 3, only the basic variables (components) of y can be positive. But a tableau for (10.18)–(10.20) has n basic variables (which is the number of constraints other than the nonnegativity constraints). Thus y has at most n positive components, corresponding to the n (or less) sets in $\mathscr{C}$.

We are now in a position to give an efficient characterization of games with nonempty cores:

10.5.10 Theorem. A necessary and sufficient condition for the n-person game v to have a nonempty core is that, for every minimal N-balanced collection $\mathscr{C} = \{S_1, \ldots, S_m\}$ with balancing vector $y = (y_1, \ldots, y_m)$

$$\sum_{j=1}^{m} y_j v(S_j) \leq v(N) \tag{10.23}$$

Proof. Necessity follows from Theorem 10.5.1 and the fact that every balancing vector must satisfy (10.19)–(10.20). To show that the condition is sufficient, we note that the maximum of program (10.18)–(10.20), which is known to exist, is to be found at an extreme point. This extreme point is the balancing vector of some minimal balanced collection. Thus, if (10.23) holds for all such vectors, the maximum of (10.18)–(10.20) must be less than or equal to $v(N)$.

Theorem 10.5.10 gives us a system of inequalities which must be satisfied if a game has a nonempty core. Now for superadditive games, the condition (10.23) will certainly hold if $\mathscr{C}$ is a partition. Thus we need concern us only with minimal balanced collections which are not partitions.

10.5.11 Example. Let $N = \{1, 2, 3\}$. Apart from the partitions there is only one minimal balanced collection, namely

$$\mathscr{C} = \{\{1, 2\}, \{1, 3\}, \{2, 3\}\}$$

with balancing vector $(\frac{1}{2}, \frac{1}{2}, \frac{1}{2})$. Thus a three-person game v has a nonempty core if and only if

$$v(\{1, 2\}) + v(\{1, 3\}) + v(\{2, 3\}) \leq 2v(N)$$

10.5.12 Example. Let $N = \{1,2,3,4\}$. The minimal N-balanced collections, other than partitions, are

(a) $\mathscr{C} = \{\{1,2,3\},\{1,2,4\},\{1,3,4\},\{2,3,4\}\}$, $y = (\frac{1}{3},\frac{1}{3},\frac{1}{3},\frac{1}{3})$
(b) $\mathscr{C} = \{\{1,2\},\{1,3\},\{1,4\},\{2,3,4\}\}$, $y = (\frac{1}{3},\frac{1}{3},\frac{1}{3},\frac{2}{3})$
(c) $\mathscr{C} = \{\{1,2\},\{1,3\},\{2,3\},\{4\}\}$, $y = (\frac{1}{2},\frac{1}{2},\frac{1}{2},1)$
(d) $\mathscr{C} = \{\{1,2\},\{1,3,4\},\{2,3,4\}\}$, $y = (\frac{1}{2},\frac{1}{2},\frac{1}{2})$

Apart from these collections, others can be obtained by permutations, giving a total of 15 such collections: one of type (a), four each of type (b) and (c), and six of type (d). Assuming the game to be in (0,1) normalization (and superadditive), this gives us a system of 15 inequalities:

(a) $v(\{1,2,3\}) + v(\{1,2,4\}) + v(\{1,3,4\}) + v(\{2,3,4\}) \le 3$
(b) $v(\{1,2\}) + v(\{1,3\}) + v(\{1,4\}) + 2v(\{2,3,4\}) \le 3$
(c) $v(\{1,2\}) + v(\{1,3\}) + v(\{2,3\}) \le 2$
(d) $v(\{1,2\}) + v(\{1,3,4\}) + v(\{2,3,4\}) \le 2$

By superadditivity, (d) implies (c). Thus we are left with 11 inequalities: (a), four of type (b), and six of type (d), which must be satisfied for v to have a nonempty core.
In case v is a symmetric four-person game, with

$$v(S) = v_s$$

where s is the cardinality of S, these inequalities will reduce to the single inequality

$$v_3 \le \frac{3}{4}$$

The proof of this is left as an exercise to the reader.

10.5.13 Example. Consider a "production game" or economy, with player set $N = \{1,2,\ldots,n\}$, each player holding a "bundle" of q commodities. More specifically, player i holds b_{i1} units of commodity C_1, b_{i2} units of $C_2,\ldots$, and b_{iq} units of C_q. The commodities are worthless in themselves, except that they can be used to produce goods $G_1,\ldots,G_m$, which can then be sold at fixed market prices. We assume a linear production process, in which one unit of G_l requires a_{1l} units of C_1, a_{2l} units of $C_2,\ldots$, and a_{ql} units of C_q, and can be sold for a price of p_l dollars.

When a coalition S forms, its members will presumably pool their resources (commodities) so as to maximize the market price of their products. Thus the characteristic function is given by the linear programs

$$v(S) = \max \sum_{l=1}^{m} p_l x_l \tag{10.24}$$

Subject to

$$\sum_{l=1}^{m} a_{kl}x_l \le b_k(S), \quad k = 1, 2, \ldots, q \tag{10.25}$$

$$x_l \ge 0, \quad l = 1, 2, \ldots, m \tag{10.26}$$

where

$$b_k(S) = \sum_{i \in S} b_{ik}$$

is the total amount of C_k that coalition S has.

We will show that v has a nonempty core. For, let $\mathcal{C} = \{S_1, S_2, \ldots, S_r\}$ be a balanced collection, with balancing vector $(y_1, y_2, \ldots, y_r)$. For each $S_i \in \mathcal{C}$, let

$$x^j = (x_1^j, \ldots, x_m^j)$$

be the optimizing vector for the program (10.24)–(10.26) defining S_j, and let

$$x^* = \sum_{j=1}^{r} y_j x^j \ge 0$$

For each k, we have

$$\sum_{l=1}^{m} a_{kl}x_l^* = \sum_{j=1}^{r} y_j \sum_{l=1}^{m} a_{kl}x_l^j$$

$$\le \sum_{j=1}^{r} y_j \sum_{i \in S_j} b_{ik} = \sum_{i \in N} b_{ik} \sum_{\substack{j \\ i \in S_j}} y_j$$

$$= \sum_{i \in N} b_{ik} = b_k(N)$$

Thus x^* satisfies (10.25)–(10.26) for the program defining $v(N)$, and therefore

$$\sum_{l=1}^{m} p_l x_l^* \le v(N)$$

Now

$$\sum_{j=1}^{r} y_j v(S_j) = \sum_{j=1}^{r} y_j \sum_{l=1}^{m} p_l x_l^j = \sum_{l=1}^{m} p_l x_l^* \le v(N)$$

Since this is true for all balanced collections $\mathcal{C}$, we conclude that the game has a nonempty core.

To find a point in the core, we proceed, as in Example 10.4.5 above, by solving the dual linear program. We leave this as an exercise for the reader (see Problem 4).

Problems

1. An $(n, n-1)$ game is a game v, with player set N, such that $v(S) = 0$ whenever $|S| \leq n-2$. Define, for each $i \in N$

$$\Delta_i = v(N) - v(N - \{i\})$$

Show that the conditions for an $(n, n-1)$ game v to be balanced reduce to a single linear inequality. Give this inequality.

2. Let v be a game with player set N. For any coalition $S \subset N$, define the *relative game on* S as the restriction of the function v to the collection 2^S. A game is said to be *totally balanced* if, for each S, the relative game on S is balanced.

   (a) Show, by a counterexample, that a balanced game need not be totally balanced.
   (b) Show that the game of Example 10.5.13 is totally balanced.

3. Let $N = R \cup L$, where R and L are disjoint sets. Assume each member of R has one right shoe, and each member of L has one left shoe. Single shoes (or several shoes for the same foot) are worthless; any pair of shoes can, however, be sold for \$1. (It is assumed that the shoes are identical except for the left-right distinction.) Let utility be identical with money.

   (a) Give the characteristic function for this game. (Hint: $v(S)$ is the number of pairs of shoes that the coalition S can form.)
   (b) Let r, l be the cardinalities R and L, respectively. Show that, if $r < l$, the core $C(v)$ consists of a single point where the members of L receive nothing. Similarly, if $r > l$, $C(v)$ has a single point where the members of R receive nothing.
   (c) Find the core in the special case $r = l$.

4. For Example 10.5.13, show that there exist "shadow prices" z_k for the several commodities such that the point $u = (u_1, u_2, \ldots, u_n)$ given by

$$u_i = \sum_k b_{ik} z_k$$

lies in the core of the game. (Hint: Obtain z by solving the dual linear program.)

5. For a game v with player set N, define a *preimputation* as a vector $x = (x_1, x_2, \ldots, x_n)$ such that

$$\sum_{i \in N} x_i = v(N)$$

For any ε, define the *strong ε-core*, denoted $C_\varepsilon(v)$, as the set of all preimputations such that, for each coalition S

$$\sum_{i \in S} x_i \geq v(S) - \varepsilon$$

(a) Show that, if $\varepsilon < \varepsilon'$, then $C_\varepsilon(v) \subseteq C_{\varepsilon'}(v)$.

(b) Show that there exist a minimal ε such that $C_\varepsilon(v) \neq \emptyset$. This particular $C_\varepsilon(v)$ is known as the least-core, $LC(v)$

(c) Find the least-core for the game v with $N = \{1, 2, 3\}$, defined by

$$v(\{i\}) = 0, \qquad v(\{1, 2\}) = 50$$

$$v(\{1, 3\}) = 80, \quad v(\{2, 3\}) = 90, \quad v(N) = 100$$

6. Let v, v' be two games with the player set N, and suppose that for some ε and ε'

$$C_\varepsilon(v) = C_{\varepsilon'}(v') \neq \emptyset$$

Then, for any $\delta > 0$

$$C_{\varepsilon - \delta}(v) = C_{\varepsilon' - \delta}(v')$$

In particular, prove that $LC(v) = LC(v')$.

Chapter 11

Stable Sets

11.1. Introduction

We saw in Chapter 10 that the core—which represents a very strong type of stability—is frequently empty. Different types of stability—not generally as strong as that represented by the core—are therefore of interest. The idea of *stable sets* was first introduced by von Neumann and Morgenstern, who called them *solutions*.

In 10.2 we gave an example of a three-person game whose characteristic function can be given most easily in the form:

$$v(S) = \begin{cases} -2 & \text{if} \quad S \text{ has 1 member} \\ 2 & \text{if} \quad S \text{ has 2 members} \\ 0 & \text{if} \quad S \text{ has 3 members} \end{cases}$$

We can leave the game in this form, or put it into $(0, 1)$ normalization. (This can be done by setting $r = \frac{1}{6}, \alpha_1 = \alpha_2 = \alpha_3 = \frac{1}{3}$.) The $(0, 1)$ normalization is a simple game in which two- and three-person coalitions win and single-player coalitions lose.

For this game in $(0, 1)$ normalization, we obtain the three imputations:

$$\alpha_{12} = (\tfrac{1}{2}, \tfrac{1}{2}, 0)$$

$$\alpha_{13} = (\tfrac{1}{2}, 0, \tfrac{1}{2})$$

$$\alpha_{23} = (0, \tfrac{1}{2}, \tfrac{1}{2})$$

as a "solution." Now, in which sense is this a solution? It is easy to see that none of these imputations dominates any one of the others. (It was this domination, in fact, which caused us to disallow the three slightly altered imputations in the subsequent variant of the game.) Of course, this is not all; any set consisting of a single imputation would have this property. This set of imputations has also, however, the following property: *any imputation other than the 3 α_{ij} is dominated by one of these three imputations.*

To see this, let $x = (x_1, x_2, x_3)$ be any imputation for this game. As the game is in $(0, 1)$ normalization, we know $x_i \geq 0$ and $x_1 + x_2 + x_3 = 1$. Thus, at most two of the components of x can be as large as $\frac{1}{2}$. If this actually happens, then both of those components must equal $\frac{1}{2}$, while the remaining component is zero. But this means x is one of the α_{ij}. If, then, x is any other imputation, then at most one of the components can be as large as $\frac{1}{2}$. At least two components, say x_i and x_j, with $i<j$, are then smaller than $\frac{1}{2}$. But it is then clear that $\alpha_{ij} \vdash x$ through the set $\{i, j\}$.

This gives us, then, a definition for our "solution concept":

11.1.1 Definition. A set $V \subset E(v)$ is said to be a *stable set* for v if

(i) $x, y \in V$, then $x \not\vdash y$
(ii) $x \notin V$, then there is some $y \in V$ such that $y \vdash x$

Thus, a stable set satisfies the two conditions of *internal stability* (no imputation in V dominates another) and *external stability* (any imputation outside V is dominated by some imputation in V).

Stable sets were first defined by von Neumann and Morgenstern; they are often called *solutions* of the game. We do not use this name, inasmuch as many other solution concepts have been formulated since then.

One of the principal difficulties with stable sets is that, from their definition, neither existence nor uniqueness is guaranteed. On the one hand, most games possess immense collections of stable sets (though there are a few with a unique stable set). On the other hand, see Section 11.4 for an example of a game with no solution.

11.1.2 Example. Let us consider once again the constant-sum three-person game in $(0, 1)$ normalization (and see Figure 11.1). As mentioned above, the set

$$V = \{(\tfrac{1}{2}, \tfrac{1}{2}, 0), (\tfrac{1}{2}, 0, \tfrac{1}{2}), (0, \tfrac{1}{2}, \tfrac{1}{2})\}$$

is stable. However, it is not the only stable set; if c is any number in the interval $[0, \tfrac{1}{2})$, it is easy to check that the set

$$V_{3,c} = \{(x_1, 1 - c - x_1, c) | 0 \le x_1 \le 1 - c\}$$

in which the third player receives the constant c, while players 1 and 2 split the remainder in all possible ways, is also stable. Internal stability follows from the fact that two such imputations x and y differ only in that, say, $x_1 > y_1$, while $x_2 < y_2$. But domination of imputations is not possible through single-player coalitions.

To show that $V_{3,c}$ is externally stable, suppose y is any imputation not in $V_{3,c}$. This means either $y_3 > c$ or $y_3 < c$.

If $y_3 > c$, say $y_3 = c + \varepsilon$. We define x by

$$x_1 = y_1 + \varepsilon/2$$
$$x_2 = y_2 + \varepsilon/2$$
$$x_3 = c$$

It is then easy to see that $x \in V_{3,c}$ and $x \vdash y$ through $\{1, 2\}$.

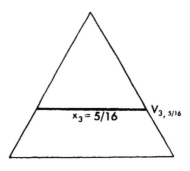

Figure 11.1.

If $y_3 < c$, we know that either $y_1 \leq \frac{1}{2}$ or $y_2 \leq \frac{1}{2}$ (or else their sum would be greater than 1). Say $y_1 \leq \frac{1}{2}$. Then let $x = (1 - c, 0, c)$. It is clear that $1 - c > \frac{1}{2} \geq y$, and so $x \vdash y$ through $\{1, 3\}$. Clearly, $x \in V_{3,c}$. If $y_2 \leq \frac{1}{2}$, we see similarly that $z \vdash y$, where $z = (0, 1 - c, c)$.

Thus it is clear that, apart from the symmetric stable set, this game has a family of solutions in which player 3 receives a fixed amount c in the interval $[0, \frac{1}{2})$. These are called *discriminatory* stable sets; player 3 is said to be discriminated. In the set $V_{3,0}$, player 3 is said to be totally discriminated, or *excluded*.

By the symmetry of the game, it is clear that there will also exist two families of stable sets, $V_{1,c}$ and $V_{2,c}$, in which players 1 and 2, respectively, are discriminated.

The preceding example shows that a game may possess a bewildering multiplicity of stable sets. It is not clear how one stable set is to be chosen from among all the sets. Nor is it clear, once a stable set is chosen, how an imputation can be chosen from this stable set. Von Neumann and Morgenstern suggest that stable sets are standards of behavior, proper to a given social order. Once a society has decided on the standard of behavior (stable set), the individual players' bargaining abilities would determine the imputation.

Unfortunately, as seen below (section 11.4) not all games have stable set solutions. Below, we consider cases in which stable sets exist, and special types of stable sets.

Let us consider the following theorems:

11.1.3 Theorem. Let v be a simple game in $(0, 1)$ normalization, and let S be a *minimal* winning coalition [i.e., a coalition such that $v(S) = 1$ but $v(T) = 0$ for all $T \subset S, T \neq S$]. Let V_s be the set of all imputations x such that $x_i = 0$ for all $i \notin S$. Then V_s is a stable set.

Proof. The proof of this theorem is the same as the proof (in Example 11.1.2) that the set $V_{3,c}$ is stable.

Theorem 11.1.3 states that every simple game has discriminatory stable sets in which a minimal winning coalition excludes all the others. In certain cases, it is possible that some of the discriminated players will receive small amounts (see Example 11.1.2). The exact values which they may receive are often quite difficult to determine; we shall not attempt to deal with this question here.

11.1.4 Example. (Symmetric Solutions to Three-Person Games) In Example 11.1.3 we gave all stable set solutions to the constant-sum three-person game. We shall see how the symmetric, three-point stable set generalizes to a symmetric stable set for other (general-sum) symmetric three-person games.

For a three-person game in $(0, 1)$ normalization, the values of the one- and three-person coalitions are known in advance; hence, it is only the values of the two-person coalitions that remain to be fixed. For symmetric games, these all have the same value, and thus there is a unique parameter, v_2, which determines these games. This parameter can vary throughout the interval $[0, 1]$; for $v_2 = 1$, we have the constant-sum game (which we have already analyzed), while for $v_2 = 0$ we have a "pure bargaining game" in which every imputation belongs to the core, so that the core is the unique stable set.

When v_2 is large (i.e., close to 1), the players may be expected to behave very much as in the constant-sum game. Thus, they will first attempt to enter into a two-player coalition. Since they are bidding against each other, they cannot ask too much of their presumptive partner, and the "natural" result may be an agreement, between the members of whichever coalition forms, to divide their winnings equally.

Once a two-player coalition has been formed, its members will bargain with the remaining player to determine the division of the remaining $1 - v_2$ units of utility. In this case there are no alternatives; there is no one else to bargain with, and so the amount which the coalition receives depends only on the bargaining abilities of the three players. Thus we may expect to find a stable set consisting of the three line-segments.

$$\left. \begin{cases} (x, x, 1 - 2x) \\ (x, 1 - 2x, x) \\ (1 - 2x, x, 2) \end{cases} \right\}, \quad v_2/2 \leq x \leq \tfrac{1}{2} \tag{11.1}$$

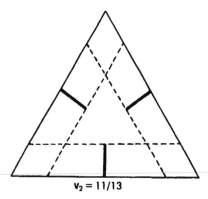

$v_2 = 11/13$

Figure 11.2.

(see Figure 11.2). It is necessary to check whether this set is, indeed, stable. If we do so, we find that the set is stable for $v_2 \geq \frac{2}{3}$.

Internal stability is proved as follows: Since $x \geq v_2/2$, we know that $2x \geq v_2$, and so an imputation of the form $(x, x, 1 - 2x)$ can only be dominated through one of the two coalitions $\{1, 3\}$ and $\{2, 3\}$. But none of the imputations given by (11.1) can dominate $(x, x, 1 - 2x)$ through $\{1, 3\}$. For, in an imputation $(y, y, 1 - 2y)$, if $y > x$, then $1 - 2x > 1 - 2y$, while in an imputation $(y, 1 - 2y, y)$, we find that, if $y > x$, then $2y > v_2$, and so the set $\{1, 3\}$ is not effective. Symmetry and similar considerations complete the proof of internal stability.

External stability is also easily proved. Since $v_2 \geq \frac{2}{3}$, we know that any imputations will have at most two components greater than or equal to $v_2/2$, with the single exception of the case $v_2 = \frac{2}{3}$, and the imputation $(\frac{1}{3}, \frac{1}{3}, \frac{1}{3})$, which will in this case satisfy (11.1). Thus any imputation which does not satisfy (11.1) will have at most two components greater than or equal to $v_2/2$. Let us suppose an imputation z has two such components; by symmetry we may assume these are the first two components, z_1 and z_2. If $z_1 = z_2$, then z satisfies (11.1). If $z_1 \neq z_2$, we may (again by symmetry) assume $z_1 = z_2 + 3\varepsilon$, with $\varepsilon > 0$. Then z is dominated by the imputation

$$(z_2 + \varepsilon, z_2 + \varepsilon, z_3 + \varepsilon)$$

through the coalition $\{2, 3\}$.

If, however, the imputation z has one component or none as great as $v_2/2$, then we may assume that both z_1 and z_2 are smaller than $v_2/2$. Then z is dominated by

$$(v_2/2, v_2/2, 1 - v_2)$$

through the coalition $\{1, 2\}$.

The above analysis loses its validity for $v_2 < \frac{2}{3}$. In fact, internal stability will be preserved, but the external stability is lost. Such games will possess a core, consisting of all imputations $y = (y_1, y_2, y_3)$ satisfying $y_i \leq 1 - v_2$ for all $i = 1, 2, 3$. It may then be

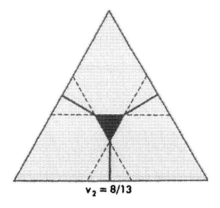

$v_2 = 8/13$

Figure 11.3.

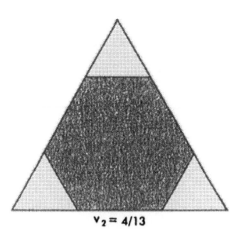

$v_2 = 4/13$

Figure 11.4.

checked that the union of the core and the three line-segments given in (11.1) is a stable set (see Figure 11.3). For $v_2 \le \frac{1}{2}$, moreover, the three line-segments will all be subsets of the core; in this case, thus, the core is the unique stable set (see Figure 11.4).

As we have seen, the set of all n-person games (in normalization) forms a compact subset of $(2^n - n - 2)$-dimensional Euclidean space. Applying Lebesgue measure, we can say that any full-dimensional subset of this set is a *positive fraction* of the set of n-person games. We give these theorems without proof:

11.1.5 Theorem. A positive fraction of all n-person games have unique stable sets consisting of the core.

11.1.6 Theorem. A positive fraction of all n-person games have stable sets which discriminate $n - 2$ players. In such sets, at least $n - 3$ of the discriminated players are excluded.

11.2. Properties of Stable Sets

In this section we study some properties of stable sets, considered from the point of view of operations on games.

11.2.1 Strategic Equivalence. We note first of all that, under strategic equivalence, the domination relation is preserved. Since stable sets are defined entirely in terms of domination, it will follow that these are preserved under strategic equivalence.

As an example, consider the two three-person games, u and v, defined by

$$u(\{i\}) = 0$$

$$u(S) = 1 \quad \text{if} \quad |S| = 2 \text{ or } 3$$

and

$$v(\{i\}) = -2$$

$$v(\{i, j\}) = +2 \quad \text{for} \quad i \neq j$$

$$v(\{1, 2, 3\}) = 0$$

It is easily checked that u and v are strategically equivalent, satisfying

$$v(S) = \beta u(S) + \sum_{i \in S} \alpha_i$$

where $\beta = 6, \alpha_i = -2$ for all i. Thus the transformation

$$y_i = 6x_i - 2$$

changes stable sets of u to stable set for v.

As we know, u has the stable set

$$V_s = \{(\tfrac{1}{2}, \tfrac{1}{2}, 0), (\tfrac{1}{2}, 0, \tfrac{1}{2}), (0, \tfrac{1}{2}, \tfrac{1}{2})\}$$

whose image

$$V'_s = \{(1, 1, -2), (1, -2, 1), (-2, 1, 1)\}$$

is stable for v. Similarly, for any $c \in [0, \tfrac{1}{2})$, u has the stable set

$$V_{3,c} = \{(t, 1 - t - c, c) | 0 \leq t \leq 1 - c\}$$

This means that v has the stable set

$$V'_{3,c} = \{(t, -t - c', c') | -2 \leq t \leq 2 - c'\}$$

for any $c' \in [-2, 1)$.

Together with similar families of sets discriminating 1 and 2, respectively, this gives all stable sets of v.

11.2.2 Composition. Let v and w be two games with disjoint player sets, M and N: we might have, e.g.,

$$M = \{1, 2, \ldots, m\}$$

$$N = \{m+1, m+2, \ldots, m+n\}$$

It is easy to see how this can be made into a "composite" game $v \oplus w$ with player set $M \cup N$. In fact, we can write

$$(v \oplus w)(S) = v(M \cap S) + w(N \cap S) \tag{11.2}$$

for any $S \subset M \cup N$.

Heuristically, Equation (11.1) tells us that there is no interaction among members of the two groups M and N. In effect, there is a juxtaposition of the two games, no more, to form an $(m+n)$-person game. Under the circumstances, it is not surprising to find that the following is true:

11.2.3 Theorem. If V is a stable set for v, and W is a stable set for w, then the Cartesian product $V \times W$ is a stable set for $v \oplus w$.

In effect, Theorem 11.2.3 simply states that the two games can be played independently of each other—there is no great surprise here. Somewhat surprising, however, is the fact that in general there are other stable sets—obtained from the games v and w by the notion of a transfer: it is, under certain circumstances, possible to "transfer" a certain amount from M to N, or vice versa. We explain this more precisely, as follows.

Given a game v with player set M and a real number q, we define $\tilde{v}^q$ by

$$\tilde{v}^q(S) = v(S) \quad \text{if} \quad S \subset M, \ S \neq M$$

$$\tilde{v}^q(M) = q$$

Thus $\tilde{v}^q$ coincides with V, except for the single set M, the "grand coalition" of the game. Of course, $\tilde{v}$ is not always, strictly speaking, a game: if q is small enough, the superadditive property will be lost. We can nevertheless think of $\tilde{v}$ as an "improper" game, and so long as

$$q \geq \sum_{i \in N} v(\{i\}) \tag{11.3}$$

we can define imputations, a core, and a domination relation among these, just as for any (proper) game. It is therefore meaningful to talk about cores and stable sets for the "game" $\tilde{v}$.

Returning now to the game $v \oplus w$, consider two numbers q and r such that

$$q + r = v(M) + w(N) \tag{11.4}$$

If x is an imputation for $\tilde{v}^q$, and y an imputation for $\tilde{w}^r$, then $(x; y)$ is clearly an imputation for $v \oplus w$. This imputation represents a *transfer* of the amount $q - v(M)$

from N to M. Assuming that this transfer is nonzero, the question is whether the game $v \oplus w$ can somehow accommodate such a transfer.

11.2.4 Definition. In the game $v \oplus w$, if $q > v(M)$, a transfer (from N to M) of the amount $q - v(M)$ is *admissible* if and only if

(a) $r = v(M) + w(N) - q \geq \sum_{i \in N} w(\{i\})$

(b) the core of the game $\tilde{v}^q$ has an empty interior

Similarly, if $q < v(M)$, the transfer (from M to N) of $v(M) - q$ is admissible if and only if

(a) $q \geq \sum_{i \in M} v(\{i\})$

(b) the core of $\tilde{w}^r$ [where $r = v(M) + w(N) - q$] has an empty interior

In Definition 11.2.4, it is easy enough to understand condition (a): a larger transfer is not admissible because there is simply no more that the "benefactors" N can give. Condition (b) is more complicated; essentially, it seems to set a limit on the amount that the "beneficiaries" M can assimilate.

11.2.5 Theorem. Let $q + r = v(M) + w(N)$ as above, and let V, W be stable sets for $\tilde{v}^q$ and $\tilde{w}^r$, respectively. Then $V \times W$ is a stable set for $v \oplus w$ if and only if the associated transfer is admissible.

Proof. Without loss of generality, we shall assume $q \geq v(M)$.

Assume first that the transfer is not admissible. This is either because (a) $r < \Sigma w(\{i\})$ or (b) the core of $\tilde{v}^q$ has a nonempty interior. In case (a), the game $\tilde{w}^r$ will have no imputations. But then W, and also $V \times W$, would be empty, and the empty set is clearly not stable for $v \oplus w$.

Suppose, instead, that $\tilde{v}^q$ has a core with nonempty interior. Thus, we can find a vector $x = (x_1, \ldots, x_m)$ such that

$$\sum_{i \in M} x_i = q$$

$$\sum_{i \in S} x_i \geq v(S) + \varepsilon \quad \text{for all} \quad S \subset M$$

where $\varepsilon > 0$. Choose, now, x' to satisfy

$$x_i' = x_i - (\varepsilon/m), \quad i \in M$$

Let $y \in W$, and choose y' by

$$y_i' = y_i + (\varepsilon/n), \quad \text{for} \quad i \in N$$

Then $(x'; y')$ is seen to be an imputation for $v \oplus w$; moreover, $x' \notin V$ and $y' \notin W$, so $(x'; y') \notin V \times W$.

If $V \times W$ is a stable set, there must be vectors $\tilde{x} \in V, \tilde{y} \in W$, such that $(\tilde{x}; \tilde{y}) \vdash (x'; y')$. Thus, there is some $S \subset M \cup N$ such that

$$\tilde{x}_i > x'_i, \quad i \in S \cap M$$

$$\tilde{y}_i > y'_i, \quad i \in S \cap N$$

$$\sum_{S \cap M} \tilde{x}_i + \sum_{S \cap N} \tilde{y}_i \leq v(S \cap M) + w(S \cap N)$$

We have, however,

$$\sum_{S \cap M} \tilde{x}_i \geq \sum_{S \cap M} x'_i \geq \sum_{S \cap M} x_i + \varepsilon \geq v(S \cap M)$$

so that

$$\sum_{S \cap M} \tilde{x}_i \geq v(S \cap M)$$

with strict inequality holding if $S \cap M \neq \emptyset$. Thus we must have $S \cap N \neq \emptyset$, and, moreover,

$$\sum_{S \cap N} \tilde{y}_i \leq w(S \cap N)$$

$$\tilde{y}_i > y'_i > y_i \quad \text{for} \quad i \in S \cap N$$

Thus $\tilde{y} \vdash y$ (in w) through the coalition $S \cap N$. But this is a contradiction as both $\tilde{y}$ and $y \in W$, which was assumed stable.

Conversely, assume that the transfer is admissible, and V, M are stable. We shall show $V \times W$ is stable for $v \oplus w$.

To prove internal stability, let $(x; y)$ and $(x'; y') \in V \times W$, and suppose $(x; y) \vdash (x'; y')$ through the coalition $S \neq \emptyset$. Then

$$x_i > x'_i, \quad i \in S \cap M$$

$$y_i > y'_i, \quad i \in S \cap N$$

$$\sum_{S \cap M} x_i + \sum_{S \cap N} y_i \leq v(S \cap M) + w(S \cap N)$$

This will mean that, either

$$\sum_{S \cap M} x_i \leq v(S \cap M), \quad S \cap M \neq \emptyset$$

or

$$\sum_{S \cap N} y_i \leq w(S \cap N), \quad S \cap N \neq \emptyset$$

Thus, either $x \vdash x'$ (in v) through $S \cap M$ or $y \vdash y'$ (in w) through $S \cap N$. Moreover, $S \cap M \neq M$ and $S \cap N \neq N$, since $\Sigma x_i = \Sigma x'_i$ and $\Sigma y_i = \Sigma y'_i$. Thus, either $x \vdash x'$ (in $\tilde{v}^q$) through $S \cap M$, or $y \vdash y'$ (in $\tilde{w}^r$) through $S \cap N$. In either case we obtain a contradiction as V and W were both assumed stable.

To prove external stability, let us suppose $(x; y)$ is an imputation for $v \oplus w$, but $(x; y) \notin V \times W$. There are three possibilities:

(a) $\sum_{i \in M} x_i = q,$ $\quad \sum_{i \in N} y_i = r$
(b) $\sum_{i \in M} x_i = q + \varepsilon,$ $\quad \sum_{i \in N} y_i = r - \varepsilon, \varepsilon > 0$
(c) $\sum_{i \in M} x_i = q - \varepsilon,$ $\quad \sum_{i \in N} y_i = r + \varepsilon, \varepsilon > 0$

In case (a), we find that either $x \notin V$ or $y \notin W$. Without loss of generality, we may assume $x \notin V$. Since x is an imputation for $\tilde{v}^q$, there is $x' \in V$ such that $x' \vdash x$ through some $S \subset N$, $\varnothing \neq S \neq N$. Then, letting $y' \in W$, it is easy to see that $(x', y') \in V \times W, (x', y') \vdash (x; y)$ through the coalition S.

In case (b), we set

$$y'_i = y_i + (\varepsilon/n) \quad \text{for} \quad i \in N$$

Clearly y' is an imputation for $\tilde{w}^r$; either $y' \in W$ or $y' \neq W$.

If $y' \in W$, we choose $x' \in V$, then $(x'; y') \in V \times W$, and it is easy to see that $(x'; y') \vdash (x; y)$ through the set N.

If $y' \in W$, there is some $\tilde{y} \in W$ such that $\tilde{y} \vdash y'$ (in $\tilde{w}$) through some $S \subset N$. Choosing $\tilde{x} \in V$, we see that $(\tilde{x}; \tilde{y}) \in V \times W$, and $(\tilde{x}; \tilde{y}) \vdash (x; y)$ through S.

In case (c), if $q = v(M)$, the situation is just as in case (b), with M and N interchanged. If $q > v(M)$, we again set

$$x'_i = x_i + (\varepsilon/m) \quad \text{for} \quad i \in M$$

and we see x' is an imputation for $\tilde{v}^q$. As the interior of $C(\tilde{v}^q)$ is empty, there must be at least one $S \subset M, S \neq M$, such that

$$\sum_{i \in S} x'_i \leq v(S)$$

If $x' \in V$, then, choosing $y' \in W$, we see that $(x'; y') \in V \times W$, and $(x'; y) \vdash (x; y)$ through the set S.

If $x' \notin V$, there is $\tilde{x} \in V$ such that $\tilde{x} \vdash x'$ through some $T \subset M$; clearly $T \neq M$. Choosing $\tilde{y} \in W$, we see $(\tilde{x}; \tilde{y}) \vdash (x, y)$ through the coalition T.

11.2.6 Example. Let $M = \{1, 2\}$ and $N = \{3, 4, 5\}$; let v and w be given by

$$v(\{i\}) = 0, \quad v(\{1, 2\}) = 100$$

$$w(\{i\}) = 0, \quad w(\{3, 4\}) = w(\{3, 5\}) = w(\{4, 5\}) = 80$$

$$w(\{3, 4, 5\}) = 100$$

It is not difficult to see that $\tilde{v}^q$ has a core with nonempty interior, for all $q > 0$. The core of w is empty, and in fact $C(\tilde{w}^r)$ will be empty for $r < 120$. For $r = 120$, $C(\tilde{w}^r)$ consists of a single point $(40, 40, 40)$, and for $r > 120$, $C(\tilde{w}^r)$ will have a nonempty interior. Thus, no positive transfer from N to M is admissible. On the other hand, a positive transfer from M to N is admissible, up to an admissible maximum of twenty units.

Consider, then, a transfer of twenty units from M to N. The game $\tilde{v}^{80}$ has a unique stable set, consisting of all imputations:

$$V = \{(x, 80 - x)|0 \leq x \leq 80\}$$

The game $\tilde{x}^{120}$ has many stable sets, including (as we saw in Example 11.1.6) one that consists of three lines from the edges of the triangle

$$\left\{ \begin{array}{ccc} (60-y, & 60-y, & 2y) \\ (60-y, & 2y, & 60-y) \\ (2y, & 60-y, & 60-y) \end{array} \middle| 0 \leq y \leq 20 \right\}$$

Then $V \times W$ is a stable set for the five-person game $v \oplus w$.

Another interesting property of stable sets deals with *composition* of simple games.

11.2.7 Definition. Let $M_1 M_2, \ldots, M_n$ be n disjoint nonempty sets of players. Let $w_1, w_2, \ldots, w_n$ be simple games in $(0, 1)$ normalization, with player sets $M_1, \ldots, M_n$, respectively. Let v be a nonnegative game over the set $N = \{1, 2, \ldots, n\}$; then the *v-composition* of $w_1, \ldots, w_n$, denoted

$$u = v[w_1, w_2, \ldots, w_n]$$

is a game with player set

$$M^* = \bigcup_{j=1}^{n} M_j$$

and characteristic function

$$u(S) = v(\{j|w_j(S \cap M_j) = 1\}) \tag{11.5}$$

for $S \subset M^*$.

Heuristically the compound game u represents a division into committees M_j. The members of M_j (the jth committee) choose an agent to represent them; the n agents then play the game v among themselves. A typical example here is a legislature, in which each set M_j represents the voters of the jth constituency. These elect a representative (Congressman, Member of Parliament, etc.) via a game w_j (usually a majority voting game); the representatives often also play a voting game.

11.2.8 Example. (The Electoral College) Let v be the 51-person weighted majority game $[270; 45, 41, \ldots, 3, 3]$. The players in v are the states of the union. For each j, M_j consists of the voters in the jth state; w_j is a simple majority game with player set M_j. The idea here, of course, is that u is a simple game with player set

$$M^* = \bigcup_{j=1}^{51} M_j$$

consisting of all voters in the United States. A set $S \subset M^*$ is winning if it contains a subset of the form

$$S' = \bigcup_{j \in T} S_j$$

where $w_j(S_j) = 1$ for all $j \in T$, and $v(T) = 1$. Thus, a coalition wins if it has at least half the popular votes [$w_j(S_j) = 1$] in states totaling at least 270 electoral votes [$v(T) = 1$].

In some cases, we find that it is possible for some of the subgames w_j to be "improper" games: the superadditivity condition is not strictly necessary, and we might weaken it to a monotonicity condition:

$$0 \leq w_j(S) \leq w_j(S') \quad \text{if} \quad S \subset S' \tag{11.6}$$

11.2.9 Example. The United Nations Security Council consists of five permanent members (United States, Soviet Union, Britain, France, and China) and ten other members. Motions must be approved by nine members, including all the permanent members. It is easy to see that we have here the voting game

$$u = v[w_1, w_2]$$

where M_1 is the set of permanent members, and

$$v_1(S) = 0 \quad \text{if} \quad S \neq M_1$$
$$w_1(M_1) = 1$$

where M_2 is the set of ten other members with

$$w_2(S) = 0 \quad \text{if} \quad |S| \leq 3$$

$$w_2(S) = 1 \quad \text{if} \quad |S| \geq 4$$

for $S \subset M_2$, and where v is given by

$$v(\{1\}) = v(\{2\}) = 0$$

$$v(\{1, 2\}) = 1$$

We note of course that w_2 is improper: two disjoint sets can both be winning in w_2. Nevertheless, u is a proper game.

For compound *simple* games, the following theorem is of importance. We shall use the notation $z^j = (z^j_i)_{i \in M_j}$ to denote imputations in w_j; $y = (y_j)_{j \in N}$ to denote imputations in v; and $x = (x_i)_{i \in M^*}$ to denote imputations in u. We shall write, moreover

$$y \otimes (z^1, z^2, \ldots, z^n)$$

to denote the vector $(x_i)_{i \in M^*}$ given by

$$x_i = y_j z^j_i \quad \text{if} \quad i \in M_j$$

With this notation, we have now

11.2.10 Theorem. Let $u, v, w_1, \ldots, w_n$ be as in Definition 11.2.7, with the added proviso that v is a simple game in $(0, 1)$ normalization.

Let $V, W_1, \ldots, W_n$ be stable sets for $v, w_1, \ldots, w_n$, respectively, and let

$$V[W_1, \ldots, W_n] = \left\{ x \, \middle| \, \begin{array}{l} x = y \otimes (z^1, \ldots, z^n) \\ y \in V \\ z^j \in W_j, \quad j = 1, \ldots, n \end{array} \right\} \tag{11.7}$$

Then $V[W_1, \ldots, W_n]$ is a stable set for the game u.

We omit the proof of Theorem 11.2.10, leaving it as an exercise for the reader. We point out merely that this proof is similar to—but lengthier than—the proof of Theorem 11.2.5, and that superadditivity of the games involved is not required for the proof.

11.2.11 Example. Let v, w_1, w_2, w_3, each be a three-person simple constant-sum game, in which the two-person coalitions win. Then $u = v[w_1, w_2, w_3]$ is a nine-person simple game, with 27 minimal winning coalitions, each of four players: two from each of two of the sets M_j. We can get a stable set for u by taking a stable set for each of v, w_1, w_2, w_3 and composing them. If, in particular, we take the symmetric stable set for each of the four games, we obtain a 27-point stable set for u, each point giving $\frac{1}{4}$ each to the members of a minimal winning coalition, and zero to the others.

If we instead compose discriminatory stable sets to each of the four games, then we obtain a stable set for u, in which the four members of a minimal winning coalition split most of the profits, while two other pairs split smaller amounts, and one player receives a small fixed amount.

There are, of course, many other possible compositions, giving rise to many other stable sets for u. Additionally, we cannot rule out the possibility of other stable sets for u, which are not composite in nature.

11.2.12 Example. Let u be a three-person game such that

$$u(\{i\}) = 0, \quad u(\{2, 3\}) = 0$$

$$u(\{1, 2\}) = u(\{1, 3\}) = u(\{1, 2, 3\}) = 1$$

It is not too difficult to see that we may write

$$u = v[w_1, w_2]$$

where $M_1 = [1]$; w_1 is given by

$$w_1(\{1\}) = 1$$

$M_2 = \{2, 3\}$; w_2 is given by

$$w_2(\{2\}) = w_2(\{3\}) = w_2(\{2, 3\}) = 1$$

and $N = \{1, 2\}$, with

$$v(\{1\}) = v(\{2\}) = 0, \quad v(\{1, 2\}) = 1$$

It is of course to be noticed that neither w_1 nor w_2 is in $(0, 1)$ normalization. Moreover w_2 is improper (i.e., not superadditive) and in fact has no imputations! Nevertheless, we will apply the composition theorem.

The game w_1 has, of course, a simple imputation, $z^1 = (1)$.

The game w_2 has no imputations; we shall, however, consider its "pseudoimputations" $z^2 = (z_1^2, z_2^2)$ satisfying $z^2 \geq 0$, $z_1^2 + z_2^2 = 1$. It may be seen that any set W_2^0, consisting of a single such pseudoimputation $(\alpha, 1 - \alpha)$ is stable in the usual sense.

Finally, v has a single stable set, consisting of all its imputations (y_1, y_2), $y_j \geq 0$, $y_1 + y_2 = 1$. We therefore combine

$$W_1 = \{(1)\}, \qquad W_2 = \{(\alpha, 1 - \alpha)\}$$

$$V = \{(1 - t, t) | 0 \leq t \leq 1\}$$

where α is a fixed number. Then, for any $0 \leq \alpha \leq 1$, the set

$$V[W_1, W_2] = \{(1 - t, t\alpha, t - t\alpha) | 0 \leq t \leq 1\}$$

will be stable for the game u.

Geometrically, this set is a line, joining the vertex $(1, 0, 0)$, to an arbitrary point $(0, \alpha, 1 - \alpha)$ on the opposite side of the triangle of imputation. But note that these are not the only stable sets.

11.3. Edgeworth Market Games—An Example

We have concerned ourselves through this chapter with some strictly mathematical ideas. We will now apply some of these to economic analysis.

Let us consider the following type of two-person nonzero sum game: Player I has a units of one commodity while player II has b units of a second commodity. Neither has any units of the other player's commodity. We represent this by saying that player I has the bundle $(a, 0)$ while II has the bundle $(0, b)$. Now, the two are to trade with each other, the idea being that both can gain some utility by changing to the bundles (x, y) and $(a - x, b - y)$, where $0 \leq x \leq a$ and $0 \leq y \leq b$.

Games such as this were first considered by Edgeworth. He gave a "solution" quite analogous to the von Neumann–Morgenstern stable sets. In fact, it is a set of distributions A such that no distribution in A is preferred by both players to any other distribution in A but, if $((x, y); (a - x, b - y))$ is any distribution not in A, there will be some distribution $((x', y'); (a - x', b - y'))$ in A which is preferred by both players. Finally, for any distribution in A, both players have at least as much utility as for the "initial" distribution $((a, 0), (0, b))$. Assuming infinite divisibility of the commodities traded, this set A will usually form a curve, and is called the *contract curve.*

The contract curve is shown graphically in Figure 11.5 by a system of *indifference curves*. These are such that any two points on one of the curves which are convex downward represent the same utility for player I, while any two points on one of the curves which are convex upward represent the same utility for player II. The contract curve will then be the locus of points of tangency of the curves from the two families, restricted to keep both players' utilities at or above the "point of no trading."

In Edgeworth's analysis he concludes that, if the number of players increases, then under certain plausible conditions the contract curve will "shrink" to a single limiting point. This is the "market price." We give below a mathematical interpretation.

Let us consider a market consisting of a set $I = M \cup N$ of traders. The members i of M have initial commodity bundles $(a_i, 0)$, while the members j of N have initial bundles $(0, b_j)$. Each player, i, has a utility function $\psi_i(x, y)$. We assume

$$\text{each } \psi_i \text{ is strictly convex,} \tag{11.8}$$

$$\lim_{x \to \infty} \psi_i(x, y) < \infty \quad \text{for each} \quad y \tag{11.9}$$

$$\lim_{y \to \infty} \psi_i(x, y) < \infty \quad \text{for each} \quad x \tag{11.10}$$

$$\text{the second partial derivatives of each } \psi_i \text{ exist} \\ \text{and are continuous every where} \tag{11.11}$$

Assume that the coalition S forms. The members of S will divide their joint bundle in such a way as to maximize their total utility, and then possibly make some side payments. We have, thus,

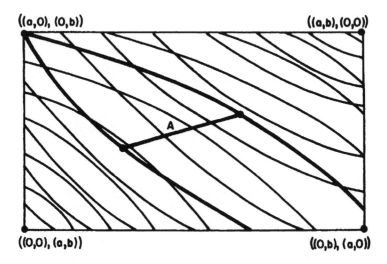

Figure 11.5.

$$v(S) = \max_{x,y} \left\{ \sum_{k \in S} \psi_k(x_k, y_k) \right\} \tag{11.12}$$

where the maximum is taken over all x, y such that $x_k \geq 0, y_k \geq 0, \Sigma_{k \in S} x_k = \Sigma_{i \in S \cap M} a_i$, and $\Sigma_{k \in S} y_k = \Sigma_{j \in S \cap N} b_j$.

We shall simplify matters by assuming that all players have the same utility function, i.e.,

$$\psi_i(x, y) = \psi(x, y) \quad \text{for all } i$$

Then, by the strict convexity of ψ, the optimal distribution of goods is obtained by distributing these equally. We assume also that each member of M has the same amount a of the first commodity, while each member of N has an amount b of the second commodity. We can now give the characteristic function implicitly; it is given by

$$v(S) = s\psi(s_m a/s, s_n b/s) \tag{11.13}$$

where s, s_m, s_n are the number of elements in S, $S \cap M$, and $S \cap N$, respectively. We shall denote a game such as this, where M has m elements and N has n elements, by $[m, n]$.

We can now look at the two-person game $[1, 1]$. The characteristic function is

$$v(\{1\}) = \psi(a, 0)$$
$$v(\{2\}) = \psi(0, b)$$
$$v(\{1, 2\}) = 2\psi(a/2, b/2)$$

This game will have a unique stable set solution consisting of all its imputations, i.e., of all vectors of the form (z_1, z_2), where

$$z_1 + z_2 = 2\psi(a/2, b/2) \tag{11.14}$$
$$z_1 \geq \psi(a, 0) \tag{11.15}$$
$$z_2 \geq \psi(0, b) \tag{11.16}$$

We will show now that for any n the game $[n, n]$ has a stable set solution analogous to that given by (11.14)–(11.16).

11.3.1 Theorem. For any n, the game $[n, n]$ has a stable set solution, V, consisting of all vectors x with $x_i = z_1$ for $i \in M$ and $x_j = z_2$ for $j \in N$, where (z_1, z_2) satisfies (11.14)–(11.16).

Proof. First, V is internally stable. In fact, domination must be through a set S such that $S \cap M \neq \varnothing$ and $S \cap N \neq \varnothing$. But, for $x, y \in V$, if $x_i > y_i$ for some $i \in M$, then $x_j < y_j$ for all $j \in N$. Hence we cannot have $x \vdash y$.

Next suppose $x \notin V$. There is some pair (i, j), with $i \in M$, $j \in N$, such that $x_i + x_j < 2\psi(a/2, b/2)$. Choose $(\hat{z}_1, \hat{z}_2)$ then so that $\hat{z}_1 > x_i$ and $\hat{z}_2 > x_j$. Then the imputation $y \in V$ corresponding to $(\hat{z}_1, \hat{z}_2)$ clearly dominates x through $\{i, j\}$. Thus V is externally stable.

Of course this shows that the stable set solutions have no tendency to "shrink" as n is increased. Such a tendency, however, is shown by the *core*. We consider two cases: monopoly (the game $[1, n]$) and pure competition (the game $[m, n]$ where both m and n are large).

11.3.2 Theorem. In the game $[1, n]$, the imputation x, such that $x_j = \psi(0, b)$ for all $j \in N$, belongs to the core. Moreover, given any $\varepsilon > 0$, there exists an $n(\varepsilon)$ such that, if $n \geq n(\varepsilon)$, no imputation y with $y_1 \leq x_1 - \varepsilon$ can belong to the core.

Proof. Domination is possible only through a set S with $1 \in S$. As x_1 is as large as possible, it follows that $x \in C(v)$.

Consider now the expression $v(I) - v(I - \{j\}) - \psi(0, b)$, where $j \in N$. If we replace v by its definition (11.13), and apply Taylor's theorem, we see that, for any $\varepsilon > 0$, if $n(\varepsilon)$ is large enough,

$$|v(I) - v(I - \{j\}) - \psi(0, b)| < \varepsilon/n$$

for all $n \geq n(\varepsilon)$. Suppose now that y is an imputation for the game $[1, n]$ such that $y_1 \leq x_1 - \varepsilon$. We must have some $j \in N$ such that $y_j \geq \psi(0, b) + \varepsilon/n$ (since we must have $\Sigma y_i = x_1 + n\psi(0, b)$). Now, it is clear that

$$\sum_{i \in I - \{j\}} y_i \leq v(I) - \psi(0, b) - \varepsilon/n < v(I - \{j\})$$

and so $y \notin C(v)$.

Theorem 11.3.2 describes what happens in the case of monopoly. For the case of pure competition, we have

11.3.3 Theorem. Suppose m' and n' are such that

$$\psi\left(\frac{m'a}{m'+n'}, \frac{n'b}{m'+n'}\right) = \max_s \psi\left(\frac{s_m a}{s}, \frac{s_n b}{s}\right) \tag{11.17}$$

Then for the game $[m, n]$, where $m = km'$ and $n = kn'$, the imputation

$$x = \left(\frac{v(I)}{m+n}, \frac{v(I)}{m+n}, \ldots, \frac{v(I)}{m+n}\right)$$

is always in the core. Moreover, for any $\varepsilon > 0$, there exists a $k(\varepsilon)$ such that, for any game $[km', kn']$ with $k \geq k(\varepsilon)$, no imputation with a component smaller than $[v(I)/(m+n)] - \varepsilon$ can belong to the core.

We skip the proof of this theorem, which is quite similar to that of 11.3.3. Equation (11.17) says that the sizes of the two sets of players, M and N, are in "optimal ratio" (e.g., the case when the amount of available capital is just enough to employ all the available labor). If M and N are in any other ratio, there will be some maladjustment (unemployment or overemployment) and some players will be disadvantaged in the core. On the other hand, if (11.17) holds, there will always be a core which shrinks to the single imputation, x, and our game can increase merrily in size with everybody happy.

11.4. A Game with No Solutions

Let us consider the ten-person game v, given by

$$v(N) = 5$$
$$v(\{1, 3, 5, 7, 9\}) = 4$$
$$v(\{3, 5, 7, 9\}) = v(\{1, 5, 7, 9\}) = v(\{1, 3, 7, 9\}) = 3$$
$$v(\{1, 4, 7, 9\}) = v(\{3, 6, 7, 9\}) = v(\{2, 5, 7, 9\}) = 2$$
$$v(\{3, 5, 7\}) = v(\{1, 5, 7\}) = v(\{1, 3, 7\}) = 2$$
$$v(\{3, 5, 9\}) = v(\{1, 5, 9\}) = v(\{1, 3, 9\}) = 2$$
$$v(\{1, 2\}) = v(\{3, 4\}) = v(\{5, 6\}) = v(\{7, 8\})$$
$$= v(\{9, 10\}) = 1$$
$$v(\{h\}) = 0 \quad \text{for all} \quad h \in N$$

and, for other $S, v(S) = 0$. The set of imputations here is

$$A = \{x | \Sigma x_h = 5, x_h \geq 0\}$$

It will be helpful to introduce the hypercube

$$B = \{x \in A | x_1 + x_2 = x_3 + x_4 = x_5 + x_6 = x_7 + x_8 = x_9 + x_{10} = 1\}$$

and also the following six vertices of B:

$$C^0 = (1, 0, 1, 0, 1, 0, 1, 0, 1, 0) \quad C^2 = (0, 1, 1, 0, 1, 0, 1, 0, 1, 0)$$

$$C^4 = (1, 0, 0, 1, 1, 0, 1, 0, 1, 0) \quad C^6 = (1, 0, 1, 0, 0, 1, 1, 0, 1, 0)$$

$$C^8 = (1, 0, 1, 0, 1, 0, 0, 1, 1, 0) \quad C^{10} = (1, 0, 1, 0, 1, 0, 1, 0, 0, 1)$$

It is easy to see that each of these imputations belongs to the core C. Thus C contains the convex hull of $C^k (k = 0, 2, 4, 6, 8, 10)$. It is also easy to check that any $x \in B$ such that

$$x_1 + x_3 + x_5 + x_7 + x_9 \geq 4$$

will lie in C. Moreover, $C \subset B$. Thus

$$C = \{x \in B | x_1 + x_3 + x_5 + x_7 + x_9 \geq 4\} \qquad (11.18)$$

To simplify our notation, we shall assume that (unless otherwise specified) the indices i, j, k, and r can have the three sets of values

$$(i, j, r, k) = (1, 3, 4, 5), \quad (3, 5, 6, 1), \quad \text{or} \quad (5, 1, 2, 3)$$

while p and q can be

$$(p, q) = (7, 9) \quad \text{or} \quad (9, 7)$$

We now define the following twelve subsets of B:

$$E_i = \{x \in B | x_j = x_k = 1, x_i < 1, x_7 + x_9 < 1\}$$

$$E = E_1 \cup E_3 \cup E_5$$

$$F_{jk} = \{x \in B | x_j = x_k = 1, x_i < 1, x_7 + x_9 \geq 1\} - C$$

$$F_p = \left\{ x \in B \left| \begin{array}{l} x_p = 1, x_q < 1, x_3 + x_5 + x_1 \geq 2 \\ x_1 + x_5 + x_q \geq 2, x_1 + x_3 + x_q \geq 2 \end{array} \right. \right\} - C$$

$$F_{79} = \{x \in B | x_7 = x_9 = 1\} - C$$

$$F_{135} = \{x \in B | x_1 = x_3 = x_5 = 1\} - C$$

$$F = F_{13} \cup F_{35} \cup F_{51} \cup F_7 \cup F_9 \cup F_{79} \cup F_{135}$$

It is easy to see $F \cap C = \varnothing$. Moreover, $E \cap C = \varnothing$. This is because, if $x \in E$, then $x_i \leq 1$ ($i = 1, 3, 5$) and $x_7 + x_9 < 1$, so that $x_1 + x_3 + x_5 + x_7 + x_9 < 4$, and so $x \notin C$. Finally, $E \cap F = \varnothing$, for clearly, $E \cap F_{135} = \varnothing$ from the definitions. If, however, $x \in F - F_{135}$, we find that $x_7 + x_9 \geq 1$, and so $x \notin E$. We see, then, that the five sets

$$A - B, B - (C \cup E \cup F), C, E, F$$

constitute a partition of A.

11.4.1 Lemma. Every $x \in A - B$ is dominated by some $y \in C$.

Proof. Let $x \in A - B$. Then $x_h \geq 0$, $\Sigma x_h = 5$, and, for some h ($h = 2, 4, 6, 8,$ or 10),

$$x_h + x_{h-1} \neq 1$$

Clearly, there will be some h ($= 2, 4, 6, 8, 10$) such that

$$x_{h-1} + x_h < 1$$

For this h, let

$$y = (x_{h-1} + \varepsilon)C^0 + (x_h + \varepsilon)C^h$$

where

$$2\varepsilon = 1 - x_{h-1} - x_h > 0$$

Then $y \in C$; moreover,

$$y_{h-1} = x_{h-1} + \varepsilon > x_{h-1}$$

$$y_h = x_h + \varepsilon > x_h$$

and so $y \vdash x$ through $\{h-1, h\}$.

11.4.2 Lemma. Every $x \in B - (C \cup E \cup F)$ is dominated by some $y \in C$.

Proof. Let $x \in B - (C \cup E \cup F)$. Then x satisfies

$$0 \leq x_h \leq 1 \quad \text{for all} \quad h \in N \tag{11.19}$$

$$x_1 + x_3 + x_5 + x_7 + x_9 < 4 \tag{11.20}$$

$$x_j < 1, x_k < 1 \quad \text{for} \quad (j,k) = (3,5),(5,1), \quad \text{or} \quad (1,3) \tag{11.21}$$

$$x_q < 1 \tag{11.22}$$

and either

$$x_p < 1 \tag{11.23}$$

or

$$x_p = 1 \quad \text{and either} \quad x_3 + x_5 + x_q < 2$$
$$\text{or} \quad x_5 + x_1 + x_q < 2$$
$$\text{or} \quad x_1 + x_3 + x_q < 2 \tag{11.24}$$

The reason for (11.19) is clear. Now, (11.20) follows from (11.18) and $x \in B$. The reason for (11.21) is that if $x_j + x_k = 1$ and $x_j < 1$, then $x \in E_i \cup F_{jk}$, whereas, if $x_1 = x_3 = x_5 = 1$, then $x \in F_{135}$. The reason for (11.22) is that, if $x_7 = x_9 = 1$, then $x \in F_{79}$. Thus either x_7 or $x_9 < 1$. Now, if $x_p < 1$ and $x_q = 1$, then (11.24) is necessary so that $x \notin F_p$.

Suppose, now, that (11.23) holds. Then $x_7 < 1, x_9 < 1$. If all three of x_1, x_3, x_5 are smaller than 1, we have

$$x_1, x_3, x_5, x_7, x_9 < 1$$

$$x_1 + x_3 + x_5 + x_7 + x_9 < 4$$

and it is easy to find $y \in C$ with $y_h > x_h$ $(h = 1, 3, 5, 7, 9)$ and $y_1 + y_3 + y_5 + y_7 + y_9 = 4$. Thus $y \vdash x$ through $\{1, 3, 5, 7, 9\}$.

Suppose, instead, $x_j < 1, x_k < 1$ but $x_7 = 1$. Then we have, by (11.19),

$$x_j + x_k + x_7 + x_9 < 3$$

and again we can find $y \in C$ such that $y \vdash x$ through $\{j, k, 7, 9\}$.

Alternatively, let us suppose (11.24) holds. If we have $x_j < 1, x_k < 1, x_q < 1$ and $x_j + x_k + x_q < 2$, then it is easy to find $y \in C, y \vdash x$ through $\{j, k, q\}$.

Suppose, however, that $x_j < 1, x_k < 1$, but $x_i = 1$, and $x_i + x_j + x_q < 2$; then $x_{i+1} = 0$, and so we have

$$x_{i+1} + x_j + x_k + x_q < 2$$

as $x_k < x_i$. Then we can find y satisfying $y_h > x_h, h = i + 1, j, k, q$, and also

$$y_{i+1} + y_j + y_k + y_q = 2$$

so that $y \vdash x$ through $\{i + 1, j, k, q\}$.

11.4.3 Lemma. If $x \in E$, there is no $y \in C$ which dominates x.

Proof. It is clear that, within A, there can be no domination through N, or through any S with $v(S) = 0$. It is also clear that, throughout B, and hence throughout E, no imputation can be dominated through any of the coalitions $\{h - 1, h\}$, where $h = 2, 4, 6, 8, 10$, since, for all $x \in B$, we have $x_{h-1} + x_h = 2 = v(\{h - 1, h\})$.

Suppose, then, that $x \in E, Y \in C, y \vdash x$. For some $i (= 1, 3, 5)$, we know $x \in E_i$. Then $x_j = x_k = 1$, but $y_j, y_k \leq 1$ if $y \in C$. Thus domination can only be through some set which includes neither j nor k, and the only possibility is $\{i, r, 7, 9\}$. Then

$$y_i + y_r + y_7 + y_9 \leq 2$$

but, since $y_r \geq 0$ and $y_r + y_{r+1} = 1$, while $y_j, y_k \leq 1$, this means

$$y_1 + y_3 + y_5 + y_7 + y_9 < 4$$

and so $y \notin C$.

11.4.4 Lemma. If $x \in E$, there is no $y \in F$ which dominates x.

Proof. Again domination can only be through $\{i, r, 7, 9\}$. If $y \vdash x$, we will have

$$y_i + y_r + y_7 + y_9 \leq 2, \quad y_r > 0$$

and these inequalities cannot both hold for $y \in F$. Thus, if $y \in F_{jk}$, we must have $y_r = 0$. If $y \in F_{ij}$ or F_{ki}, we have $y_i + y_7 + y_9 \geq 2$, and this gives us $y_i + y_r + y_7 + y_9 > 2$, etc. Thus no $y \in F$ can dominate x.

11.4.5 Corollary. If V is a stable set solution for v, then any $x \in E - V$ must be dominated by some $y \in E \cap V$.

Proof. We know $C \subset V$ for all games. By internal stability, no x dominated by C can lie in V; this eliminates all $x \in A - B$ or $x \in B - (C \cup E \cup F)$. Thus we must have

$$V \subset C \cup E \cup F$$

Suppose $x \in E - V$. By external stability, there must be $y \in V$ which dominates x. But $y \notin C \cup F$. Thus $y \in E$, and so $y \in V \cap E$.

11.4.6 Lemma. If $x \in E_i$, then there is no $y \in E_i \cup E_k$ which dominates x, but there is some $y \in E_j$ which dominates x through $\{i, r, 7, 9\}$.

Proof. If $x \in E_i$, it can only be dominated through $\{i, r, 7, 9\}$. Now if $y \in E_i \cup E_k$, then $y_r = 0$, and hence no domination is possible.

Since $x \in E_i$ we have

$$x_i < 1, \quad x_r = 0, \quad x_7 + x_9 < 1$$

and so it is possible to find $y \in E_j$ satisfying

$$y_7 > x_7, \quad y_9 > x_9$$

$$y_i + y_r + y_7 + y_9 \le 2$$

It suffices to choose $y_7 = x_7 + \varepsilon, y_9 = x_9 + \varepsilon, y_r = \varepsilon$, where $3\varepsilon = 1 - x_7 - x_9$.

11.4.7 Theorem. The game v has no stable sets.

Proof. By Corollary 11.4.5, $V \cap E$ is not empty. Choose $x \in V \cap E$; then $x \in E_i$. Choose $y \in E_j, y \vdash x$; clearly (by internal stability) $y \notin V$. As $y \in E - V$ there is $z \in E \cap V, z \vdash y$, and clearly $z \in E_k$. Choose $w \in E_i, w \vdash z$. Then $w \notin V$. Then there is $u \in E_j \cap V, u \vdash w$. We have then

$$u_7 > w_7 > z_7 > y_7 > x_7$$

$$u_9 > w_9 > z_9 > y_9 > x_9$$

$$u_i = 1 \ge x_i$$

$$u_r > 0 = x_r$$

and also

$$u_i + u_r + u_7 + u_9 < 2$$

But this means

$$u \vdash x$$

through $\{i, r, 7, 9\}$. This is a contradiction as u, x both belong to V.

It will be noticed that the game v, as defined above, is not superadditive. We may, however, define a game v' which is superadditive, and whose domination structure is the same as that of v: this is given by

$$v'(S) = \max\{v(T_1) + \ldots + v(T_k)\} \tag{11.25}$$

where the maximum is taken over all partitions $\{T_1, \ldots, T_k\}$ of S. Now,

$$v'(N) = 5, \quad v'(\{h\}) = 0 \tag{11.26}$$

as may be easily checked; thus $A(v)$ and $A(v')$ are the same. The reader is invited to prove that, if $y \vdash x$ in v, then $y \vdash x$ in v', and conversely. Thus v and v' will have the same stable sets, and in this case, v' has no stable sets.

Problems

1. (a) The core of any n-person game v is always a subset of any stable set.
   (b) The core has a nonempty intersection with each face $x_i = v(\{i\})$ of the simplex of imputations, whenever it is the (unique) stable set solution of the game v.
   (c) A positive (full-dimensional) fraction of n-person games [in $(0, 1)$ normalization] has a unique stable set solution (Theorem 11.1.5).

2. For a game v, define a semi-imputation as a vector $x = (x_1, \ldots, x_n)$ such that $x_i \geq v(\{i\})$ and $\Sigma_{i \in N} x_i \leq v(N)$. Then, if V is any stable set solution to v and x is a semi-imputation but $x \notin V$, there exists $y \in V$ such that $y \vdash x$.

3. For a game v, define the number b_i, as

$$b_i = \max_{S \subset N - \{i\}} [v(S \cup \{i\}) - v(S)]$$

Show that an imputation x which does not satisfy $x_i \leq b_i$ cannot belong to the core or to any stable set.

4. Show that the symmetric stable set solution V of the three-person game considered in 11.1.2 generalizes, as follows, to the *main simple solution* of a simple game.
   Let v be a simple, constant-sum n-person game in $(0, 1)$ normalization, and suppose there exists a vector $\alpha = (\alpha_1, \ldots, \alpha_n)$ such that $\alpha_i \geq 0$ and $\Sigma_{i \in S} \alpha_i = 1$ for every minimal winning coalition S. For such a coalition, let x^S be the vector defined by $x_i^S = \alpha_i$, if $i \in S$, $x_i^S = 0$ otherwise. Then x^S is an imputation, and the set $V = \{x^S | S$ is a minimal winning coalition$\}$ is a stable set solution of v, called a *main simple solution*.

5. Consider the simple n-person game in $(0, 1)$ normalization in which the winning coalitions consist of (a) player 1 and at least one other player, (b) the set $\{2, 3, \ldots, n\}$.
   (a) Find the main simple solution of this game.
   (b) For $S = \{1, j\}$, where $j \neq 1$, let V_s consist of all the imputations such that $x_1 + x_j = 1$. For $S = \{2, 3, \ldots, n\}$, let V_s consist of all the imputations such that $x_1 = 0$. In either case, V_s is a stable set solution.

(c) For $S = \{2, 3, \ldots, n\}$ and $0 \le c \le 1$, let $V_s(c)$ consist of all the imputations such that $x_1 = c$. Then $V_s(c)$ is a stable set for $0 \le c < (n-2)/(n-1)$, but not for $c \ge (n-2)/(n-1)$.

(d) For $S = \{1, j\}$ the only stable set solution which discriminates the members of $N - S$ is the set V_s mentioned in (b) above.

6. Let v be a simple game for a denumerable infinity of players, for which a set is winning if and only if its complement is finite. If v is in $(0, 1)$ normalization, the set $E(v)$ consists of all sequences $(x_1, x_2, \ldots)$ of nonnegative numbers such that $\Sigma x_i = 1$.
Then v has no stable set solutions. (Show that no imputation can belong to a stable set.)

7. Let v be the simple four-person game in $(0, 1)$ normalization with winning coalitions $\{1, 2, 4\}, \{1, 3, 4\}, \{2, 3, 4\}$ and $\{1, 2, 3, 4\}$. Let C be any closed subset of $[0, 1)$, and let J be the set of all vectors $x = (0, (1-u)/2, (1-u)/2, u)$ for $y \in C$. Then v has a stable set solution V, such that $J \subset V$, and the two sets $J, V - J$ are closed and disjoint (and therefore separated).

(a) Let $\rho(u, C) = \min_{v \in C} |u - v|$. Then let K be the set of all imputations of the form
$$\left(\frac{1 - u - \rho(u, C)}{2}, x, y, u \right)$$
for all $u \in [0, 1]$. Show K is closed and $K \cap J = \varnothing$.

(b) Let H consist of all those imputations in K which are dominated by some imputation in J. Then
$$V = J \cup (K - H)$$
is the desired stable set.

8. Prove Theorem 11.3.3.

9. Let v be an $(n, n-1)$ game (see Chapter 10, Problem 1). For a given $i \in N$, let V_i be the set of all imputations x such that $x_i = A_i$. Under what conditions will V_i be a von Neumann–Morgenstern solution (stable set)?

Chapter 12

Indices of Power

12.1. The Shapley Value

The fact that no general existence theorem has been given for stable sets of n-person games has led mathematicians to look for other solution concepts. One such concept is the Shapley value.

Shapley approaches his value axiomatically. We give two definitions first.

12.1.1 Definition. A *carrier* for a game v is a coalition T such that, for any S, $v(S) = v(S \cap T)$.

Intuitively, Definition 12.1.1 states that any player who does not belong to a carrier is a *dummy* — i.e., can contribute nothing to any coalition.

12.1.2 Definition. Let v be an n-person game, and let π be any permutation of the set N. Then, by πv we mean the game u such that, for any $S = \{i_1, i_2, \ldots, i_s\}$

$$u(\{\pi(i_1), \pi(i_2), \ldots, \pi(i_s)\}) = v(S)$$

Effectively, the game πv is nothing other than the game v, with the roles of the players interchanged by the permutation π.

With these two definitions, it is possible to give an axiomatic treatment. We point out merely that, as games are essentially real-valued functions, it is possible to talk of the sum of two or more games, or of a number times a game.

12.1.3 Axioms (Shapley). By the value of a game v, we shall mean an n-vector, $\varphi[v]$, satisfying

S1. If S is any carrier of v, then

$$\sum_S \varphi_i[v] = v(S)$$

S2. For any permutation π, and $i \in N$

$$\varphi_{\pi(i)}[\pi v] = \varphi_i[v]$$

S3. If u and v are any games

$$\varphi_i[u + v] = \varphi_i[u] + \varphi_i[v]$$

These are Shapley's axioms. It is a remarkable fact that these axioms are sufficient to determine a value φ uniquely, for all games.

12.1.4 Theorem. There is a unique function φ, defined on all games, satisfying Axioms S1–S3.

The proof of 12.1.4 is given by the following sequence of lemmas.

12.1.5 Lemma. For any coalition S, let the game w_S be defined by

$$w_S(T) = \begin{cases} 0 & \text{if} \quad S \not\subset T \\ 1 & \text{if} \quad S \subset T \end{cases}$$

Then, if s is the number of players in S

$$\varphi_i[w_S] = \begin{cases} 1/s & \text{if} \quad i \in S \\ 0 & \text{if} \quad i \notin S \end{cases}$$

Proof. It is clear that S is a carrier for w_S, as is any superset T of S. By Axiom S2, then, it follows that

$$\sum_T \varphi_i[w_S] = 1 \quad \text{if} \quad S \subset T$$

But this means (as $S = T$ is also a possibility) that $\varphi_i[w_S] = 0$ for $i \notin S$.

Now, if π is any permutation which carries S into itself, it is clear that $\pi w_S = w_S$. Hence, by S2, $\varphi_i[w_S] = \varphi_j[w_S]$ for any $i, j \in S$. As there are s of these terms, and their sum is 1, it follows that $\varphi_i[w_S] = 1/s$ if $i \in S$.

12.1.6 Corollary. If $c > 0$, then

$$\varphi_i[cw_S] = \begin{cases} c/s & \text{if} \quad i \in S \\ 0 & \text{if} \quad i \notin S \end{cases}$$

12.1.7 Lemma. If v is any game, then there exist $2^n - 1$ real numbers c_s for $S \subset N$ such that

$$v = \sum_{S \subset N} c_S w_S$$

where w_S is as in 12.1.5.

Proof. Let

$$c_S = \sum_{T \subset S} (-1)^{s-t} v(T) \tag{12.1}$$

(where t is the number of elements in T). We shall show that these c_s satisfy the lemma. In fact, if U is any coalition

$$\sum_{S \subset N} c_S w_S(U) = \sum_{S \subset U} c_S$$

$$= \sum_{S \subset U} \left(\sum_{T \subset S} (-1)^{s-t} v(T) \right)$$

$$= \sum_{T \subset U} \left(\sum_{\substack{S \subset U \\ S \supset T}} (-1)^{s-t} \right) v(T)$$

Now, consider the inner parenthesis in this last expression. For every value of s between t and u, there will be $\binom{u-t}{u-s}$ sets S with s elements such that $T \subset S \subset U$. Hence the inner parenthesis may be replaced by

$$\sum_{s=t}^{u} \binom{u-t}{u-s} (-1)^{s-1}$$

But this is precisely the binomial expansion of $(1-1)^{u-t}$. Hence this will be zero for all $t < u$, and 1 for $t = u$. Hence we have

$$\sum_{S \subset N} c_S w_S(U) = v(U)$$

for all $U \subset N$.

We can now proceed to prove Theorem 12.1.4. In fact, Lemma 12.1.7 shows that any game can be written as a linear combination of games w_S. By Lemma 12.1.5, the function φ is uniquely defined for such games. Now, some of the coefficients c_s are negative; however, it is easy to see from Axiom S3 that if u, v, and $u - v$ are all games, then $\varphi[u - v] = \varphi[u] - \varphi[v]$. By S3, then, the function φ is uniquely defined for all games v. We can also give the exact form of the function φ. We know that

$$v = \sum_{S \subset N} c_S w_S$$

and so

$$\varphi_i[v] = \sum_{S \subset N} c_S \varphi_i[w_S]$$

$$= \sum_{\substack{S \subset N \\ i \in S}} c_S \cdot \frac{1}{s}$$

But c_s is defined by (12.1); substituting this here, we obtain

$$\varphi_i[v] = \sum_{\substack{S \subset N \\ i \in S}} \frac{1}{s} \left\{ \sum_{T \subset S} (-1)^{s-t} v(T) \right\}$$

and so

$$\varphi_i[v] = \sum_{T\subset N}\left\{\sum_{\substack{S\subset N \\ T\cup\{i\}\subset S}} (-1)^{s-t}\frac{1}{s}v(T)\right\} \tag{12.2}$$

From (12.2) let us write

$$\gamma_i(T) = \sum_{\substack{S\subset N \\ T\cup\{i\}\subset S}} (-1)^{s-t}\frac{1}{s} \tag{12.3}$$

It is easy to see that, if $i \notin T'$ and $T = T' \cup \{i\}$, then $\gamma_i(T') = -\gamma_i(T)$. In fact, all the terms in the right side of (12.3) will be the same in both cases, except that $t = t' + 1$, and so there is a change of sign throughout. This means we will have

$$\varphi_i[v] = \sum_{\substack{T\subset N \\ i\in T}} \gamma_i(T)\{v(T) - v(T - \{i\})\}$$

Now, if $i \in T$, we see there are exactly $\binom{n-t}{s-t}$ coalitions S with s elements such that $T \subset S$. Thus we have

$$\begin{aligned}
\gamma_i(T) &= \sum_{s=t}^{n}(-1)^{s-t}\binom{n-t}{s-t}\frac{1}{s} \\
&= \sum_{s=t}^{n}(-1)^{s-t}\binom{n-t}{s-t}\int_0^1 x^{s-1}\,dx \\
&= \int_0^1\sum_{s=t}^{n}(-1)^{s-t}\binom{n-t}{s-t}x^{s-1}\,dx \\
&= \int_0^1 x^{t-1}\sum_{s=t}^{n}(-1)^{s-t}\binom{n-t}{s-t}x^{s-t}\,dx \\
&= \int_0^1 x^{t-1}(1-x)^{n-t}\,dx
\end{aligned}$$

But this is a well-known definite integral; we have, thus

$$\gamma_i(T) = \frac{(t-1)!(n-t)!}{n!}$$

so that

$$\varphi_i[v] = \sum_{\substack{T\subset N \\ i\in T}} \frac{(t-1)!(n-t)!}{n!}[v(T) - v(T - \{i\})] \tag{12.4}$$

Formula (12.4) gives the Shapley value, explicitly. It is then easy to check that this expression will satisfy Axioms S1–S3. Moreover, it can be seen that the sum of the coefficients $\gamma_i(T)$ is equal to 1. For the numerator is equal to the number of permutations of N in which i is preceded exactly by the elements of T, while the numerator is the total number of permutations. By superadditivity, the bracket is always at least equal to $v(\{i\})$. Hence we see that

$$\varphi_i[v] \geq v(\{i\}) \tag{12.5}$$

and so $\varphi[v]$ is always an imputation.

Apart from the axiomatic treatment, the Shapley value, as defined by (12.4), can be given the following heuristic explanation: Suppose the players (the elements of N) agree to meet at a specified place and time. Naturally, because of random fluctuations, all will arrive at different times; it is assumed, however, that all orders of arrival (permutations of the players) have the same probability: viz., $1/n!$. Suppose that, if a player, i, arrives, and finds the members of the coalition $T - \{i\}$ (and no others) already there, he receives the amount $v(T) - v(T - \{i\})$, i.e., the marginal amount which he contributes to the coalition, as payoff. Then the Shapley value $\varphi_i[v]$ is the expected payoff to player i under this randomization scheme.

In case the game v is simple, the formula for the Shapley value becomes especially simple. In fact, the term $v(T) - v(T - \{i\})$ will always have value 0 or 1, taking the value 1 whenever T is a winning coalition but $T - \{i\}$ is not. Hence we would have

$$\varphi_i[v] = \sum_T \frac{(t-1)!(n-t)!}{n!}$$

where the summation is taken over all winning coalitions T such that $T - \{i\}$ is not winning.

12.1.8 Example. Consider a corporation with four stockholders, having respectively 10, 20, 30, and 40 shares of stock. It is assumed that any decision can be settled by approval of stockholders holding a simple majority of the shares. This can be treated as a simple four-person game, in which the winning coalitions are: $\{2, 4\}$, $\{3, 4\}$, $\{1, 2, 3\}$, $\{1, 2, 4\}$, $\{1, 3, 4\}$, $\{2, 3, 4\}$, and $\{1, 2, 3, 4\}$. We would like to find the Shapley value of this game.

To find φ_1, we note that the only winning coalition T such that $T - \{1\}$ is not winning, is $\{1, 2, 3\}$. Hence, as $t = 3$

$$\varphi_1 = \frac{2!1!}{4!} = \frac{1}{12}$$

Similarly, we find that $\{2, 4\}$, $\{1, 2, 3\}$, and $\{1, 2, 4\}$ are winning coalitions which are not winning if 2 is removed. Hence

$$\varphi_2 = \tfrac{1}{12} + \tfrac{1}{12} + \tfrac{1}{12} = \tfrac{1}{4}$$

In the same way, we find $\varphi_3 = \frac{1}{4}$, while $\varphi_4 = \frac{5}{12}$. Thus the Shapley value is the vector $\left(\frac{1}{12}, \frac{1}{4}, \frac{1}{4}, \frac{5}{12}\right)$. This contrasts with the "vote vector," which would be $\left(\frac{1}{10}, \frac{1}{5}, \frac{3}{10}, \frac{2}{5}\right)$. Note that the value is the same for players 2 and 3, even though player 3 has many more shares. This is not surprising, since player 3 has no greater opportunity than 2 to form a winning coalition; the game (thought of as a characteristic function) treats these two players symmetrically.

In a similar way, it is clear that player 4's strength is greater than his share of the stock, while player 1's strength is less.

12.1.9 Example. Consider a game similar to that of Example 12.1.8, except that the players have 10, 30, 30, and 40 shares respectively. It may be seen that any two of players 2, 3, 4 form a winning coalition, while player 1 is a dummy: he cannot contribute anything to any coalition. Thus the value vector is $\left(0, \frac{1}{3}, \frac{1}{3}, \frac{1}{3}\right)$: player 1's shares are useless, while player 4's extra shares give him no advantage over players 2 and 3.

12.2. Multilinear Extensions

One of the principal difficulties with the Shapley value is that its computation generally requires the sum of a very large number of terms. Thus, even when the characteristic function is easy to define (e.g., a weighted majority game), evaluation may require a prohibitive amount of work. Recourse is therefore frequently had to the *multilinear extension* (MLE) of the game.

As we have studied it, an *n*-person game (in characteristic function form) is merely a real-valued function v whose domain is 2^N — the set of subsets of $N = \{1, \ldots, n\}$. However, we can also interpret 2^N as the set of all vectors $(x_1, x_2, \ldots, x_n)$ whose components are either 0 or 1. Thus

$$2^N = \{0, 1\}^N$$

is the set of all corners of the unit cube in *n*-dimensional space. The function v, then, is a real-valued function defined on the corners of the cube, and it seems reasonable to try to extend it throughout the cube

$$[0, 1]^N = \{(x_1, x_2, \ldots, x_n) | 0 \le x_i \le 1\}$$

There are, of course, many ways of extending the function. We shall extend it in a manner that is linear in each variable (though not in all variables simultaneously). This is the multilinear extension of v.

12.2.1 Definition. Let v be an *n*-person game with carrier $N = \{1, \ldots, n\}$. The *multilinear extension* of v is a function f defined by

$$f(x_1, x_2, \ldots, x_n) = \sum_{S \subset N} \left\{ \prod_{i \in S} x_i \prod_{i \notin S} (1 - x_i) \right\} v(S) \tag{12.6}$$

for $0 \le x_i \le 1$, $i = 1, \ldots, n$.

12.2.2 Example. Let v be the three-person simple majority game in $(0,1)$ normalization. Its multilinear extension is

$$f(x_1, x_2, x_3) = x_1 x_2 (1 - x_3) + x_1 x_3 (1 - x_2) + x_2 x_3 (1 - x_1) + x_1 x_2 x_3$$

or, equivalently

$$f(x_1, x_2, x_3) = x_1 x_2 + x_1 x_3 + x_2 x_3 - 2 x_1 x_2 x_3$$

12.2.3 Example. Let v be the four-person simple game in which a coalition S wins if (a) S has two players, and $1 \in S$ or (b) S has three or four players. Then the MLE of v is

$$\begin{aligned}
&x_1 x_2 (1 - x_3)(1 - x_4) + x_1 x_3 (1 - x_2)(1 - x_4) \\
&+ x_1 x_4 (1 - x_2)(1 - x_3) + x_2 x_3 x_4 (1 - x_1) + x_1 x_2 x_3 (1 - x_4) \\
&+ x_1 x_3 x_4 (1 - x_2) + x_1 x_2 x_4 (1 - x_3) + x_1 x_2 x_3 x_4
\end{aligned}$$

Some simplification and collection of terms gives us

$$\begin{aligned}
f(x_1, x_2, x_3, x_4) = {} & x_1 x_2 + x_1 x_3 + x_1 x_4 + x_2 x_3 x_4 \\
& - x_1 x_2 x_3 - x_1 x_2 x_4 - x_1 x_3 x_4
\end{aligned}$$

To justify the name we have given the function f, we shall show, first, that f is multilinear (i.e., linear in each x_i); second, that it coincides with v where v is defined; third, that it is unique with respect to those properties.

It is easy enough to see that f is multilinear. For, each of the terms $\Pi x_i \; \Pi (1 - x_i) v(S)$ is linear in each variable, and f, as the sum of such terms, is also linear in each variable, hence multilinear.

We show next that f is an extension of v. For $S \subset N$, we let α^S be the S-corner of the cube, i.e.,

$$\alpha_i^S = \begin{cases} 1 & \text{if} \quad i \in S \\ 0 & \text{if} \quad i \notin S \end{cases}$$

Then

$$f(\alpha^S) = \sum_{T \subset N} \left\{ \prod_{i \in T} \alpha_i^S \prod_{j \notin T} (1 - \alpha_i^S) \right\} v(T) \tag{12.7}$$

It is easy to see that, if $T \neq S$, then the term in braces in (12.7) vanishes. For, if there is some $i \in T - S$, the corresponding $\alpha_i^S = 0$; if there is some $i \in S - T$, the corresponding $1 - \alpha_i^S = 0$. In either case, at least one of the factors vanishes, and so the entire product vanishes. If, however, $T = S$, then the term in braces is equal to 1. Thus

$$f(\alpha^S) = v(S)$$

and f is indeed an extension of v.

To show uniqueness, we point out that a multilinear function of n variables, being a polynomial, has the form

$$f(x_1, \ldots, x_n) = \sum_{T \subset N} C_T \prod_{j \in T} x_j \tag{12.8}$$

and so depends on the 2^n constants C_T.

The equation (12.8) reduces to

$$f(\alpha^S) = \sum_{T \subset S} C_T$$

so that the condition $f(\alpha^S) = v(S)$ reduces to

$$\sum_{T \subset S} C_T = v(S) \quad \text{for all} \quad S \subset N$$

This is a system of 2^n linear equations in the 2^n unknowns C_T. We know that it can be solved — in fact (12.6) will give us a solution, which after expansion reduces to (12.1) — for any choice of $v(S)$. But for a linear system, this means that the corresponding matrix of coefficients is nonsingular, and so the system always has a unique solution. We conclude that v has a unique MLE f, given by (12.6). An alternative form is (12.8), where the C_T are given by (12.1).

Heuristically, we can give a probabilistic interpretation to the multilinear extension f of the game v. Let us suppose that a coalition is to be formed at random. Let $\mathfrak{S}$ be this random coalition, and let the event

$$A_i : \{i \in \mathfrak{S}\}$$

have probability x_i $(i = 1, \ldots, n)$. Assuming that the events A_i are independent, then, for any fixed $S \subset N$

$$\text{Prob}\{\mathfrak{S} = S\} = \prod_{i \in S} x_i \prod_{i \notin S} (1 - x_i)$$

and so

$$E[v(\mathfrak{S})] = f(x_1, \ldots, x_n)$$

so that $f(x)$ can be thought of as the mathematical expectation of $v(\mathfrak{S})$, under the given randomization scheme. Thus we might say: if each player $i \in N$ is x_i-committed to joining, then the coalition $\mathfrak{S}$ can, in expectation, command the amount $f(x)$ of utility.

We give, next, some interesting properties of MLEs.

12.2.4 Theorem. Let v and w be games with the same player set N, and let f and g be the corresponding multilinear extensions. Then, for any scalars α and β, the game $\alpha v + \beta w$ has the MLE $\alpha f + \beta g$. (But note that, strictly speaking, this need not be a game unless the scalars are positive.)

Proof. Immediate from the definition of MLE.

12.2.5 Theorem. Let v and w be games with disjoint player sets M and N, and let $v \oplus w$ be the von Neumann–Morgenstern sum (composition) of these games. If f and g are the MLEs of v and w, respectively, then the MLE of $v \oplus w$ is the function $f \oplus g$, whose domain is $[0, 1]^{M \cup N}$, defined by

$$f \oplus g(x_1, \ldots, x_m; y_1, \ldots, y_n) = f(x_1, \ldots, x_m) + g(y_1, \ldots, y_n)$$

12.2.6 Theorem. Let v be a constant-sum game, and f its MLE. Then for any x

$$f(1 - x_1, 1 - x_2, \ldots, 1 - x_n) = v(N) - f(x_1, x_2, \ldots, x_n)$$

Proof. We use the probabilistic interpretation of the MLE. Then it is not too difficult to see that

$$f(1 - x_1, 1 - x_2, \ldots, 1 - x_n) = E[v(N - \mathfrak{S})]$$

where $\mathfrak{S}$ is a random coalition as above. For, the probability that $N - S$ form, given the probabilities $1 - x_i$, is the same as that of coalition S forming, given the probabilities x_i. Then

$$f(1 - x_1, 1 - x_2, \ldots, 1 - x_n) + f(x_1, x_2, \ldots, x_n)$$
$$= E[v(N - \mathfrak{S}) + v(\mathfrak{S})]$$

Now, v is constant-sum, so that $v(N - S) + v(S) = v(N)$ for all $S \subset N$. Thus

$$f(1 - x_1, 1 - x_2, \ldots, 1 - x_n) + f(x_1, x_2, \ldots, x_n) = v(N)$$

as desired.

The above properties of MLEs, though interesting, are rather trivial. The following property, which is not nearly so obvious, relates the MLE to the Shapley value, and is the real reason for studying MLEs.

Let, now, f_i be the ith partial derivative of f. We have

$$f_i(x) = \sum_{\substack{T \subset N \\ i \in T}} \prod_{\substack{j \in T \\ j \neq i}} x_j \prod_{j \notin T} (1 - x_j) v(T) - \sum_{\substack{S \subset N \\ i \notin S}} \prod_{j \in S} x_j \prod_{\substack{j \notin S \\ j \neq i}} (1 - x_j) v(S)$$

Letting $T = S \cup \{i\}$ above, this will reduce to

$$f_i(x) = \sum_{\substack{S \subset N \\ i \notin S}} \prod_{j \in S} x_j \prod_{\substack{j \notin S \\ j \neq i}} (1 - x_j) [v(S \cup \{i\}) - v(s)] \tag{12.9}$$

In particular, we can set $x = (t, t, \ldots, t)$. Then

$$f_i(t, t, \ldots, t) = \sum_{\substack{S \subseteq N \\ i \notin S}} t^s (1-t)^{n-s-1}[v(S \cup \{i\}) - v(S)]$$

where s is the cardinality of S. Integrating this, we obtain

$$\int f_i(t, t, \ldots, t) dt = \sum_{\substack{S \subseteq N \\ i \notin S}} \left\{ \int t^s (1-t)^{n-s-1} \ dt \right\} [v(S \cup \{i\}) - v(S)]$$

or

$$\int_0^1 f_i(t, t, \ldots t) dt = \varphi_i[v] \tag{12.10}$$

where $\varphi[v]$ is the Shapley value of v.

We have thus obtained the Shapley value in a totally different way, by integrating the partial derivatives of f along the main diagonal $x_1 = x_2 = \cdots = x_n$ of the cube $[0, 1]^N$.

12.2.7 Example. Consider the four-person game of Example 12.2.3, whose MLE is

$$f(x) = x_1 x_2 + x_1 x_3 + x_1 x_4 + x_2 x_3 x_4 - x_1 x_2 x_3 - x_1 x_2 x_3 - x_1 x_3 x_4$$

We have here

$$f_1(x) = x_2 + x_3 + x_4 - x_2 x_3 - x_2 x_4 - x_3 x_4$$

so that

$$f_1(t, t, \ldots, t) = 3t - 3t^2$$

and so

$$\int f_1(t, t, \ldots, t) \ dt = \left[\tfrac{3}{2} t^2 - t^3 \right]_0^1 = \tfrac{1}{2}$$

Similarly

$$f_2(x) = x_1 + x_3 x_4 - x_1 x_3 - x_1 x_4$$
$$f_2(t, t, \ldots, t) = t - t^2$$

and so

$$\int f_2(t, t, \ldots, t) \ dt = \tfrac{1}{6}$$

Thus we will have $\varphi_1 = \tfrac{1}{2}, \varphi_2 = \tfrac{1}{6}$, and by symmetry, $\varphi_3 = \varphi_4 = \tfrac{1}{6}$. This is then the value of v.

At this point, we find that we have replaced evaluation of the sum (12.4) by that of the sum (12.9), which contains as many terms, plus an integration. Thus no great simplification has been made. The point is that the MLE approach has two advantages. The first is due to the fact that, thanks to our probabilistic interpretation, limit theorems of probability theory (in particular, the central limit theorem) can be applied. The second deals with the behavior of multilinear extensions under composition of games.

To understand the first point, let us consider the equation (12.9). The coefficient of $v(S \cup \{i\}) - v(S)$ is

$$\prod_{j \in S} x_j \prod_{\substack{j \notin S \\ j \neq i}} (1 - x_j)$$

which can be interpreted as the probability that $\mathfrak{S} = S$, where $\mathfrak{S}$ is a random subset of $N - \{i\}$, whose distribution is given by the fact that, for each $j \neq i$

$$\text{Prob}\{j \in \mathfrak{S}\} = x_j$$

and that all of these are independent. Thus, we can write

$$f_i(x) = E[v(\mathfrak{S} \cup \{i\}) - v(\mathfrak{S})]$$

under the corresponding randomization scheme.

Assuming, now, that v is a simple game in $(0, 1)$ normalization, we note that $v(\mathfrak{S} \cup \{i\}) - v(\mathfrak{S})$ is equal to 1 if $\mathfrak{S}$ loses but $\mathfrak{S} \cup \{i\}$ wins, and vanishes otherwise. Thus

$$f_i(x) = \text{Prob}\{\mathfrak{S} \text{ loses, } \mathfrak{S} \cup \{i\} \text{ wins}\}$$

Suppose, further, that v is a weighted majority game, with symbol $[q; w_1, \ldots, w_n]$. We have then

$$f_i(x) = \text{Prob}\{q - w_i \le Y < q\}$$

where Y is the random variable

$$Y = \sum_{j \in \mathfrak{S}} w_j$$

or, equivalently

$$Y = \sum_{\substack{j \in N \\ j \neq i}} Z_j$$

where Z_j is a random variable, equal to w_j if $j \in \mathfrak{S}$ and 0 otherwise. It is easy to see that Z_j has mean $w_j x_j$ and variance $w_j^2 x_j (1 - x_j)$. Since the Z_j are independent, Y has mean and variance

$$\mu(Y) = \sum_{j \neq i} w_j x_j \tag{12.11}$$

$$\sigma^2(Y) = \sum_{j \neq i} w_j^2 x_j (1 - x_j) \tag{12.12}$$

Now the distribution of Y, which is the sum of $n-1$ random variables, is not, in general, easy to obtain exactly. Under certain circumstances, however (mainly, that n be large and that none of the Z_j have a variance considerably larger than the others), Y can be approximated by a normal random variable. We can then conclude that $f_i(x)$ is approximately equal to the probability that a normal random variable $\tilde{Y}$, with mean and variance given by (12.11)–(12.12), satisfy

$$q - w_i - \tfrac{1}{2} \leq \tilde{Y} \leq q - \tfrac{1}{2}$$

(The $\tfrac{1}{2}$ here is standard procedure, used whenever an integer-valued random variable is approximated by a continuous variable.) In case $x = (t, t, \ldots, t)$, the mean and variance take the simpler forms

$$\mu(Y) = t \sum_{j \neq i} w_j \tag{12.13}$$

$$\sigma^2(Y) = t(1 - t) \sum_{j \neq i} w_j^2 \tag{12.14}$$

Numerical integration of this probability will then yield an approximation to the Shapley value.

An interesting example of this is treated in some detail in Section 12.4. The following example is slightly different, but also uses the central limit theorem.

12.2.8 Example. Let L and R be disjoint sets of players with l and r elements, respectively, and let $N = L \cup R$. Each member of L has one left-handed glove; each member of R has one right-handed glove. We assume that a single glove is worthless, but a pair (one right-handed and one left-handed glove) can be sold for $1. Identifying utility with money, we have then

$$v(S) = \min\{|S \cap R|, |S \cap L|\}$$

for any $S \subset N$, as this is the total number of pairs which can be formed by the members of S.

Assume, now, that $i \in R$. It is easy to see that, for $i \notin S$

$$v(S \cup \{i\}) - v(S)$$

will be equal to 1 if $s_l > s_r$ (where $s_l = |S \cap L|$, $s_r = |S \cap R|$) and 0 otherwise, Thus

$$f_i(x) = \text{Prob}\{s_l > s_r\}$$

where s_l and s_r are the cardinalities of the random coalitions $\mathfrak{S} \cap L$ and $\mathfrak{S} \cap R$, respectively.

Letting $x = (t, t, \ldots, t)$, we see that s_l is a binomial random variable with parameters l and t, respectively, while s_r is a binomial variable with parameters $r - 1$ and t. Assuming l and r are large, we can approximate them by normal variables with the same mean and variance. Their difference Y then has the mean and variance

$$\mu(Y) = \mu(s_l) - \mu(s_r) = (l - r + 1)t \tag{12.15}$$

$$\sigma^2(Y) = \sigma^2(s_l) + \sigma^2(s_r) = (l + r - 1)t(1 - t) \tag{12.16}$$

and so we have the approximation

$$f_i(t, t, \ldots, t) \cong \text{Prob}\{Y \geq \tfrac{1}{2}\}$$

where Y is a normal random variable with (12.15)–(12.16). This can then be integrated (numerically) over $0 \leq t \leq 1$ to obtain the value.

In this fashion, we evaluated a game with $r = 96$, $l = 101$. The slight surplus of left-handed gloves suggests that the members of R should be at a slight advantage. In fact, the value obtained was $\varphi_i = 0.600$ for $i \in R$, $\varphi_j = 0.380$ for $j \in L$. Thus, the advantage to the members of R was really more than would have been expected from the mere comparison of the two numbers l and r.

We consider, next, the behavior of MLEs under composition of games. For simple games $w_1, \ldots, w_n$ in $(0,1)$ normalization and disjoint carriers $M_1, \ldots, M_n$, and a nonnegative game v with carrier $N = \{1, \ldots, n\}$, we gave in Chapter 11 the formula

$$u(S) = v(\{j | w_j(S \cap M_j) = 1\}) \tag{12.17}$$

for $S \subset M^* = \cup_j M_j$ to define the compound game $u = v[w_1, \ldots, w_n]$. Consider, however, the formula

$$u(S) = \sum_{t \subset N} \prod_{j \in T} w_j(S_j) \prod_{j \notin T} [1 - w_j(S_j)] v(T) \tag{12.18}$$

or, equivalently

$$u(S) = f(w_1(S_1), \ldots, w_n(S_n)) \tag{12.19}$$

where $S_j = S \cap M_j$ and f is the multilinear extension of v. In case the w_j are simple, we see that

$$(w_1(S_1), w_2(S_2), \ldots, w_n(S_n)) = \alpha^T$$

where $T \subset N$ is given by

$$T = \{j | w_j(S_j) = 1|\}$$

and so

$$u(S) = f(\alpha^T) = v(T)$$

so that for simple games, formulas (12.4) and (12.18) coincide. The advantage of (12.18) is that it is meaningful also for nonsimple games, subject to the conditions that

$$w_j(S) \geq 0 \quad \text{for all} \quad S \subset M_j \tag{12.20}$$

$$w_j(M_j) = 1 \quad \text{for all} \quad j \tag{12.21}$$

Thus (12.18) is a generalization of (12.4).

Let g_j, $j = 1, \ldots, n$, be the MLE of w_j. Then g_j is a function mapping the cube $[0, 1]^{M_j}$ into the reals. In fact, because of conditions (12.20)–(12.21), the image of g_j the unit interval $[0, 1]$. The cartesian product

$$g = g_1 \times g_2 \times \ldots \times g_n$$

is a function whose domain is

$$[0, 1]^{M_1} \times [0, 1]^{M_2} \times \cdots \times [0, 1]^{M_n} = [0, 1]^{M^*}$$

and whose image is $[0, 1]^N$. The function g is given by

$$g(x) = (g_1(x^1), g_2(x^2), \ldots, g_n(x^n))$$

where x^j is the restriction of the vector x to indices $i \in M_j$.

Now f, the multilinear extension of v, has the domain $[0, 1]^N$. We can therefore form the composite function

$$h(x) = f(g_1(x^1), g_2(x^2), \ldots, g_n(x^n))$$

defined on the cube $[0, 1]^{M^*}$.

It may be seen, now, that h is a multilinear function of the variables x_i, $i \in M^*$. For, suppose $i \in M_j$. By hypothesis the sets M_k are disjoint, and so $g_k(x^k)$, $k \neq j$, do not depend on x_i. Therefore

$$\frac{\partial h}{\partial x_i} = \frac{\partial f}{\partial y_j} \frac{\partial g_j}{\partial x_i}$$

or, using subscripts to denote differentiation

$$h_i(x) = f_j(g(x)) g_{ji}(x^j)$$

Now, since g_j is multilinear, its derivative g_{ji} does not depend on x_i. Since f is multilinear, its derivative f_j depends only on $g_k(x^k)$, $k \neq j$, and therefore does not depend on x_i. Thus $h_i(x)$ does not depend on x_i, and we conclude that h is linear in x_i. As this is true for each $i \in M^*$, we see that h is multilinear.

Since h is multilinear, it is the extension of some real-valued function, defined on the corners of the cube. In fact, it is the extension of the compound game u.

12.2.9 Theorem. Let v be a nonnegative n-person game, let $w_1, \ldots, w_n$ be as in (12.20)–(12.21), and let $u = v[w_1, \ldots, w_n]$. Let f, $g_1, \ldots, g_n$ be the MLEs of v, $w_1, \ldots, w_n$, respectively, and let $h = f \circ g$. Then h is the multilinear extension of u.

Proof. We evaluate h at the S-corner, α^S of the cube $[0,1]^{M^*}$. We have, for any $S \subset M^*$

$$h(\alpha^S) = f(g_1(\alpha^{S_1}), g_2(\alpha^{S_2}), \ldots, g_n(\alpha^{S_n}))$$

where, as before, $S_j = S \cap M_j$, and α^{S_j} is the restriction of α^S to indices $i \in M_j$. We know that $g_j(\alpha^{S_j}) = w_j(S_j)$, and so

$$h(\alpha^S) = f(w_1(S_1), w_2(S_2), \ldots, w_n(S_n))$$

or, by (12.19)

$$h(\alpha^S) = u(S)$$

Thus h coincides with u on the corners of the cube, and is the MLE of u.

Theorem 12.2.9 tells us that composition of games corresponds to composition of their MLEs. Since the MLE is closely related to the Shapley value, one would hope that the value would also compose, so that, say, we would have $\varphi_i[u] = \varphi_i[w_j]\varphi_j[v]$ for $i \in M_j$. In fact, this is generally not true, though in certain special cases it may happen to be true. To see why, in general, the value does not compose, we remember that for $i \in M_j$ we have

$$h_i(x) = f_j(g(x))g_{ji}(x)$$

and, setting $x = (t, t, \ldots, t)$

$$h_i(t, t, \ldots, t) = f_j(y(t))g_{ji}(t, t, \ldots, t)$$

where

$$y_k(t) = g_k(t, t, \ldots, t) \tag{12.22}$$

Thus

$$\varphi_i[u] = \int_0^1 f_j(y(t))g_{ji}(t, t, \ldots, t)dt \tag{12.23}$$

whereas

$$\varphi_j[v] = \int_0^1 f_j(t, t, \ldots, t)dt$$

and

$$\varphi_i[w_j] = \int_0^1 g_{ji}(t, t, \ldots, t)dt$$

Thus, we see two reasons why the value does not, in general, compose: in the first place, $y(t)$ is generally not equal to $(t, t, \ldots, t)$; in the second place, even if $y_k(t)$ is equal to t for all k, the product of two integrals is generally not equal to the integral of the product of their two integrands.

Formula (12.23) allows us to compute the value for compound games. The related formula,

$$\sum_{i \in M_j} \varphi_i[u] = \int_0^1 f_j(y(t)) \sum_{i \in M_j} g_{ji}(t, t, \ldots, t) dt$$

reduces, by (12.22), to

$$\sum_{i \in M_j} \varphi_i[u] = \int_0^1 f_j(y(t)) \frac{dy_j}{dt} dt \tag{12.24}$$

which is also useful.

12.2.10 Example. We saw above that the United Nations Security Council can be represented as a compound game $u = v[w_1, w_2]$, where w_1 is a five-person simple game in which only the five-person coalition wins, w_2 is an improper ten-person simple game in which any coalition with four or more players wins, and v is a two-person simple game in which the two-person coalition wins.

For w_1, we have the multilinear extension

$$g_1(x) = x_1 x_2 x_3 x_4 x_5$$

and so

$$y_1(t) = t^5$$

Rather than give $g_2(x)$ for general x, we shall give g_{2i} at points $x = (t, t, t, \ldots, t)$. In fact, we have, for any $i \in M_2$

$$g_{2i}(t, t, \ldots, t) = \sum_S t^s (1 - t)^{9 - s}$$

where the sum is taken over all sets $S \subset M_2 - \{i\}$, such that S loses but $S \cup \{i\}$ wins, and s is the cardinality of S. Thus, the sum is taken over all $S \subset M_2 - \{i\}$ which have exactly three elements, and there are $\binom{9}{3} = 84$ of these. Therefore

$$g_{2i}(t, t, \ldots, t) = 84 t^3 (1 - t)^6$$

Finally, v has the extension

$$f(y_1, y_2) = y_1 y_2$$

We have, then, for any $i \in M_2$

$$\varphi_i[u] = \int f_2(y(t))g_{2i}(t, t, \dots, t)dt$$

$$= \int 84t^5t^3(1-t)^6 dt = \int_0^1 84t^8(1-t)^6 dt$$

or

$$\varphi_i[u] = 84\frac{8!6!}{15!} = \frac{4}{2145}$$

Since there are 10 elements in M_2, this leaves a total of $1 - 40/2145 = 2105/2145$ for the members of M_1. By symmetry, each of them receives one fifth of this, or $421/2145$. Thus the game has the value

$$\varphi_i[u] = \begin{cases} 421/2145 = 0.1963 & \text{if } i \text{ is a permanent member} \\ 4/2145 = 0.00186 & \text{for other members} \end{cases}$$

We conclude from this that the permanent members of the Security Council have the lion's share of the power.

12.2.11 Example. The Victoria Proposal suggests that Amendments to the Canadian Constitution, to be ratified, must be approved by at least

1. Ontario and Quebec,
2. two of the four Maritime Provinces,
3. either British Columbia, and one of the Prairie Provinces, or all three of the Prairie Provinces.

Setting this as a simple game, we note that we have here a natural division of the Provinces into three subsets. There are the large provinces

$$M_1 = \{\text{Ontario, Quebec}\} = \{1, 2\}$$

the Maritime Provinces

$$M_2 = \{\text{New Brunswick, Nova Scotia, Newfoundland,}$$
$$\text{Prince Edward Island}\}$$
$$= \{3, 4, 5, 6\}$$

and the Western Provinces

$$M_3 = \{\text{British Columbia, Alberta, Saskatchewan, Manitoba}\}$$
$$= \{7, 8, 9, 10\}$$

and we find that the game can be represented as

$$u = v[w_1, w_2, w_3]$$

where w_1 is a two-player game in which $\{1, 2\}$ is the only winning coalition; w_2 is a four-player improper game in which any two-player coalition (or larger) wins; and w_3 is a four-player simple game similar to that studied in Examples 12.2.3 and 12.2.7. Finally, v is a three-person simple game in which only the three-person coalition wins.

For w_1 here, we have the MLE

$$g_1(x_1, x_2) = x_1 x_2$$

so that

$$y_1(t) = t^2$$

For w_2, we evaluate the MLE on the points (t, t, t, t)

$$g_2(t, t, t, t) = 6t^2(1-t)^2 + 4t^3(1-t) + t^4$$

or

$$y_2(t) = 6t^2 - 8t^3 + 3t^4$$

Next, for w_3 we have (from Example 12.2.3)

$$g_3(x) = x_7 x_8 + x_7 x_9 + x_7 x_{10} + x_8 x_9 x_{10} - x_7 x_8 x_9 - x_7 x_8 x_{10} - x_7 x_9 x_{10}$$

so that

$$y_3(t) = 3t^2 - 2t^3$$

$$g_{37}(t, t, t, t) = 3t - 3t^2$$

$$g_{38}(t, t, t, t) = t - t^2$$

and so on. Finally, v has the extension

$$f(y) = y_1 y_2 y_3$$

We have then

$$\varphi_1[u] + \varphi_2[u] = \int (6t^2 - 8t^3 + 3t^4)(3t^2 - 2t^3)2t \, dt$$

$$= \int_0^1 (36t^5 - 72t^6 + 50t^7 - 12t^8)dt = 106/168$$

and, by symmetry

$$\varphi_1 = \varphi_2 = 53/168 = 0.3155$$

Next, we have

$$\sum_{i \in M_2} \varphi_i = \int t^2(3t^2 - 2t^3)(12t - 24t^2 + 12t^3)\,dt$$

so that

$$\varphi_3 + \varphi_4 + \varphi_5 + \varphi_6 = 20/168$$

and, again by symmetry

$$\varphi_3 = \varphi_4 = \varphi_5 = \varphi_6 = 5/168 = 0.0298$$

For player 7 (British Columbia), we have

$$\varphi_7 = \int f_3(y)g_{37}(t, t, t, t)\,dt$$
$$= \int t^2(6t^2 - 8t^3 + 3t^4)(3t - 3t^2)\,dt$$
$$= 21/168 = 0.0125.$$

Finally, for the Prairie Provinces, we will have

$$\varphi_8 = \varphi_9 = \varphi_{10} = \int t^2(6t^2 - 8t^3 + 3t^4)(t - t^2)\,dt = 7/168 = 0.0417.$$

We therefore have the values

$$\varphi = 0.3155 \quad \text{for Ontario and Quebec,}$$
$$\varphi = 0.0298 \quad \text{for the Maritime Provinces,}$$
$$\varphi = 0.1250 \quad \text{for British Columbia,}$$
$$\varphi = 0.0417 \quad \text{for the Prairie Provinces.}$$

12.3.   The Banzhaf–Coleman Index of Power

A second index of power has been suggested by Banzhaf [12.1] and Coleman [12.3]. Assuming v to be a simple game [in $(0,1)$ normalization], a *swing* for player i is defined as a set $S \subset N$ such that $i \in S$, S wins, and $S - \{i\}$ loses.

Letting θ_i be the number of swings for player i, we can then write

$$\beta_i[v] = \frac{\theta_i}{\sum\limits_{j=1}^{n} \theta_j} \tag{12.25}$$

This is the *normalized Banzhaf–Coleman index*.

12.3.1 Example. Consider the four-person game of Examples 12.2.3 and 12.2.7. Player 1 has six swings, namely, the sets $\{1, 2\}, \{1, 3\}, \{1, 4\}, \{1, 2, 3\}, \{1, 2, 4\}$, and $\{1, 3, 4\}$. Player 2 has two swings, namely, $\{1, 2\}$ and $\{2, 3, 4\}$. Thus $\theta_1 = 6$, and, by symmetry, $\theta_2 = \theta_3 = \theta_4 = 2$. Therefore

$$\beta[v] = (\tfrac{1}{2}, \tfrac{1}{6}, \tfrac{1}{6}, \tfrac{1}{6})$$

is the normalized power index for this game. Note that, in this case, it coincides with the Shapley value. This does not generally happen.

12.3.2 Example. Let v be the three-person simple game in which $\{1, 2\}, \{1, 3\}$, and $\{1, 2, 3\}$ are the only winning coalitions. All of these are swings for player 1. On the other hand, player 2 has only the swing $\{1, 2\}$. Thus $\theta_1 = 3$, and $\theta_2 = \theta_3 = 1$. The game, therefore, has the normalized index

$$\beta = (\tfrac{3}{5}, \tfrac{1}{5}, \tfrac{1}{5})$$

which is clearly different from the Shapley value

$$\varphi = (\tfrac{2}{3}, \tfrac{1}{6}, \tfrac{1}{6})$$

To obtain a coherent theory of this index, we shall set

$$\theta_i[v] = \sum_{\substack{S \subseteq N \\ i \in S}} [v(S) - v(S - \{i\})] \tag{12.26}$$

which coincides with our original definition of θ_i. For, if v is a simple game, we note that $v(S) - v(S - \{i\})$ will vanish, except when S is a swing for player i. The advantage of (12.26), of course, is that it is meaningful for games in general (even when not simple). Let us also set

$$\psi[v] = 2^{1-n} \theta[v]$$

or

$$\psi[v] = \sum_{\substack{S \subset N \\ i \in S}} (\tfrac{1}{2})^{n-1} [v(S) - v(S - \{i\})] \tag{12.27}$$

A comparison of this with (12.4) shows a certain relation between the Shapley value and the Banzhaf–Coleman index: both give averages of player i's marginal contributions $v(S) - v(S - \{i\})$. The difference lies in the weighting coefficients used: for the value, these varied according to the size of S; for the present index, they are all equal. It is easy to see, in any case, that there are 2^{n-1} terms on the sum; thus the sum of the coefficients — all equal to 2^{1-n} — is indeed unity.

An alternative form of (12.27) is

$$\psi_i[v] = f_i(\tfrac{1}{2}, \tfrac{1}{2}, \dots, \tfrac{1}{2}) \tag{12.28}$$

where, as usual, f is the MLE of v and f_i is the ith partial derivative of f. The following properties of the index are of interest now.

12.3.3 Theorem. If v and w are games with player set N, and α, γ are scalars, then

$$\psi[\alpha v + \gamma w] = \alpha\psi[v] + \gamma\psi[w]$$

Proof. Immediate.

12.3.4 Theorem. If v, w are games with disjoint player sets M and N and $v \oplus w$ is the von Neumann–Morgenstern composition (sum), then, for any $i \in M$ or $j \in N$

$$\psi_i[v \oplus w] = \psi_i[v]$$
$$\psi_j[v \oplus w] = \psi_j[w]$$

Proof. This follows immediately from (12.28) and Theorem 12.2.5.

In particular, w can be a null game, so that $v \oplus w$ is merely v with some dummies adjoined. Thus we see that the present index is dummy-invariant, i.e., $\psi_i[v] = \psi_i[v']$ if v' is obtained by the adjunction of dummies to v.

12.3.5 Theorem. If i is a dummy for v, then $\psi_i[v] = 0$.

Proof. If i is a dummy, then $v(S) = v(S - \{i\})$ for all S. Thus, from (12.27), $\psi_i[v] = 0$.

12.3.6 Theorem. Let $w_1, w_2, \ldots, w_n$ be games with disjoint player sets $M_1, M_2, \ldots, M_n$, respectively, satisfying $w_j \geq 0$, $w_j(M_j) = 1$, let v be a nonnegative game with player set $N = \{1, 2, \ldots, n\}$, and let

$$u = v[w_1, w_2, \ldots w_n]$$

Then, for each $j \in N$, there is a constant $\lambda_j \geq 0$ such that, if $i \in M_j$, then

$$\psi_i[u] = \lambda_j \psi_i[w_j]$$

(In other words, the members of M_j have a power in u which is proportional to their power in w_j.)

Proof. Letting $f, g_1, \ldots, g_n$ and h be the MLEs of $v, w_1, \ldots, w_n$, and u, respectively, we have

$$h_i(\tfrac{1}{2}, \tfrac{1}{2}, \ldots, \tfrac{1}{2}) = f_j(y(\tfrac{1}{2}))g_{ji}(\tfrac{1}{2}, \ldots, \tfrac{1}{2})$$

where, as always

$$y_k(t) = g_k(t, t, \ldots, t)$$

Thus, from (12.28), we have

$$\psi_i[u] = f_j(y(\tfrac{1}{2}))\psi_i[w_j]$$

As v is nonnegative, all the partial derivatives of f are nonnegative. Thus the theorem is proved, with

$$\lambda_j = f_j(y(\tfrac{1}{2}))$$

12.3.7 Theorem. Let $u, v, w_1, \ldots, w_n$ be as in Theorem 12.3.6, and assume, moreover, that all w_j are constant-sum games. Then, if $i \in M_j$

$$\psi_i[u] = \psi_j[v]\psi_i[w_j]$$

Proof. From Theorem 12.3.6, we have

$$\lambda_j = f_j(y(\tfrac{1}{2}))$$

Each w_k is constant-sum, and so from Theorem 12.2.6, we have

$$g_k(\tfrac{1}{2}, \tfrac{1}{2}, \ldots, \tfrac{1}{2}) = \tfrac{1}{2}w_k(M_k)$$

Now, $w_k(M_k) = 1$, and so $y_k(\tfrac{1}{2}) = \tfrac{1}{2}$. Thus

$$\lambda_j = f_j(\tfrac{1}{2}, \ldots, \tfrac{1}{2}) = \psi_j[v]$$

and we have, from 12.3.7

$$\psi_i[u] = \psi_j[v]\psi_i[w_j]$$

12.3.8 Example. Let u be the Security Council game of Example 12.2.10. We have, as before

$$u = v[w_1, w_2]$$

where M_1 is the set of permanent members, and M_2 is the set of other members.
Using the notation of 12.13, we have here

$$f_1(y) = y_2, \quad f_2(y) = y_1$$

$$y_1(\tfrac{1}{2}) = (\tfrac{1}{2})^5 = \frac{1}{32}$$

and, for all $i \in M_1$

$$g_{1i}(\tfrac{1}{2}, \tfrac{1}{2}, \ldots, \tfrac{1}{2}) = \frac{1}{16} \quad \text{for all} \quad i \in M_1$$

$$y_2(\tfrac{1}{2}) = \frac{53}{64}$$

$$g_{2i}(\tfrac{1}{2}, \tfrac{1}{2}, \ldots, \tfrac{1}{2}) = \frac{21}{128} \quad \text{for all} \quad i \in M_2$$

Thus, for $i \in M_1$

$$\psi_i[u] = \left(\frac{53}{64}\right) \cdot \left(\frac{1}{16}\right) = \frac{53}{1024}$$

and, for $i \in M_2$

$$\psi_i[u] = \left(\frac{21}{128}\right) \cdot \left(\frac{1}{32}\right) = \frac{21}{4096}$$

We note that the power index for permanent members is about 10 times that for the other members. As against this, the Shapley value also gave most of the power to the permanent members, but the ratio there was much greater (about 100 to 1).

12.3.9 Example. Again we consider the Victoria Proposal to the Canadian Constitution. Using the notation of Example 12.2.11, we have

$$g_{1i}\left(\tfrac{1}{2}, \tfrac{1}{2}\right) = \tfrac{1}{2} \quad i = 1, 2$$
$$y_1\left(\tfrac{1}{2}\right) = \tfrac{1}{4}$$
$$g_{2i}\left(\tfrac{1}{2}, \ldots, \tfrac{1}{2}\right) = \tfrac{3}{8} \quad i = 3, 4, 5, 6$$
$$y_2\left(\tfrac{1}{2}\right) = \tfrac{11}{16}$$
$$g_{37}\left(\tfrac{1}{2}, \ldots, \tfrac{1}{2}\right) = \tfrac{3}{4}$$
$$g_{3i}\left(\tfrac{1}{2}, \ldots, \tfrac{1}{2}\right) = \tfrac{1}{4} \quad i = 8, 9, 10$$
$$y_3\left(\tfrac{1}{2}\right) = \tfrac{1}{2}$$
$$f_1(y) = y_2 y_3$$
$$f_2(y) = y_1 y_3$$
$$f_3(y) = y_1 y_2$$

Thus

$$\psi_i[u] = \left(\tfrac{1}{2}\right) \cdot \left(\tfrac{11}{16}\right) \cdot \left(\tfrac{1}{2}\right) = \tfrac{11}{64} \quad i = 1, 2$$
$$\psi_i[u] = \left(\tfrac{3}{8}\right) \cdot \left(\tfrac{1}{4}\right) \cdot \left(\tfrac{1}{2}\right) = \tfrac{3}{64} \quad i = 3, 4, 5, 6$$
$$\psi_j[u] = \left(\tfrac{3}{4}\right) \cdot \left(\tfrac{1}{4}\right) \cdot \left(\tfrac{11}{16}\right) = \tfrac{33}{256}$$
$$\psi_i[u] = \left(\tfrac{1}{4}\right) \cdot \left(\tfrac{1}{4}\right) \cdot \left(\tfrac{11}{16}\right) = \tfrac{11}{256} \quad i = 8, 9, 10$$

and we have the power index

0.1718   for Ontario and Quebec,
0.0469   for the Maritime Provinces,
0.1289   for British Columbia,
0.0430   for the Prairie Provinces.

It is of interest to note that the Banzhaf–Coleman index assigns greater power to the Maritime Provinces than to the Prairie Provinces. By contrast, the Shapley value favored the Prairie Provinces more than the Maritimes.

The above properties of the index can now be used to develop it axiomatically. As in Section 12.1 (when we introduced the Shapley value), we are looking for a mapping ψ which assigns to each n-person game v an n-vector $\psi[v] = (\psi_1, \psi_2, \ldots, \psi_n)$. We have, then, the following system of axioms:

B1. If i is a dummy in v, then

$$\psi_i[v] = 0$$

B2. If v is a game with player set N, and v' is obtained from v by adjoining a dummy, then

$$\psi_i[v'] = \psi_i[v] \quad \text{for all} \quad i \in N$$

B3. If π is a permutation of N, and $i \in N$, then

$$\psi_{\pi(i)}[\pi v] = \psi_i[v]$$

B4. If u and v are games, and α and γ are positive scalars, then

$$\psi[\alpha u + \gamma v] = \alpha \psi[u] + \gamma \psi[v]$$

B5. Let $u = v[w_1, w_2, \ldots, w_n]$, suppose $w_1, \ldots, w_n$ are constant-sum games. Then, for $i \in M_j$

$$\psi_i[u] = \psi_j[v]\psi_i[w_j]$$

We have, of course, proved that the Banzhaf–Coleman index ψ satisfies B1–B5. On the other hand, it is not the only power index to satisfy the axioms. The *null index*

$$\lambda[v] = 0 \quad \text{for all} \quad v \tag{12.29}$$

clearly satisfies them also. Moreover, so does the *dictatorial index*

$$\rho_i[v] = v(\{i\}) \tag{12.30}$$

For, ρ clearly satisfies B1–B4. To prove B5, we note that, for any game

$$v(\{i\}) = f_i(0, 0, \ldots, 0)$$

where f is the MLE of v. Then, if $u = v[w_1, \ldots, w_n]$, and $i \in M_j$

$$h_i(0, \ldots, 0) = f_i(y(0))g_{ji}(0, \ldots, 0)$$

But $y_k(0) = 0$ for all k, and so

$$h_i(0, \ldots, 0) = g_j(0, \ldots, 0)g_{ji}(0, \ldots, 0)$$

or

$$\rho_i[u] = \rho_j[v]\rho_i[w_j]$$

We will now prove, more or less, that the Banzhaf–Coleman index is the only nontrivial index satisfying B1–B5. An exact statement of this will be given below, as Theorem 12.3.11. We will start by considering, first, constant-sum games; the proof will later be extended to general-sum games.

Essentially, Axioms B1–B5 refer to four different operations which may be performed on games: adjunction of a dummy; permutation of the players; linear combinations; and composition. It is of interest to note that the space $\mathfrak{C}$ of constant-sum games is closed under all these operations. This is obviously true for dummy adjunction, permutation, and linear combinations. For composition, we give the following lemma:

12.3.10 Lemma. If $v, w_1, \ldots, w_n$ are constant-sum games, and $u = v[w_1, \ldots, w_n]$, then u is constant-sum.

Proof. For any $S \subset M^*$, we have, by (12.19)

$$u(M^* - S) = f(w_1(M_1 - S_1), \ldots, w_n(M_n - S_n))$$

where, as before, $S_j = S \cap M_j$, and f is the MLE of v. By 12.11 and the constant-sum properties of v and w_j this is

$$u(M^* - S) = f(1 - w_1(S_1), \ldots, 1 - w_n(S_n))$$
$$= v(N) - f(w_1(S_1), \ldots, w_n(S_n))$$
$$= v(N) - u(S)$$

so that

$$u(M^* - S) = u(M^*) - u(S)$$

and u is constant-sum.

Since $\mathfrak{C}$ is closed under these operations, we can effectively restrict the Axioms B1–B5 to $\mathfrak{C}$. Then we have the following:

12.3.11 Theorem. There are only three indices which satisfy the Axioms B1–B5 over the space of all constant-sum n-person games. These are the Banzhaf–Coleman index (12.27), the null index (12.29), and the dictatorial index (12.30).

To prove this, we will show, broadly speaking, that there are two fundamental games which can be used as "basic building blocks" for the space $\mathfrak{C}$. These are the one-person dictator game $\mathscr{E}_i$ and the three-person majority game, $\mathscr{M}_{ijk}$. These are defined as follows.

For any player i, the game $\mathscr{E}_i$ has the characteristic function

$$\mathscr{E}_i(S) = \begin{cases} 1 & \text{if } i \in S \\ 0 & \text{if } i \notin S \end{cases}$$

Thus, i is a dictator for the game $\mathscr{E}_i$. All other players are dummies.

For distinct i, j, and k, we define $\mathcal{M}_{ijk}$ by

$$\mathcal{M}_{ijk}(S) = \begin{cases} 1 & \text{if} \quad |\{i, j, k\} \cap S| \geq 2 \\ 0 & \text{otherwise} \end{cases}$$

Thus i, j, and k are the three decision makers for the game $\mathcal{M}_{ijk}$; any two of these, acting together, can win, and all other players are dummies.

We can think of $\mathcal{E}_i$ as a one-person game, and of $\mathcal{M}_{ijk}$ as a three-person game. We can also, however, consider them as games with larger sets of players, in which all but one or three of the players are dummies. Essentially, Axioms B1–B2 tell us that it does not matter which we do. Moreover, if we think of $\mathcal{E}_i$ as a one-person game, or of $\mathcal{M}_{ijk}$ as a three-person game, we might omit the subscripts when no confusion is possible.

For these two fundamental games, we shall show that there exist only three possibilities for an index satisfying the axioms. Now, all constant-sum games can be obtained from $\mathcal{E}$ and $\mathcal{M}$ by the four operations mentioned in the axioms; thus, the three possibilities for $\mathcal{E}$ and $\mathcal{M}$ exhaust all possibilities for the entire space $\mathfrak{C}$.

By B1–B3, we will have

$$\psi_j[\mathcal{E}_i] = \begin{cases} p & \text{if} \quad j = i \\ 0 & \text{if} \quad j \neq i \end{cases} \tag{12.31}$$

and

$$\psi_l[\mathcal{M}_{ijk}] = \begin{cases} q & \text{if} \quad l = i, j, \text{ or } k \\ 0 & \text{otherwise} \end{cases} \tag{12.32}$$

where p, q are constants (to be determined) which, due to B3, do not depend on i, j, or k.

We have, then, the following:

12.3.12 Lemma. For the null index, $p = 0$. Otherwise, $p = 1$.

Proof. The first assertion is obvious. Assuming ψ is an index, other than the null index, satisfying B1–B5, there must be a game v and a player i such that $\psi_i[v] \neq 0$. Now let

$$u = v[\mathcal{E}_1, \mathcal{E}_2, \ldots, \mathcal{E}_n]$$

It is clear that $u = v$. Then by B5, we must have

$$\psi_i[v] = \psi_i[u] = \psi_i[v]\psi_i[\mathcal{E}_i] = p\psi_i[v]$$

and, since $\psi_i[v] \neq 0$, this means $p = 1$.

12.3.13 Lemma. In (12.3.8), q must be either 0 or $\frac{1}{2}$.

Proof. If $p = 0$, we have the null index, and clearly $q = 0$. Otherwise, we will have $p = 1$. Consider, then, the two games

$$u = \mathcal{M}[\mathscr{E}_1, \mathscr{E}_2, \mathcal{M}_{345}]$$
$$u' = \mathcal{M}[\mathcal{M}_{123}, \mathscr{E}_4, \mathscr{E}_5]$$

We must have, by B5

$$\psi[u] = (q, q, q^2, q^2, q^2)$$
$$\psi[u'] = (q^2, q^2, q^2, q, q)$$

Both u and u' are games with player set (or carrier) $\{1, 2, 3, 4, 5\}$. It is not too difficult to verify that

$$2u - 2u' = \mathcal{M}_{124} + \mathcal{M}_{125} - \mathcal{M}_{145} - \mathcal{M}_{245}$$

and an application of B4 gives us

$$(2q - 2q^2, 2q - 2q^2, 0, 2q^2 - 2q, 2q^2 - 2q) = (q, q, 0, -q, -q)$$

which reduces to the single equation

$$2q - 2q^2 = q$$

with the two solutions $q = 0$ and $q = \frac{1}{2}$.

We show, next, that any index satisfying B1–B5 is effectively determined by the two numbers p and q. We shall define the *degree* of a game v as the polynomial degree of its MLE f. It is clear that if the coalition S with s players is a carrier for v, then the degree of v is not greater than s. For f is a polynomial in the variables x_i, $i \in S$, and the degree of a multilinear function cannot be greater than the number of variables.

There are games for which the degree is equal to the cardinality of the minimal carrier; for example, $\mathscr{E}_i$ and $\mathcal{M}_{ijk}$ have the MLEs

$$f = x_i \tag{12.33}$$

and

$$f = x_i x_j + x_i x_k + x_j x_k - 2 x_i x_j x_k \tag{12.34}$$

respectively, so that their degrees are 1 and 3, respectively. On the other hand, this is not true for all games; the four-person simple game of Example 12.2.3 has the MLE

$$f = x_1 x_2 + x_1 x_3 + x_1 x_4 + x_2 x_3 x_4 - x_1 x_2 x_3 - x_1 x_2 x_4 - x_1 x_3 x_4$$

so that its degree is 3, although its minimal carrier has four elements. In fact, no four-person constant-sum game has a degree greater than 3 (see Problem 6).

12.3.14 Lemma. Let S have s elements, s odd. Then there is a constant-sum game v, with carrier S and degree s, such that $\psi[v]$ is determined entirely by p and q.

Proof. By induction on s. We have already proved this for $s = 1$ and $s = 3$ (the games are $\mathscr{E}$ and $\mathscr{M}$, respectively). Assume, then, that this is true for $s - 2$. Let $S = T \cup \{i, j\}$, where T has $s - 2$ elements, and let w be the corresponding game for the carrier T. Let

$$v = \mathscr{M}[w, \mathscr{E}_i, \mathscr{E}_j]$$

then v has the carrier $T \cup \{i, j\} = S$, and the MLE

$$h = f(g, x_i, x_j)$$

where f is as in (10.3.10), and g is the MLE of w. Thus

$$h = gx_i + gx_j + x_i x_j - 2gx_i x_j$$

Since g has degree $s - 2$, it follows that h has degree s. By the induction hypothesis, the two numbers p and q determine the index on the games $\mathscr{M}$, w, $\mathscr{E}_i$, and $\mathscr{E}_j$. Thus by B5, p and q determine the index on the game v.

12.3.15 Lemma. Any index satisfying B1–B5 is entirely determined on $\mathfrak{C}$ by p and q.

Proof. Essentially, we shall show that the games guaranteed by Lemma 12.3.14 form a basis for the space.

Letting $N = \{1, 2, \ldots, n\}$, we see that a constant-sum game v is essentially determined by 2^{n-1} numbers: the numbers $v(S)$ for all S which contain player 1. Thus $\mathfrak{C}$ is contained in a Euclidean space of dimension 2^{n-1}. Now, N has 2^n subsets, of which exactly one-half, or 2^{n-1}, have an odd number of elements. This is best proved by noting that the number of sets with an even number of elements, minus that with an odd number of elements, is

$$\binom{n}{0} - \binom{n}{1} + \binom{n}{2} - \cdots \pm \binom{n}{n-1} \mp \binom{n}{n}$$

or

$$\sum_{k=0}^{n} \binom{n}{k}(-1)^k$$

and this is easily recognized as the binomial expansion of $(1 - 1)^n$, which is 0 if $n \geq 1$. Thus exactly half the subsets of N have odd cardinality.

Let $S_1, S_2, \ldots, S_{2^{n-1}}$ be the odd-sized subsets of N, and let $v_1, \ldots, v_{2^{n-1}}$ be the corresponding games guaranteed by 12.38. We claim now that these 2^{n-1} games (treated as vectors) are linearly independent. For, suppose they satisfy a relation of the form

$$\sum_{k=1}^{2^{n-1}} \alpha_k v_k = 0$$

where not all $\alpha_k = 0$. Let S_l be a coalition of maximal size for which the corresponding coefficient $\alpha_l \neq 0$. Then we can write

$$v_l = \sum_k \beta_k v_k$$

where, on the right-hand side, each v_k either has degree smaller than s_l (the cardinality of S_l), or has degree equal to s_l, with a carrier S_k of size s_l but different from S_l. A similar relation holds clearly among the MLEs

$$g_l = \sum_k \beta_k g_k$$

Now, the function g_l has degree s_l, and so the monomial $\Pi_{S_l} x_i$, with a nonzero coefficient, must appear on the left-hand side of this equation. On the other hand, the only terms of degree s_l on the right-hand side are the monomials $\Pi_{S_k} x_i$ with $S_k \neq S_l$. This contradiction proves that the games v_k are linearly independent. Since there are 2^{n-1} of them, they span a space of dimension 2^{n-1}, and thus form a basis for the space $\mathfrak{C}$ of n-person constant-sum games; i.e., every constant-sum game can be expressed as a linear combination of these games. By Lemma 12.3.14, the index is uniquely determined on the basic games by p and q, and so by axiom B4, p and q uniquely determine the index throughout the entire space $\mathfrak{C}$.

We have seen, then, that any index satisfying B1–B5 is entirely determined, for all constant-sum games, by the numbers p and q. Now, by 12.36 and 12.37, there are only three possibilities for these numbers. These three possibilities correspond to the three indices discussed above:

(a) $p = 0$, $q = 0$ (null index),
(b) $p = 1$, $q = 0$ (dictatorial index),
(c) $p = 1$, $q = \frac{1}{2}$ (Banzhaf–Coleman index),

which are therefore the only three indices possible.

For general-sum games, the situation is somewhat more complicated. In fact, we are looking for an extension to $\mathfrak{S}$, the space of all n-person games, of the three mappings ψ, λ, and ρ defined on the space $\mathfrak{C}$. It turns out that ψ and λ can only be extended in the obvious unique manner; the mapping ρ, however, has more than one extension.

Essentially, we find that the space $\mathfrak{S}$ of n-person games can be built up from three fundamental games. These are the games $\mathscr{E}$ and $\mathscr{M}$ discussed above, and also the game $\mathscr{B}_{ij}$ defined by

$$\mathscr{B}_{ij}(S) = \begin{cases} 1 & \text{if } \{i, j\} \subset S \\ 0 & \text{otherwise} \end{cases}$$

It is clear once again that, for any index satisfying B1–B5

$$\psi_k[\mathscr{B}_{ij}] = \begin{cases} r & \text{if } k = i \text{ or } j \\ 0 & \text{otherwise} \end{cases} \tag{12.35}$$

where once again r is a constant, independent of i or j.

Much as in the constant-sum case, we shall show that the index is entirely determined by p, q, and r.

12.3.16 Lemma. For any nonempty set S with s elements, there exists a game v with carrier S and degree s such that $\psi[v]$ is entirely determined by p, q, and r.

Proof. We have already proved this for all odd values of s. For $s = 2$, we have $S = \{i, j\}$. Then $\mathscr{B}_{ij}$ is the desired game, for it has the MLE

$$f = x_i x_j \tag{12.36}$$

which is of degree 2.

For arbitrary even s, let S have s elements, and let $S = T \cup \{i\}$, where T has $s - 1$ elements. By 12.38, there exists a constant-sum game w with carrier T and degree $s - 1$, having the desired properties. Now set

$$v = \mathscr{B}[w, \mathscr{E}_i]$$

Then v has the carrier $T \cup \{i\} = S$ and the MLE

$$f = g x_i$$

where g is the MLE of w. Since g has degree $s - 1$, we see f, and hence v, has the degree s as desired. Since both w and $\mathscr{E}_i$ are constant-sum games, we can apply B5, and we see that the index is determined on v by the three numbers p, q, and r.

12.3.17 Lemma. Any index satisfying B1–B5 for the space of all n-person games is uniquely determined by the three numbers p, q, and r.

Proof. We know that the space of n-person games with player set N has the dimension $2^n - 1$. Now N has $2^n - 1$ nonempty subsets $S_1, \ldots, S_{2^n - 1}$, with corresponding games $v_1, \ldots, v_{2^n - 1}$ guarantees by Lemma 12.3.16. Much as in Lemma 12.3.15, we see that these $2^n - 1$ games form a basis for the space $\mathfrak{S}$, and an application of B4 tells us that the index is determined throughout the space by p, q, and r.

It remains to determine the permissible values of p, q, and r. Consider, then, the four-person games

$$u = \mathscr{B}[\mathscr{E}_1, \mathscr{M}_{234}]$$
$$u' = \mathscr{B}[\mathscr{M}_{123}, \mathscr{E}_4]$$

with the indices

$$\psi[u] = (pr, qr, qr, qr)$$
$$\psi[u'] = (qr, qr, qr, pr)$$

It can be seen that

$$2(u - u') = \mathscr{B}_{12} + \mathscr{B}_{13} - \mathscr{B}_{24} - \mathscr{B}_{34} - \mathscr{M}_{123} + \mathscr{M}_{234}$$

and, applying B4, we have

$$2(pr - qr, 0, 0, qr - pr) = (2r - q, 0, 0, q - 2r)$$

which reduces to

$$2pr - 2qr = 2r - q$$

or

$$r = \frac{q}{2 + 2q - 2p}$$

Now for $p = 0$, $q = 0$, this gives us $r = 0$. If $p = 1$ and $q = \frac{1}{2}$, we obtain $r = \frac{1}{2}$. If, however, $p = 1$ and $q = 0$, the equation reduces to $0 = 0$, and so any value of r will work. We thus obtain the following theorem:

12.3.18 Theorem. There exists a unique extension of the null index λ from the space $\mathfrak{C}$ to the space $\mathfrak{S}$, and it is the null index on $\mathfrak{S}$. There exists a unique extension of the Banzhaf–Coleman index ψ from $\mathfrak{C}$ to $\mathfrak{S}$, and it is the Banzhaf–Coleman index on $\mathfrak{S}$. There exists a one-parameter family of extension of the dictatorial index ρ from $\mathfrak{C}$ to $\mathfrak{S}$, and it includes the dictatorial index ρ, the marginal index

$$\mu_i[v] = v(N) - v(N - \{i\}) \tag{12.37}$$

and all combinations of the form $t\rho + (1 - t)\mu$.

Proof. We have seen above that there are the following possibilities for p, q, and r:

(a) $p = 0$, $q = 0$, $r = 0$;
(b) $p = 1$, $q = \frac{1}{2}$, $r = \frac{1}{2}$;
(c) $p = 1$, $q = 0$, r arbitrary.

Case (a) gives us the null index; case (b) is the Banzhaf–Coleman index. For case (c), $r = 0$ gives us the dictatorial index, while $r = 1$ gives us the marginal index.

We know already that λ, ψ, and ρ satisfy the axioms. To show that μ, also satisfies B1–B5, we note that

$$\mu_i[v] = f_i(1, 1, \ldots, 1) \tag{12.38}$$

where f is the MLE of v. The proof then proceeds as for the dictatorial index ρ, using the fact that, in composition, $y_k(1) = g_k(1, 1, \ldots, 1) = w_k(M_k) = 1$.

Strictly speaking, we should prove that all combinations $t\rho + (1 - t)\mu$ satisfy the axioms as well. We leave the details of this to the reader, pointing out that an important element of the proof is the fact that, for constant-sum games, ρ and μ, and all their combinations, coincide (see Problem 8).

Let us now consider the following axiom:

B6. If $u = v[w_1, \ldots, w_n]$, then there is a vector $\gamma = (\gamma_1, \ldots, \gamma_n)$ such that, for $i \in M_j$

$$\psi_i[u] = \gamma_j \psi_i[w_j]$$

We have already seen that the Banzhaf–Coleman index ψ satisfies B6, with $\gamma_j = f_j(v(\frac{1}{2}))$. Moreover, the null index satisfies B6 trivially, while the dictatorial index

ρ satisfies B6 with $\gamma_j = \rho_j[v]$, and the marginal index μ satisfies B6 with $\gamma_j = \mu_j[v]$. We have, then, a final theorem:

12.3.19 Theorem. There exist exactly four indices satisfying B1–B6 over the space of all n-person games. These are the Banzhaf–Coleman index (12.27), the null index (12.29), the dictatorial index (12.30), and the marginal index (12.36).

Proof. We have seen that these four indices satisfy the axioms. To show that these are the only such indices, we must, in effect, eliminate all values of r other than 0 or 1 for the case $p = 1$, $q = 0$.

Let us assume $p = 1$, $q = 0$ and set

$$u = \mathscr{B}[w, \mathscr{E}_3]$$

where

$$w = \frac{1}{2}\mathscr{E}_1 + \frac{1}{2}\mathscr{B}_{12}$$

Here, w has the index

$$\psi[w] = \left(\frac{1+r}{2}, \frac{r}{2}\right)$$

and so, by B6, u must have the index

$$\psi[u] = \left(\frac{\gamma + \gamma r}{2}, \frac{\gamma r}{2}, \gamma'\right)$$

where γ and γ' are to be determined. We note that

$$4u = 3\mathscr{B}_{13} + \mathscr{B}_{12} + \mathscr{B}_{23} - \mathscr{M}_{123}$$

and so by B4,

$$(2\gamma + 2\gamma r, 2\gamma r, 4\gamma') = (4r, 2r, 4r)$$

which has the solutions $\gamma = \gamma' = r$, $r = 0$ or 1. Thus we conclude that, for $p = 1$ and $q = 0$, we must have either $r = 0$ or $r = 1$. There are no other possibilities, and hence the dictatorial and marginal indices are the only possible extensions of ρ.

12.4. The Presidential Election "Game"

The present method of choosing a president for the United States of America presents us with a very interesting game-theoretic problem. As is well known, the voters within each state elect "great electors," or members of the Electoral College, who in turn vote for the president. It is generally assumed (though in practice there are occasional exceptions to this rule) that all of a given state's electors will vote for whichever candidate is preferred by a majority of the state's voters; thus a very thin majority in a large state (i.e., one with a large number of electoral votes) will more than annul large majorities in several small states. It is clear that such a "game" is not symmetric, i.e., not all the minimal winning coalitions are of equal size, and certain coalitions with less than half the voters can win. What is not obvious is whether the game is fair (in some sense of the word), or, if unfair, which voters the game favors. We shall analyze the game, here, from the point of view of both the Shapley value and the Banzhaf–Coleman index of power.

We have here, in effect, a compound simple game

$$u = v[w_1, w_2, \ldots, w_{51}]$$

where u is the Presidential Election Game (i.e., the game among all voters in the United States), v is the Electoral College Game (i.e., the voting game among the great electors), and $w_1, \ldots, w_{51}$ are the state games, i.e., w_j is the game played among the voters in the jth state.

Let us consider, first, the Electoral College Game v. Assuming (as we have) that all electors from a given state vote together (the unit rule), this is a 51-person weighted majority game with the symbol

$$[270; p_1, p_2, \ldots, p_{51}]$$

where p_j is the number of electoral votes from the jth state (ranging, as of 1977, from a maximum of 45 for California to a minimum of 3 for the smallest states and the District of Columbia).

For the jth state, we will have here the Shapley value

$$\varphi_j[v] = \sum_S \frac{s!(50-s)!}{51!}$$

where the summation is taken over all $S \subset N - \{j\}$ such that S loses, but $S \cup \{j\}$ wins. This is equivalent to saying that

$$270 - p_j \le \sum_{k \in S} p_k \le 269$$

and we can write this summation in the form

$$\varphi_j = \sum_{s=1}^{50} \sum_{k=270-p_j}^{269} \frac{s!(50-s)!}{51!} C_{sk}^j \tag{12.39}$$

where C_{sk}^j is the number of subsets $S \subset N - \{j\}$, having exactly s elements and exactly k electoral votes.

As may be seen, the number of terms in the sum (12.39) is not terribly large; the principal difficulty lies in evaluating the numbers C_{sk}^j. To see how this can best be done, let us use the probabilistic interpretation of the MLE

$$f_j(t, t, \ldots, t) = \text{Prob}\{270 - p_j \le Y^j \le 269\}$$

where, as we saw before

$$Y^j = \sum_{q \ne j} Z_q$$

the Z_q being independent random variables with the possible values p_q and 0, and probabilities t and $1 - t$, respectively. Each such Z_q has the probability-generating function

$$G_{Z_q}(\lambda) = 1 - t + t\lambda^{p_q} \tag{12.40}$$

and so Y^j has the generating function

$$G_{Y^j}(\lambda) = \prod_{q \ne j} [(1 - t) + t\lambda^{p_q}]$$

This is, of course, a polynomial in t and λ, and it is not difficult to see that it can be written as

$$G_{Y^j}(\lambda) = \sum_s \sum_k C_{sk}^j (1 - t)^{50 - s} t^s \lambda^k$$

where the C_{sk}^j are as above.

To simply things somewhat, we can use the slightly different generating function

$$\Gamma_j(\lambda) = \prod_{q \ne j} (1 + t\lambda^{p_q}) \tag{12.41}$$

and then C_{sk}^j is simply the coefficient of $t^s \lambda^k$ in the expansion of $\Gamma_j(\lambda)$.

It is relatively straightforward, by the use of a small computer, to expand the several products $\Gamma_j(\lambda)$. This allows us to calculate the C_{sk}^j; computation of $\varphi_j[v]$ then follows from equation (12.39).

To compute the Banzhaf–Coleman index for v, we will have

$$\psi_j[v] = 2^{-50} \sum_{s=1}^{50} \sum_{k=270 - p_j}^{269} C_{sk}^j \tag{12.42}$$

where the C_{sk}^j are as above. Thus the Banzhaf–Coleman index is somewhat easier to calculate than the Shapley value.

The above method will give us an exact computation of both the value and the index for the game v. For the game u, unfortunately, the necessary computations are clearly impractical, as the number of players is of the order of 10^8. Let us therefore approach this from the point of view of multilinear extensions.

As before, we write

$$f_j(t, t, \ldots, t) = \text{Prob}\{270 - p_j \le Y^j \le 269\}$$

Rather than attempting an exact evaluation, we note that Y^j is approximately normal, and has, moreover, a mean

$$\mu_j = t \sum_{q \ne j} p_q = (538 - p_j)t \qquad (12.43)$$

and a variance

$$\sigma_j^2 = t(1 - t) \sum_{q \ne j} p_q^2 = (9902 - p_j^2)t(1 - t) \qquad (12.44)$$

Thus we obtain the approximation

$$f_j(t, t, \ldots, t) \approx \text{Prob}\{269.5 - p_j < Y^j < 269.5\}$$

where Y^j is a normal random variable with mean (12.43) and variance (12.44). This can also be written as

$$f_j(t, t, \ldots, t) \approx \Phi\left(\frac{269.5 - \mu}{\sigma}\right) - \Phi\left(\frac{269.5 - p - \mu}{\sigma}\right) \qquad (12.45)$$

where Φ is the cumulative normal distribution function

$$\Phi(s) = \int_{-\infty}^{s} \frac{1}{\sqrt{2\pi}} e^{-(1/2)r^2} dr$$

We can now integrate (12.7) (numerically) to obtain

$$\varphi_j[v] = \int f_j(t, t, \ldots, t) dt \qquad (12.46)$$

or evaluate it at $t = \frac{1}{2}$ to obtain

$$\psi_j[v] = f_j(\tfrac{1}{2}, \ldots, \tfrac{1}{2}) \qquad (12.47)$$

Table 12.1 gives the exact values of both φ and ψ as well as the approximations obtained by using (12.7)–(12.47) for the value and index, respectively. It may be seen that the error, in general, is very small.

We next consider the state games $w_1, \ldots, w_{51}$. For a given j, w_j is a simple majority game among m_j players, where m_j is the number of voters (potential rather than actual) in the jth state. Thus, for $S \subset M_j$, we have

$$w_j(S) = \begin{cases} 1 & \text{if } s > m_j/2 \\ 0 & \text{if } s \le m_j/2 \end{cases}$$

Table 12.1: Shapley value φ and Banzhaf index ψ for the electoral college game (1970 Census).

Electoral votes	Number of states	Exact value φ	Approximation φ'	Approximation ψ'
45	1	0.08831	0.08852	0.38694
41	1	0.07973	0.07976	0.34806
27	1	0.05096	0.05113	0.22151
26	2	0.04898	0.04915	0.21291
25	1	0.04700	0.04716	0.20435
21	1	0.03917	0.03928	0.17056
17	2	0.03147	0.03157	0.13736
14	1	0.02577	0.02586	0.11276
13	2	0.02388	0.02396	0.10462
12	3	0.02200	0.02207	0.09648
11	1	0.02013	0.02019	0.08838
10	4	0.01827	0.01833	0.08028
9	4	0.01641	0.01647	0.07221
8	2	0.01456	0.01461	0.06414
7	4	0.01272	0.01276	0.05610
6	4	0.01088	0.01092	0.04806
5	1	0.009053	0.009078	0.04004
4	9	0.007230	0.007243	0.03202
3	7	0.005412	0.005431	0.02402

It is not too difficult to see that, if g_j is the MLE of w_j, we have

$$g_j(t, t, \ldots, t) = \text{Prob}\{X > m_j/2\}$$

where X is a binomial random variable with parameters m_j and t. We can approximate this (with truly negligible error, if m_j is large) by assuming that X is normally distributed, with mean tm_j and variance $t(1-t)m_j$. Thus we will obtain

$$y_j(t) = \Phi\left(\frac{t - \frac{1}{2}}{\sqrt{t(1-t)}}\sqrt{m_j}\right) \tag{12.48}$$

Making the substitution

$$\tau = \frac{t - \frac{1}{2}}{\sqrt{t(1-t)}}$$

this becomes

$$y_j = \Phi(\tau\sqrt{m_j}) \tag{12.49}$$

and

$$\frac{dy_j}{dt} = \frac{\sqrt{m_j}}{\sqrt{2\pi}} e^{-(1/2)m_j\tau^2} \frac{d\tau}{dt} \tag{12.50}$$

In particular, for $t = \frac{1}{2}$, we will have $\tau = 0$, $d\tau/dt = 2$, and so

$$y_j\left(\frac{1}{2}\right) = \Phi(0) = \frac{1}{2} \tag{12.51}$$

$$\frac{dy_j}{dt}\left(\frac{1}{2}\right) = \frac{\sqrt{2m_j}}{\sqrt{\pi}} \tag{12.52}$$

By symmetry, it is not difficult to see, that, for each $i \in M_j$

$$g_{ji}(t, t, \ldots, t) = \frac{1}{m_j}\frac{dy_j}{dt} = \frac{1}{\sqrt{2\pi m_j}} e^{-(1/2)m_j\tau^2} \frac{d\tau}{dt} \tag{12.53}$$

and, in particular

$$g_{ji}(\tfrac{1}{2}, \tfrac{1}{2}, \ldots, \tfrac{1}{2}) = \frac{\sqrt{2}}{\sqrt{\pi m_j}} \tag{12.54}$$

Thus we have

$$\psi_i[w_j] = \frac{\sqrt{2}}{\sqrt{\pi m_j}} \tag{12.55}$$

which is an important result, and also

$$\varphi_i[w_j] = \frac{1}{m_j}$$

which is not important.

We proceed, now, to an analysis of the compound game, u. We have, by (12.24)

$$\sum_{i \in M_j} \varphi_i[u] = \int f_j(y(t)) \frac{dy_j}{dt} dt$$

which can most easily be evaluated by substituting τ, to give us

$$\sum_{i \in M_j} \varphi_i[u] = \int_{-\infty}^{\infty} \frac{\sqrt{m_j}}{\sqrt{2\pi}} f_j(y(t)) e^{-(1/2)m_j\tau^2} d\tau \tag{12.56}$$

To evaluate $f_i(y(t))$, we note that, as before

$$f_j(y(t)) = \text{Prob}\{270 - p_j \leq Y^j \leq 269\}$$

where Y^j is the sum of several independent random variables

$$Y^j = \sum_{k \neq j} Z_k$$

Each Z_k can have the values 0 and p_k with probabilities $1 - y_k$ and y_k, respectively. Thus Y^j will have the mean and variance

$$\mu(Y^j) = \sum_{k \neq j} p_k y_k \tag{12.57}$$

$$\sigma^2(Y^j) = \sum_{k \neq j} p_k^2 y_k (1 - y_k) \tag{12.58}$$

where y_k is given by (12.49). Since the number of variables Z_k is large, we can use the approximation

$$f_i(y) \approx \text{Prob}\{269.5 - p_j < Y^j < 269.5\}$$

where Y^j is a normal variable with the mean (12.57) and variance (12.58). This value can then be substituted in (12.56), or in the equivalent

$$\varphi_i[u] = \int_{-\infty}^{\infty} \frac{1}{\sqrt{2\pi m_j}} f_j(y) e^{-(1/2)m_j \tau^2} \, d\tau \tag{12.59}$$

to obtain the value for an individual voter in the jth state.

One obvious difficulty arises here, and this lies in the fact that m_j, the number of players in w_j, is not well known. In fact, m_j is the number of people who are eligible to vote in the jth state for the presidential election, and, quite apart from the fact that this number changes every day, no accurate count of this is ever taken. Census figures tell us the total population of the state, or even the total population of voting age, while voting statistics tell us the number of people who actually vote, but no figures are available on the potential voters.

If we consider the expression

$$\sum_{i \in M_j} \varphi_i[u] = \int_{-\infty}^{\infty} f_j(y) \frac{dy_i}{d\tau} \, d\tau$$

then it may be seen that this is homogeneous, of degree 0, in the vector $(m_1, \ldots, m_{51})$. For, in fact, we have

$$y_k(\tau; m_k) = \Phi(\tau \sqrt{m_k})$$

Let us suppose that all the m_k are multiplied by the same positive constant c; we will then have

$$y_k(\tau; cm_k) = y_k(\tau\sqrt{c}; m_k)$$

and, more generally

$$y(\tau; cm) = y(\tau\sqrt{c}; m)$$

so that, making the substitution

$$T = \sqrt{c}\tau$$

we have

$$\int_{-\infty}^{\infty} f_j(y(\tau; cm)) \frac{dy_j}{d\tau} d\tau = \int_{-\infty}^{\infty} f_j(y(T; m)) \frac{dy_j}{dT} dT$$

and, as the τ or T is merely a dummy variable, we see that the expression

$$\sum_{i\in M_j} \varphi_i[u]$$

is unchanged. Thus, in order to calculate this, it suffices to know, not the exact values of the m_j, but only their relative ratios. As a reasonable approximation, we shall assume that the number of potential voters is proportional to the total population, i.e.

$$m_j = cq_j$$

where q_j is the population of the state, obtained from census data, and c is a constant independent of j. The actual value of c is probably about 0.7, but is not really important, for our desire was to compute the discrepancies in the value or power index among the voters of the several states, and this will be independent of c. In any case, we can calculate the sum (12.56) by assuming $m_j = q_j$ (the total census population). Then we will have, for $l \in M_j$

$$\varphi_l[u] = \frac{1}{cq_j} \sum_{i\in M_j} \varphi_i[u]$$

and, for citizens of two different states

$$\frac{\varphi_l[u]}{\varphi_{l'}[u]} = \frac{q_{j'}}{q_j} \frac{\sum_{i\in M_j} \varphi_i[u]}{\sum_{i\in M_{j'}} \varphi_i[u]}$$

which does not depend on c.

For the Banzhaf–Coleman index, the situation is considerably simpler. We have, for each k

$$y_k\left(\tfrac{1}{2}\right) = \Phi(0) = \tfrac{1}{2}$$

and so we will have, for $i \in M_j$

$$\psi_i[u] = \psi_j[v]\psi_i[w_j]$$

or, by (12.55)

$$\psi_i[u] = \frac{\sqrt{2}}{\sqrt{\pi m_j}}\psi_j[v] \tag{12.60}$$

Since we have already computed $\psi_j[v]$ this gives us $\psi_i[u]$ directly. Once again, we note that, if we assume $m_j = cq_j$ we will have, for voters in different states,

$$\frac{\psi_i[u]}{\psi_{i'}[u]} = \frac{\sqrt{q_{j'}}}{\sqrt{q_j}}\frac{\psi_j[v]}{\psi_{j'}[v]} \tag{12.61}$$

which is also independent of c. Thus, as far as the discrepancies in value or index are concerned, we note that we can base our calculations directly on the census figures q_j, without worrying about the value of the constant c. Table 12.2 gives the results obtained in this manner for the several states, based on the 1970 population. The discrepancies are obtained by giving the value and index for an individual voter in each state, divided by the corresponding number of voters in the least advantaged state (actually, the District of Columbia). Both the value and the index suggest that a voter in California has approximately three times as much power as a voter in the District of Columbia — an observation which is doubtless of interest in the present discussion on the abolition of the Electoral College.

12.5. The Coalition Value

As we have seen, the values described above — both the Shapley value and the Banzhaf–Coleman index of power — depend only on the characteristic function of the game. This is quite proper if we wish to know, in the abstract and with no further information, what an individual player may reasonably expect to win in the game. Quite frequently, however, there is additional information available. If nothing else, the players' identities are known; in many cases some of the coalitions, while not certain, are very likely to form, while others are seemingly out of the question due to the players' previous relations with each other.

In such cases, a realistic appraisal should take these prior situations into account.

In the simplest of cases, we might imagine a partition of the set N of players into *a priori unions*

$$\mathbf{T} = \{T_1, \ldots, T_m\}$$

where the coalitions (unions) T_k are pairwise disjoint, and their union is N.

We can think of these unions as prior agreements among the several players: perhaps they are firms. The members of each of these unions (i.e., the owners,

Table 12.2: Shapley value φ and Banzhaf–Coleman index ψ for the presidential election (composed) game, for the 1970 Census[a].

State	Electoral votes	Population	Shapley value $\varphi_i(\times 10^{-9})$	Disparity $\varphi_i/\varphi_{\text{D.C.}}$	Banzhaf index $\psi_i(\times 10^{-5})$	Disparity $\psi_i/\psi_{\text{D.C.}}$
Alabama	9	3,444,165	3.4849	1.406	3.8909	1.409
Alaska	3	302,173	3.9253	1.584	4.3696	1.582
Arizona	6	1,772,482	3.2385	1.307	3.6099	1.307
Arkansas	6	1,923,295	3.1079	1.254	3.4655	1.255
California	45	19,953,134	7.8476	3.166	8.6624	3.137
Colorado	7	2,207,259	3.3861	1.366	3.7760	1.367
Connecticut	9	3,032,217	3.7171	1.500	4.1468	1.502
Delaware	3	548,104	2.9131	1.175	3.2445	1.175
District of Columbia	3	765,510	2.4783	1.000	2.7616	1.000
Florida	17	6,789,433	4.7326	1.906	5.2716	1.909
Georgia	12	4,589,575	4.0341	1.628	4.5035	1.631
Hawaii	4	769,913	3.2804	1.324	3.6524	1.323
Idaho	4	713,008	3.4073	1.375	3.7932	1.374
Illinois	26	11,113,976	5.7332	2.313	6.3865	2.313
Indiana	13	5,193,669	4.1106	1.659	4.5907	1.662
Iowa	8	2,825,041	3.4202	1.380	3.8161	1.382
Kansas	7	2,249,071	3.3542	1.353	3.7408	1.355
Kentucky	9	3,219,311	3.6062	1.455	4.0245	1.457
Louisiana	10	3,643,180	3.7686	1.521	4.2060	1.523
Maine	4	993,663	2.8860	1.164	3.2148	1.164
Maryland	10	3,922,399	3.6300	1.464	4.0535	1.468
Massachusetts	14	5,689,170	4.2333	1.708	4.7275	1.712
Michigan	21	8,875,083	5.1322	2.071	5.7252	2.073
Minnesota	10	3,805,069	3.6864	1.487	4.1155	1.490
Mississippi	7	2,216,912	3.3787	1.363	3.7678	1.364
Missouri	12	4,677,399	3.9954	1.612	4.4610	1.615

Table 12.2: *(Continued)*

State	Electoral votes	Population	Shapley value $\varphi_i(\times 10^{-9})$	Disparity $\varphi_i/\varphi_{D.C.}$	Banzhaf index $\psi_i(\times 10^{-5})$	Disparity $\psi_i/\psi_{D.C.}$
Montana	4	694,409	3.4516	1.393	3.8407	1.391
Nebraska	5	1,483,791	2.9488	1.190	3.2871	1.190
Nevada	3	488,738	3.0854	1.245	3.4359	1.244
New Hampshire	4	737,681	3.3486	1.351	3.7281	1.350
New Jersey	17	7,168,164	4.5946	1.854	5.1305	1.858
New Mexico	4	1,016,000	2.8516	1.151	3.1767	1.150
New York	41	18,190,740	7.3756	2.976	8.1607	2.955
North Carolina	13	5,082,059	4.1563	1.677	4.6408	1.680
North Dakota	3	617,761	2.7435	1.107	3.0561	1.107
Ohio	25	10,652,017	5.6191	2.267	6.2612	2.267
Oklahoma	7	2,559,253	3.1423	1.268	3.5068	1.270
Oregon	6	2,091,385	2.9793	1.202	3.3233	1.203
Pennsylvania	27	11,793,909	5.7910	2.337	6.4501	2.336
Rhode Island	4	949,723	2.9498	1.190	3.2857	1.190
South Carolina	8	2,590,516	3.5734	1.442	3.9851	1.443
South Dakota	4	666,257	3.5260	1.423	3.9250	1.421
Tennessee	10	3,924,164	3.6292	1.464	4.0526	1.467
Texas	26	11,196,730	5.7115	2.305	6.3628	2.304
Utah	4	1,059,273	2.7924	1.127	3.1111	1.127
Vermont	3	444,732	3.2362	1.306	2.6035	1.305
Virginia	12	4,648,494	4.0080	1.617	4.4749	1.620
Washington	9	3,409,169	3.5030	1.413	3.9109	1.416
West Virginia	6	1,744,237	3.2648	1.317	3.6390	1.318
Wisconsin	11	4,417,933	3.7645	1.519	4.2048	1.523
Wyoming	3	332,416	3.7423	1.510	4.1661	1.509

[a]From Owen, G. (1975). American Political Science Review, 69, 951 and Owen, G. (1975). Naval Research Log Quarterly, 22, 749.

management, and employees of the firms) have agreed to cooperate, and in fact they will do so in most cases, though they are in no way forced to do so. Thus members of each firm can always threaten to defect, and this threat must be taken into account by the remaining members. We find, then, that a two-part process takes place: on the one hand, the firms deal with each other to determine how much each one receives; on the other, the members of each firm bargain within the firm to determine how these receipts are to be divided.

Given the partition **T**, we can define a *quotient game* $u = v/\mathbf{T}$ with player set $M = \{1, 2, \ldots, m\}$, by

$$u(H) = v\left(\bigcup_{h \in H} T_h\right)$$

for $H \subseteq M$. Effectively, this is the game played by the firms; player h in this game is really firm T_h in the original game. In such a case, $u(\{h\})$ is the amount that firm T_h can earn in the game v, acting as a coalition, while $u(H)$ is the amount that all the firms T_h, $h \in H$, can earn (in v) by acting together.

12.5.1 Example. Let v be the four-person game of Example 12.1.8, where the players have 10, 20, 30, and 40 shares of stock, respectively, with 51 shares (or more) necessary for a winning coalition. Suppose players 1 and 2 agree to work together (perhaps they marry each other). In this case **T** would be the partition $\{\{1, 2\}, \{3\}, \{4\}\}$. Thus u would be a 3-player game between the "unions" $\{1, 2\}$, $\{3\}$, and $\{4\}$, holding 30, 30, and 40 shares respectively. It is easily checked that $u(H) = 1$ if H has at least two members and 0 otherwise.

12.5.2 Example. Consider a parliament of n members; if a majority (consisting of q members) is needed to pass a bill, then v will be a symmetric simple game given by

$$v(S) = \begin{cases} 1 & \text{if} \quad |S| \geq q \\ 0 & \text{if} \quad |S| < q \end{cases}$$

Suppose, now, that the parliament is divided into m parties, $T_1, \ldots, T_m$, with $p_1, \ldots, p_m$ members, respectively. In this case u is a weighted majority game $[q, p_1, \ldots, p_m]$ as in definition 10.3.9. Here, then, the quotient game is the game played among the parties *assuming some reasonably strong party discipline*. Note that, if one of the parties holds a majority (not an uncommon event in some countries), then that party is a dictator, and all other parts are dummies, in game u.

12.5.3 Example. More generally, let v be a weighted majority game $[q; p_1, \ldots, p_n]$, and let **T** be the partition $\{T_1, \ldots, T_m\}$. Then u is the weighted majority game $[q; p(T_1), \ldots, p(T_m)]$, where $p(S)$ is simply the sum $\sum_{i \in S} p_i$.

For a given set $S \subseteq N$, and given a partition **T** of N, we now define the set

$$S/\mathbf{T} = \{j | S \cap T_j \neq \phi\}$$

The following two lemmas will be useful:

12.5.4 Lemma. Let S be a carrier for the game v. Then $S/\mathbf{T}$ is a carrier for the game $u = v/\mathbf{T}$.

Proof. Let $S^* = \cup(t_j | S \cap T_j \neq \phi)$. It is clear that $S \subset S^*$, and, since S is a carrier, so is S^*.

Let $H \subset M$, and let $K = \cup_{j \in H} T_j$. We know $v(K) = v(K \cap S^*)$, and so $u(H) = v(K) = v(K \cap S^*) = u(H \cap S/\mathbf{T})$ Thus $S/\mathbf{T}$ is a carrier for u.

12.5.5 Lemma. Let w_s be the unanimity game with carrier S, as defined in Lemma 12.1.5. Then the quotient game $w_S/\mathbf{T}$ is the same as the game $w_{S/T}$.

We can now look for a modified value, which will be a mapping from the space of all pairs $\langle v, \mathbf{T} \rangle$, where v is a game with player set N, and $\mathbf{T}$ is a partition of N, into the space of n-tuples of real numbers (payoff vectors). Let us call this mapping Ω. As in Section 12.1, we will approach this axiomatically. The following axioms seem reasonable:

12.5.6 Axioms. The coalition value, Ω, satisfies:

C1. If K is any carrier of v, then

$$\sum_{i \in K} \Omega_i[v, \mathbf{T}] = v(K)$$

C2. For any $T_j \in \mathbf{T}$, the quantity $\sum_{i \in T_j} \Omega_i[v, \mathbf{T}]$ depends only on $v/\mathbf{T}$.
C3. Let σ be any permutation of the set $\{1, \ldots, m\}$. Then for all $i \in N$

$$\Omega_i[v; \{T_1, \ldots, T_m\}] = \Omega_i[v; \{T_{\sigma(1)}, \ldots, T_{\sigma(m)}\}]$$

C4. Let π be any permutation of N which leaves the sets $T_1, \ldots, T_m$ invariant. Then

$$\Omega_{\pi(i)}[\pi^* v; \mathbf{T}] = \Omega_i[v; \mathbf{T}]$$

(where $\pi^* v$ is an in 12.2).
C5. For any two games v and w, and any partition $\mathbf{T}$

$$\Omega[v + w; \mathbf{T}] = \Omega[v; \mathbf{T}] + \Omega[w; \mathbf{T}]$$

We can, perhaps, best understand these axioms by comparison with Shapley's axioms, A1–A3, as given in 12.2. C1 is, of course, Axiom A1. C4 and C5 are restricted version of A2 and A3, saying in effect that symmetry of treatment applies only within a given union (e.g., two equivalent workers will be paid quite different salaries if one works for a profitable firm while another works for a struggling small business), and that the value is additive only if the same coalition structure is kept throughout.

Axiom C2 says, more or less, that the unions (members of **T**) do, in effect, play the game $v/\mathbf{T}$ among themselves. C3 is a different symmetry axiom, and says that the outcome of the game depends on the unions formed, but not on the labels given them.

As before, we are able to give a neat existence/uniqueness theorem:

12.5.7 Theorem. There exists a unique mapping satisfying Axioms C1–C5. This mapping is given by, for $i \in T_j \in \mathbf{T}$

$$\Omega_i[v; \mathbf{T}] = \sum_{\substack{H \subset M \\ j \notin H}} \sum_{\substack{S \subset T_j \\ i \notin S}} \frac{h!(m-h-1)!s!(t_j-s-1)!}{m!t_j!}$$

$$\times [v(Q \cup S \cup \{i\}) - v(Q \cup S)]$$

(12.62)

where h, s, and t_j are the cardinalities of H, S, and T_j, respectively, and where $Q = \cup_{k \in H} T_k$.

Proof. As with the Shapley value, the proof of uniqueness is based on the unanimity games w_s, $S \subset N$, which, as was seen in 12.1, form a basis (in the linear algebraic sense) for the space of all games with player set N.

For the game w_s, we know S is a carrier. Hence by Axiom C1

$$\sum_{i \in S} \Omega_i[w_S; \mathbf{T}] = 1$$

and by the same argument as in 12.1, $\Omega_i[w_S; \mathbf{T}] = 0$ for all $i \notin S$.

Let j and k be two elements of $S/\mathbf{T}$, and let σ be that permutation of M which interchanges j and k and leaves all other elements of M unchanged. By Lemma 12.5.3, $w_S/\mathbf{T} = w_{S/\mathbf{T}}$, and so $\sigma^*(w_S/\mathbf{T}) = w_S/\mathbf{T}$. By Axiom C2, it follows that

$$\sum_{i \in T_j} \Omega_i[w_S; \mathbf{T}] = \sum_{i \in T_k} \Omega_i[w_S; \mathbf{T}]$$

Let $h(S)$ be the cardinality of $S/\mathbf{T}$, i.e., the number of j such that $T_j \cap S \neq \phi$. Since all such T_j receive equal shares, and all together receive one unit, we must have

$$\sum_{i \in T_j} \Omega_i[w_S; \mathbf{T}] = \frac{1}{h(S)} \quad \text{if} \quad j \in S/\mathbf{T}$$

Let i and l be two elements of $S \cap T_j$ for some j. Let π be the permutation of N which interchanges i and l and leaves all other elements fixed. Clearly $\pi^* v = v$, and otherwise π satisfies the conditions of Axiom C4. Hence $\Omega_i[w_S; T] = \Omega_l[w_S; T]$.

Let s_j, now, be the number of elements of $S \cap T_j$. Since all players receive the same amount, and together they receive $1/h(S)$, we find, finally

$$\Omega_i[w_S; \mathbf{T}] = \begin{cases} 1/s_j h(S) & \text{if} \quad i \in S \cap Tj, \ T_j \in \mathbf{T} \\ 0 & \text{if} \qquad i \notin S \end{cases}$$

We see, then, that there is at most one value satisfying Axioms C1–C4 on the unanimity games w_S. Since these games form a basis for the game space, then by Axiom C5 there is at most one value on the space of all these pairs $\langle v, \mathbf{T} \rangle$.

To prove existence, it suffices to show that the value given by equation (12.5.1) satisfies the axioms. We leave this straightforward exercise to the reader.

12.5.8 Example. We consider once again the game of Example 12.5.1, using once again the partition $\{\{1, 2), \{3\}, \{4\}\}$. As we saw there, the quotient game $v/\mathbf{T}$ is a symmetric three-person game. As $v(N) = 1$, it follows that each of the three unions receives $1/3$ unit. Thus individuals 3 and 4 each receive $1/3$; individuals 1 and 2, together, receive $1/3$ unit. The question is how 1 and 2 will divide their share.

To compute Ω_1, we note that, in sum 12.5.1, there will be 8 summands, but of these, only the one corresponding to $S = \{2\}$, $H = \{2\}$ will be different from zero. For this term we have $m = 3$, $t_j = 2$, $s = 1$, $h = 1$, and so the nonzero term is

$$\frac{1!0!1!1!}{2!3!}[v((\{1, 2, 3\}) - v(\{2, 3\})]$$

We see thus that $\Omega_1 = 1/12$, and so $\Omega_2 = 1/3 - 1/12 = 1/4$. Thus we have the value for this particular game and partition: it is the vector $(1/12, 1/4, 1/3, 1/3)$. We may compare this with the value found in 12.1.8 to see how the coalition formation has affected the value. In fact, the two members of the new coalition, $\{1, 2\}$, see no change in their expectations; rather, it is player 3 who gains at the expense of player 4. The reason for this is that, in the original game, 4 could hope to form the winning coalition $\{2, 4\}$; now, he cannot hope to win with 2's help unless 1 also collaborates.

An alternative approach is here possible. We can think of the several coalitions $T_1, \ldots, T_m$ as forming; if they are familiar with the arguments for the Shapley value (and accept them as valid), then each of them feels that it can "count on obtaining" the amount $\Phi_j[v/\mathbf{T}]$, its Shapley value in the quotient game. Each of these unions now convenes to determine how it should divide its share. We may think of the members of one of these unions, say T_j, as playing a *reduced game*, w_j, in which each coalition $S \subset T_j$ bargains with its partners $T_j - S$ by considering what it can do *with no help from $T_j - s$ but with some reasonable degree of cooperation from the remaining players, $N - T_j$.*

In that case, S may think of itself as playing an m-player game, in which S is one player, and the remaining $m - 1$ unions $T_k(k \neq j)$ are the others players. If $u = v/\mathbf{T}$ is the quotient game, we then introduce the $T_j : S$ *altered game* (with player set M), $u_{j,s}$, by

$$u_{j,\,s}(K) = \begin{cases} v\left(\bigcup_{k \in H} T_k\right) & \text{if } j \notin H \\ v\left(\bigcup_{k \in H-j} T_k \cup S\right) & \text{if } j \in H \end{cases} \qquad (12.63)$$

In this altered game, player $k(k \neq j)$ is actually union T_k; player j, however, is the coalition S.

12.5.9 Example. As in Example 12.5.3, let v be the weighted voting game $[q; p_1, \ldots, p_n]$. Then the quotient game u is the voting game $[q; p(T_1), \ldots, p(T_m)]$. The $T_j : S$ altered game is the weighted voting game $[q; p(T_1), \ldots, p(S), \ldots, p(T_m)]$. Note that q is the same for all three games. The difference between the game u and the altered game $u_{j,S}$ is that $p(T_j)$ is replaced by $p(S)$.

What amount, then, can S reasonably expect to obtain by playing this game with $N - T_j$? This depends, naturally, on the type of solution concept used. Assume, however, that the Shapley value is accepted as reasonable. In this case, S may feel that it can hope for the amount $\Phi_j[u_{j,S}]$, the Shapley value (to player j) of this altered game. Thus, in bargaining within the union T_j, coalition S will feel that it can count on this amount.

We define, therefore, the reduced game, w_j, with player set T_j: it is given by

$$w_j(S) = \Phi_j[u_{j,S}] \quad \text{for} \quad S \subset T_j \tag{12.64}$$

The members of T_j, then, will bargain among themselves, feeling that they are playing the game w_j. In such a case, player $i \in T_j$ can reasonably expect to gain the amount $\Phi_i[w_j]$. We shall call this the two-step value.

12.5.10 Example. We consider once again the game of Example 12.5.1, with the unions $\{1, 2\}, \{3\}, \{4\}$. To compute the reduced game w_1, we note that (as mentioned above), the reduced game u is a three-person game in which the three players have 30, 30, and 40 shares respectively, with 51 shares needed for a winning coalition: Its value is $\Phi[u] = (1/3, 1/3, 1/3)$. This is the same as the altered game $u_{1,\{1,2\}}$. Thus $w_1(\{1, 2\}) = \Phi_1[u] = 1/3$.

Next, consider $S = \{1\}$. The altered game $u_{1,\{1\}}$ is a three-player game in which the three players have 10, 30, and 40 shares respectively. A winning coalition is still any coalition with 51 or more shares. It may be seen that player 1 is a dummy in this game; its value is $\Phi[u_{1,\{1\}}] = (0, 1/2, 1/2)$. Thus $w_1(\{1\}) = 0$.

Next, let $S = \{2\}$. The altered game is a 3-player game with 20, 30, and 40 shares respectively. Since 51 shares are need to win, we see that the last player has a veto and can win with the help of either one of the others. The value here is $\Phi[u_{1,\{2\}}] = (1/6, 1/6, 2/3)$. Hence $w_1(\{2\}) = 1/6$.

Finally, let $S = \phi$. In the altered game $u_{1,\phi}$, the players have 0, 30, and 40 shares. Clearly, 1 is a dummy. The value is, once again, $\Phi[u_{1,\phi}] = (0, 1/2, 1/2)$. Therefore $w_1(\phi) = 0$.

We have thus obtained the reduced game w_1. It is given by

$$w_1(\phi) = 0; \quad w_1(\{1\}) = 0; \quad w_1(\{2\}) = 1/6; \quad w_1(\{1, 2\}) = 1/3$$

It is easy, now, to calculate the Shapley value of this game; player 2 receives $1/6$, with the remaining $1/6$ split evenly between the two players. Thus $\Phi[w_1] = (1/12, 1/4)$.

The other two "unions" in this example are single-player sets. The reduced games will be 1-player games; the two players, 3, and 4, receive the whole amount that their "unions," T_2 and T_3, respectively, receive in the quotient game u. Thus the two-step value here is $(1/12, 1/4, 1/3, 1/3)$.

It may be seen that in the example above, the reduced game w_1 is actually a game, in the sense that it satisfies the conditions (10.1–10.2). More interesting still, the two-step value coincides with the coalitional value Ω. The obvious question is as to whether this is always so.

12.5.11 Theorem. The reduced game function, w_j, defined by (12.64), satisfies condition (10.1) but not necessarily (10.2).

Proof. To see that $w_j(\phi) = 0$, we note that, by (12.63), player j is a dummy in game $u_{j, \phi}$. It follows that $\Phi_j[u_{j, \phi}] = 0$. Hence $w_j(\phi) = 0$.

12.5.12 Example. To see that the reduced game need not be super-additive, consider the voting game [4; 3, 1, 1, 1], and let **T** be the partition $\{\{1\}, \{2\}, \{3, 4\}\}$. Then the quotient game u is the voting game [4; 3, 1, 2], with value $\Phi[u] = (2/3, 1/6, 1/6)$. The altered quotient game $u_{3,\{3\}}$ is the voting game [4; 3, 1, 1], as is $u_{3, \{4\}}$. This game also has value $(2/3, 1/6, 1/6)$. Finally, $u_{3, \phi}$ is the voting game [4; 3, 1, 0], with value $(1/2, 1/2, 0)$. Thus the reduced game w_3 is given by $w_3(\phi) = 0$, $w_3(\{3\}) = w_3(\{4\}) = w_3(\{3,4\}) = 1/6$. This reduced game, which is not superadditive, has value $\Phi[w_3] = (1/12, 1/12)$. Thus the two-step value is $\Omega = (2/3, 1/6, 1/12, 1/12)$.

Functions such as that of Example 12.5.11, which satisfy (10.1) but not (10.2), are not, strictly speaking, games. Nevertheless, they will frequently appear in reductions such as this, much in the way that imaginary numbers appear when solving algebraic equations. We will call such functions *improper* games. It may be seen that most of our solution concepts (in particular the value) do not require superadditivity for their validity. Thus there is no problem in taking the value of such improper games — as is, in fact, done when we calculate the two-step value. The value of the reduced game w_3 in this example is $(1/12, 1/12)$, which is not even an imputation — but Shapley's axioms do not require the value to be an imputation. (Of course, the value $\Phi[v]$ will be an imputation *whenever v is a proper game* — but that is simply a consequence of the axioms.)

In general, then, even for improper games, the value can be defined, and is in fact given by the formula (12.4).

With this in mind, we can now state the principal result of this section:

12.5.13 Theorem. On the space of all pairs $\langle v, \mathbf{T} \rangle$, the two-step value coincides with the coalitional value Ω.

We omit the proof, as it is quite straightforward and consists merely in substituting the reduced game into the formula (12.4).

12.5.14 Example. Let v be the weighted voting game [3; 2, 1, 1, 1], and let $\mathbf{T} = \{\{1, 2\}, \{3\}, \{4\}\}$. The quotient game u is the voting game [3; 3, 1, 1], and player 1 (i.e., union $\{1, 2\}$) is a dictator here, so that $\Phi[u] = (1, 0, 0)$.

The altered game $u_{1, \{1\}}$ is the voting game $[3; 2, 1, 1]$, which has value $(2/3, 1/3, 1/3)$, while the altered game $u_{1, \{2\}}$ the voting game $[3; 1, 1, 1]$, with value $(1/3, 1/3, 1/3)$. Thus the reduced game w_1 is given by $w_1(\{1\}) = 2/3$, $w_1(\{2\}) = 1/3$, $w_1(\{1, 2\}) = 1$. This game has value $(2/3, 1/3)$. The two-step value is therefore $(2/3, 1/3, 0, 0)$. It will follow that this is also the coalitional value Ω.

Problems

1. If v is a constant-sum game, then the Shapley value φ is given by

$$\varphi_i[v] = 2 \sum_{\substack{S \subseteq N \\ i \in S}} \left[\frac{(n-s)!(s-1)!}{n!} v(S) \right] - v(N)$$

2. A game v is said to be convex if, for every S and T

$$v(S \cup T) + v(S \cap T) \geq v(S) + v(T)$$

Prove that a convex game has a nonempty core, and, in fact

$$\phi[v] \in C(v)$$

3. Let $w = (w_1, w_2, \ldots, w_n)$ be a nonnegative vector, and define a function v by

$$v(S) = \left(\sum_{i \in S} w_i \right)^2$$

   (a) Show that v is a convex game (see Problem 2).
   (b) Find the MLE of the game v. (*Hint*: Use the probabilistic interpretation.)
   (c) Use the MLE to compute both $\phi[v]$ and $\psi[v]$.

4. Compute the Shapley value and the Banzhaf–Coleman index for the six-person weighted majority game $[8; 4, 3, 3, 2, 2, 1]$.
5. Let

$$\bar{\beta}[v] = 2^{n-1} \sum_{i \in N} \psi_i[v]$$

be the generalization, to arbitrary games, of the number of swings in the game v. (This was defined originally only for simple games.) Show that, if $0 \leq v(S) \leq 1$, for all s, then

$$\bar{\beta}[v] \leq m \binom{n}{m}$$

where $m = [n/2] + 1$, i.e., m is the smallest integer larger than $n/2$.

(a) Prove first that

$$\bar{\beta}[v] = \frac{d}{dt}f(t, t, \ldots, t)$$

where f is the MLE of v.

(b) Show that $\bar{\beta}[v]$ is maximized if $v(S) = 0$ whenever $s \le n/2$, and $v(S) = 1$ whenever $s > n/2$.

(c) Prove that, in this maximizing case, equality will be obtained. (Compute $\psi_i[v]$ directly in this case.)

6. Compute both $\phi[u]$ and $\psi[u]$, where u is the compound simple game

$$u = v[w_1, w_2, w_3]$$

and where v is a three-person simple game in which the winning coalitions are $\{1, 2\}$, $\{1, 3\}$ and $\{1, 2, 3\}$; w_1 is a game with carrier $\{1, 2\}$ in which $\{1, 2\}$ is the only winning coalition; w_2 is a three-person game with carrier $\{3, 4, 5\}$ in which any coalition of two or more players wins; and w_3 is a five-person game in which any coalition of four or more players wins.

7. Prove that the degree of any constant-sum game is even. (*Hint*: Assume $v(N) = 1$, let f be the MLE of v, and consider the polynomial

$$g(s_1 s_2, \ldots, s_n) = f(s_1 + \tfrac{1}{2}, s_2 + \tfrac{1}{2}, \ldots, s_n + \tfrac{1}{2}) - \tfrac{1}{2}$$

Show that g is an even function.)

8. Show that for any real t, the value

$$\gamma[v] = t\rho[v] + (1 - t)\mu[v]$$

where ρ and μ are the dictatorial and marginal indices, respectively, satisfies the axioms B1–B5. (*Hint*: B1–B4 are easy to verify. For B5, prove that ρ and μ, and hence γ as well, coincide for constant-sum games. This will simplify the proof.)

9. For the game of Problem 4, above, compute the coalitional value given the coalition structures

(a) $\{\{1, 2\}, \{3, 4\}, \{5\}, \{6\}\}$
(b) $\{\{1\}, \{2\}, \{3, 4, 5, 6\}\}$
(c) $\{\{1, 2\}, \{3, 6\}, \{4\}, \{5\}\}$

10. Prove Theorem 12.5.13. (*Hint*: Essentially, what is needed is to obtain a formula for the characteristic function of the reduced game — which is of course obtained by the use of formula (12.4). Once this is done, a second application of (12.4) will give the desired result.)

Chapter 13

The Bargaining Set and Related Concepts

13.1. The Bargaining Set

One of the difficulties with the concepts that we have previously studied is that, in general, they do not seem to explain what happens in any play of the game. Stable sets are given as "standards of behavior"; the Shapley value gives some type of expectation. The *bargaining set* is obtained by considering the discussion that may actually take place during a play of the game. Thus we shall consider the possible threats and counterthreats made by the several players.

13.1.1 Definition. By a *coalition structure* for an *n*-person game we mean a partition

$$\mathcal{T} = \{T_1, T_2, \ldots, T_m\}$$

of *N*.

Heuristically, a coalition structure represents the "breaking up" of the set *N* into mutually disjoint coalitions. Suppose such a structure is reached. We assume that each of the coalitions T_k that forms will receive the amount $v(T_k)$ to be divided among its members. But how will this amount be divided?

13.1.2 Definition. A *payoff configuration* is a pair

$$(x; \mathcal{T}) = (x_1, \ldots, x_n; T_1, \ldots, T_m)$$

where $\mathcal{T}$ is a coalition structure and *x* is an *n*-vector satisfying

$$\sum_{i \in T_k} x_i = v(T_k) \tag{13.1}$$

for $k = 1, \ldots, m$.

We wish, then, to know which payoff configurations will be reached. An obvious requirement will be that of individual rationality: i.e., we may demand

$$x_i \geq v(\{i\}) \quad \text{for all} \quad i \in N \tag{13.2}$$

A further possible requirement may be that no coalition T_k will form if one of its subcoalitions can obtain more than the payoff vector *x* gives it. Thus

$$\sum_{i \in S} x_i \geq v(S) \quad \text{for} \quad S \subset T_k \in \mathcal{T} \tag{13.3}$$

It is clear, of course, that (13.3) is stronger than (13.2). On the other hand, the necessity for (13.3) is not as clear as that for (13.2). Without going into the pros and cons of (13.3), we shall merely say here that a payoff configuration satisfying (13.3) will be called *coalitionally rational* (c.r.p.c.), whereas if it satisfies only (13.2) it will be called *individually rational* (i.r.p.c.). We shall deal, first of all, with c.r.p.c.'s.

Suppose, then, that some c.r.p.c. has been reached. When will this be stable? It is clear that if the c.r.p.c. is an imputation in the core of the game, it will be stable in the sense that no coalition has both the power and the inclination to change it. Since, however, the core is often empty, it follows that we cannot be as demanding as this.

Consider, e.g., the symmetric three-person game previously treated. If the coalition $\{1,2\}$ forms, we say that the payoff vector $\left(\frac{1}{2},\frac{1}{2},0\right)$ is quite likely. Nevertheless, either of the two players in the coalition might demand more of the other, and threaten to join player 3 if his demand is not met. Thus player 1 might threaten to form the coalition $\{1,3\}$, obtaining, say, the payoff $\left(\frac{3}{4},0,\frac{1}{4}\right)$. Player 2, however, can counter this threat by pointing out that, in such a case, he will try to form the coalition $\{2,3\}$, offering 3 the amount $\frac{1}{2}$, to obtain the vector $\left(0,\frac{1}{2},\frac{1}{2}\right)$. Thus player 2 has a counterthreat that "protects" his share of $\frac{1}{2}$.

To formulate this idea mathematically, we begin by defining the idea of partners.

13.1.3 Definition. Let $\mathcal{T} = \{T_1,\ldots,T_m\}$ be a coalition structure, and let K be a coalition. Then, by *the partners* of K in $\mathcal{T}$, we mean the set

$$P(K;\mathcal{T}) = \{i | i \in T_k, T_k \cap K \neq \varnothing\}$$

Thus the player i is a partner of K in $\mathcal{T}$ if he belongs to the same coalition T_k as some member of K. (Note that each member of K is also a partner of K.) The idea is that, for the members of K to get their share of the c.r.p.c. $(x;\mathcal{T})$, they need the consent of their partners and nothing else.

13.1.4 Definition. Let $(x;\mathcal{T})$ be a c.r.p.c. for a game v. Let K and L be nonempty disjoint subsets of some $T_k \in \mathcal{T}$. Then an *objection* of K against L is a c.r.p.c. $(y;\mathcal{U})$, satisfying

$$P(K;\mathcal{U}) \cap L = \varnothing \tag{13.4}$$

$$y_i > x_i \quad \text{for all} \quad i \in K \tag{13.5}$$

$$y_i \geq x_i \quad \text{for all} \quad i \in P(K;\mathcal{U}) \tag{13.6}$$

13.1.5 Definition. Let $(x;\mathcal{T})$ be a c.r.p.c., let K, L be as in Definition 11.1.4, and let $(y;\mathcal{U})$ be an objection of K against L. Then a *counterobjection* of L against K is a c.r.p.c. $(z;\mathcal{V})$ satisfying

$$K \not\subset P(L;\mathcal{V}) \tag{13.7}$$

$$z_i \geq x_i \quad \text{for} \quad i \in P(L;\mathcal{V}) \tag{13.8}$$

$$z_i \geq y_i \quad \text{for} \quad i \in P(L;\mathcal{V}) \cap P(K;\mathcal{U}) \tag{13.9}$$

Briefly, the members of K, in their objection against L, claim that they can obtain more by changing to a new c.r.p.c., and that their new partners will agree to this. The members of L can counterobject if it is possible for them to find a third c.r.p.c. in which they and all their partners receive at least their original shares. If they need some of K's partners for this, they give these players at least as much as in the objection c.r.p.c. Note that it may be necessary for L to use some members of K as partners; they may not, however, use them all.

13.1.6 Definition. A c.r.p.c. $(x; \mathcal{T})$ is called *stable* if for every objection of a K against an L, L has a counterobjection. The *bargaining set* $\mathcal{M}$ is the set of all stable c.r.p.c.'s.

Now, the form of the bargaining set is open to some question. Certain modifications might, indeed, be made. For example, we might insist that only single players be allowed to make objections, or that it should be sufficient for one member of L to possess a counterobjection.

13.1.7 Definition. The bargaining set $\mathcal{M}_1$ is the set of all c.r.p.c.'s $(x; \mathcal{T})$ such that, whenever any set K has an objection against a set L, at least one member of L has a counterobjection.

13.1.8 Definition. The bargaining set $\mathcal{M}_2$ is the set of all c.r.p.c.'s $(x; \mathcal{T})$ such that, if any single player, i, has an objection against a set L, then L has a counterobjection against i.

It is easy to see that $\mathcal{M} \subset \mathcal{M}_1$ and $\mathcal{M} \subset \mathcal{M}_2$. The relationship between $\mathcal{M}_1$ and $\mathcal{M}_2$ is not clear.

One final variation might be suggested for the bargaining set. It is that, instead of c.r.p.c.'s, we deal with i.r.p.c.'s. This will give rise to the three sets, $\mathcal{M}^{(i)}$, $\mathcal{M}_1^{(i)}$, $\mathcal{M}_2^{(i)}$, derived from $\mathcal{M}$, $\mathcal{M}_1$, $\mathcal{M}_2$, respectively.

In contrast with the core, which is so often empty, it may be seen that none of the bargaining sets mentioned above is empty. In fact (assuming the game to be in $(0,1)$ normalization) the c.r.p.c. $(x; \mathcal{T})$, where $x = (0,0,\ldots,0)$ and $\mathcal{T} = \{\{1\}, \{2\}, \ldots, \{n\}\}$, is obviously stable. On the other hand, it represents complete failure to cooperate. Thus we would like to find elements in the bargaining set for some of the more interesting coalition structures. We have, in fact, the following valuable result:

13.1.9 Theorem. Let v be an n-person game, and let $\mathcal{T}$ be any coalition structure. Then there is at least one vector x such that $(x; \mathcal{T}) \in \mathcal{M}_1^{(i)}$.

We shall assume the game is in $(0,1)$ normalization. For a given coalition structure, $\mathcal{T}$, we shall let $X(\mathcal{T})$ denote the set of all vectors x such that $(x; \mathcal{T})$ is an i.r.p.c. The following lemma is that of B. Peleg:

13.1.10 Lemma. Let $c_1(x), c_2(x), \ldots, c_n(x)$ be nonnegative continuous real-valued functions defined for $x \in X(\mathcal{T})$. If, for each $x \in X(\mathcal{T})$, and each $S_j \in \mathcal{T}$, there is a player $i \in S_j$ such that $c_i(x) \geq x_i$ then there exists a point $\xi = (\xi_1, \ldots, \xi_n) \in X(\mathcal{T})$ such that $c_i(\xi) \geq \xi_i$ for $i = 1, 2, \ldots, n$.

Proof of Lemma. For $x \in X(\mathcal{T})$, and $i \in N$, we set

$$d_i = \begin{cases} x_i - c_i(x) & \text{if} \quad x_i \geq c_i(x) \\ 0 & \text{if} \quad x_i \leq c_i(x) \end{cases}$$

and, if $i \in S_j$

$$y_i = x_i - d_i + \frac{1}{s_j} \sum_{k \in S_j} d_k$$

where s_j is the number of elements in S_j.

It is clear that y is a continuous function of x. Moreover, we can see that $y_i \geq 0$ as x_i, d_i and $c_i(s)$ are all non-negative. Finally, we have

$$\sum_{i \in S_j} y_i = \sum_{i \in S_j} x_i = v(S_j)$$

and so $y \in X(\mathcal{T})$.

Suppose now that $x_i > c_i(x)$. This means $d_i > 0$. However, the hypothesis is that, for some $k \in S_j$, $x_k \leq c_k(x)$, and so $d_k = 0$. Hence

$$y_k \geq x_k + (d_i/s_j) > x_k$$

and so x is not fixed under this mapping. As $X(\mathcal{T})$ is convex, however, the Brouwer fixed point theorem applies, and so there must be some ξ such that $y(\xi) = \xi$. We have just seen that this ξ must satisfy $\xi_i \leq c_i(\xi)$ for all $i = 1, 2, \ldots, n$.

13.1.11 Definition. We shall say a player i is *stronger* than k in $(x; \mathcal{T})$ if i has an objection against k, which cannot be countered. We denote this as $i \gg k$. We say i and k are *equal*, denoted $i \sim k$, if neither $i \gg k$ nor $k \gg i$.

13.1.12 Definition. Let $(x; \mathcal{T})$ be an i.r.p.c., and let C be a coalition. Then the *excess* of C is $e(C) = v(C) - \sum_{i \in C} x_i$.

13.1.13 Lemma. Let $(x; \mathcal{T})$ be an i.r.p.c. Then the relation $\gg$ is acyclic.

Proof. It is clear that if i and k are in different coalitions, then $i \sim k$. Suppose, then, that a coalition, S_j, contains the players $1, 2, \ldots, t$, and that $1 \gg 2 \gg 3 \gg \cdots \gg t \gg 1$. Thus player $i(i = 1, \ldots, t)$ has an objection, which cannot be countered, against player $i + 1 \pmod{t}$ through the coalition C_i. Let C_{i_0} be that coalition (among $C_1, C_2, \ldots, C_t$) which has maximal excess. Then we claim that i_0 can counterobject against $i_0 - 1 \pmod{t}$ through the coalition C_{i_0}. Indeed, $i_0 - 1 \pmod{t}$ has only the amount $e(C_{i_0 - 1})$ at his disposal to form the objecting coalition; player i_0, having the amount $e(C_{i_0}) \geq e(C_{i_0 - 1})$ at his disposal, can always counterobject unless $i_0 - 1 \pmod{t} \in C_{i_0}$. Repeating this argument, we must have $i_0 - 2 \pmod{t} \in C_{i_0}$, etc. and eventually $i_0 + 1 \pmod{t} \in C_{i_0}$. But this is obviously impossible.

13.1.14 Proof of Theorem 13.1.9. Let $(x; \mathcal{T})$ be an i.r.p.c. We shall denote by $(y^{S_j}, x^{N-S_j}; \mathcal{T})$ the i.r.p.c., which is obtained by keeping x_i fixed for $i \in N - S_j$, and replacing x_k by y_k for $k \in S_j$, where $y_k \geq 0$ and $\sum_{k \in S_j} y_k = v(S_j)$.

Let $E_j^i(x)$ be the set of points y^{S_j} such that, in the i.r.p.c. $(y^{S_j}, x^{N-S_j}; T)$, player $i(i \in S_j)$ is *not weaker* than any other player. The set $E_j^i(x)$ is closed and contains the set $x_i = 0$ (since, if $x_i = 0$, i can always counterobject as a one-player coalition). We then define the function

$$c_i(x) = x_i + \max_{y^{S_j} \in E_j^i(x)} \min_{k \in S_j} (x_k - y_k)$$

where S_j is the coalition in T, which contains i. Now, $c_i(x)$ is easily seen to be continuous; moreover, it is non-negative, as can be seen by showing that if $y_i = 0$, and $y_k \geq x_k$ for all $k \in S_j$, $k \neq i$, then $y^{S_j} \in E_j^i(x)$.

By Lemma 13.1.13, for any $x \in X(\mathcal{T})$ and any $S_j \in \mathcal{T}$, there is some $i \in S_j$ such that i is *not weaker* than any $k \in S_j$. Thus, $x^{S_j} \in E_j^i(x)$, and so $c_i(x) \geq x_i$. Hence, by Lemma 13.1.10, there is some ξ such that $c_i(\xi) \geq \xi_i$ for all $i \in N$.

Now it is clear from the fact that

$$\sum_{k \in S_j} x_k = \sum_{k \in S_j} y_k$$

that $c_i(x) \leq x_i$ for all i. Hence, we actually have $c_i(\xi) = \xi_i$ for all i. But this means that there is some $y \in E_j^i(\xi)$, for each i, such that $y_k = \xi_k$. But this in turn means $\xi^{S_j} \in E_j^i(\xi)$ for each i and j. Thus, in $(\xi; \mathcal{T})$, *no player is stronger than another*. This means, of course, that $(\xi; \mathcal{T}) \in \mathcal{M}_1^{(i)}$.

13.1.15 Example. Consider the five-person simple game in $(0,1)$ normalization in which the minimal winning coalitions are $\{1, 2\}, \{1, 3\}, \{1, 4\}, \{1, 5\}$, and $\{2, 3, 4, 5\}$. We shall consider two types of coalition structures: one in which $\{1, 2\}$ forms, and a second one in which $\{2, 3, 4, 5\}$ forms.

Suppose that $\{1, 2\}$ forms. It does not matter what the other players do, but we shall assume they do not form any coalition. Hence $\mathcal{T} = \{\{1,2\}, \{3\}, \{4\}, \{5\}\}$. The i.r.p.c.'s. will consist of all $(x; \mathcal{T})$ where x is a vector with $x_1 + x_2 = 1$, $x_3 = x_4 = x_5 = 0$, $x_1 \geq 0$, $x_2 \geq 0$. Now, if $x_1 > \frac{3}{4}$, we must have $x_2 < \frac{1}{4}$, and so player 2 can object with the imputation $(0, \frac{1}{4}, \frac{1}{4}, \frac{1}{4}, \frac{1}{4})$. It is easy to see that 1 will have no counterobjection. If, on the other hand, $x_1 \leq \frac{3}{4}$, we find that any objection of 2 will give one of his partners an amount less than $\frac{1}{4}$. Say $x_3 < \frac{1}{4}$. Then 1 can counterobject with $(\frac{3}{4}, 0, \frac{1}{4}, 0, 0)$. Similarly, if $x_1 < \frac{1}{2}$, then 1 can object with, say, $(\frac{1}{2}, 0, \frac{1}{2}, 0, 0)$ and 2 has no counterobjection. But if $x_1 \geq \frac{1}{2}$, then any objection y by 1 will give the players 3, 4, and 5 a total of less than $\frac{1}{2}$. Thus 2 can counterobject with $z = (0, \frac{1}{2}, y_3 + \varepsilon_3, y_4 + \varepsilon_4, y_5 + \varepsilon_5)$. Thus, for the given $\mathcal{T}$, $(x; \mathcal{T})$ will belong to $\mathcal{M}_1^{(i)}$ and only if x satisfies

$$x_1 + x_2 = 1 \tag{13.10}$$

$$x_1 \geq \tfrac{1}{2} \tag{13.11}$$

$$x_2 \geq \tfrac{1}{4} \tag{13.12}$$

$$x_3 = x_4 = x_5 = 0 \tag{13.13}$$

By symmetry, similar results will hold if the coalitions $\{1, 3\}$, $\{1, 4\}$, or $\{1, 5\}$ form.

Suppose, instead that the coalition $\{2, 3, 4, 5\}$ forms. We must consider all i.r.p.c.'s $(x; \mathcal{T})$ where x is a non-negative vector of sum 1 with $x_1 = 0$, and $\mathcal{T} = \{\{1\}, \{2, 3, 4, 5\}\}$. Suppose we have $x_2 > x_3$. Then 3 can object against 2 with the imputation $(1 - x_3 - \varepsilon, 0, x_3 + \varepsilon, 0, 0)$ where $0 < \varepsilon < x_2 - x_3$. It is easy to see that 2 has no counter-objection. By symmetry, it follows that the only x such that $(x; \mathcal{T}) \in \mathcal{M}_1^{(i)}$ for this $\mathcal{T}$ is

$$x_1 = 0 \tag{13.14}$$
$$x_2 = x_3 = x_4 = x_5 = \tfrac{1}{4} \tag{13.15}$$

13.2.   The Kernel

Closely related to the bargaining set is the concept of the kernel. This is mainly based on two ideas: those of *excess* and *surplus*.

13.2.1 Definition. For the *n*-person game v, let S be a coalition and $x = (x_1, \ldots, x_n)$ a payoff vector (not necessarily an imputation). Then the excess of S with respect to x is

$$e(S, x) = v(S) - \sum_{i \in S} x_i$$

13.2.2 Definition. For the *n*-person game v, let $i \neq j$ be players, and $x = (x_1, \ldots, x_n)$. Then the *surplus of i against j* is

$$s_{ij}(x) = \max \; e(S; x)$$

where the maximum is taken over all coalitions S such that $i \in S$ and $j \notin S$.

Thus, s_{ij} represents the most that player i could—under the best of circumstances—hope to gain without the cooperation of j.

13.2.3 Definition. Let $\langle x; \mathcal{T} \rangle$ be an individually rational payoff configuration, and let i, j be distinct members of some $T_k \in \mathcal{T}$. We shall say i outweighs j (notation, $i \succ\succ j$) if and only if

$$s_{ij}(x) > s_{ji}(x) \quad \text{and} \quad x_j > v(\{j\})$$

Essentially, it seems that, if $i \succ\succ j$, then i can make a demand on j, which, in some sense, j cannot contest. Thus, if $i \succ\succ j$, there is a certain instability. We now define the kernel as the set of i.r.p.c.'s for which no such instability occurs.

13.2.4 Definition. The kernel of a game v is the set $\mathscr{K}$ of all i.r.p.c.s $\langle x, \mathscr{T} \rangle$ such that, for $T_k \in \mathscr{T}$, there are no $i, j \in T_k$ with $i \succ\succ j$.

13.2.5 Example. Consider the three-person simple game in which each two-person coalition wins. For $\mathscr{T} = \{\{1,2\},\{3\}\}$, it is not too difficult to see that $x = \left(\frac{1}{2},\frac{1}{2},0\right)$ is the only payoff vector such that $\langle x, \mathscr{T} \rangle \in \mathscr{K}$. On the other hand, if $\mathscr{T} = \{N\}$, then $x = \left(\frac{1}{3},\frac{1}{3},\frac{1}{3}\right)$ is the only x with $\langle x; \mathscr{T} \rangle \in \mathscr{K}$.

13.2.6 Example. Consider the three-person simple game in which $\{1, 2\}$, $\{1, 3\}$, and $\{1, 2, 3\}$ win. Then, for $\mathscr{T} = \{\{1,2\},\{3\}\}$ or $\mathscr{T} = \{N\}$, we find $\langle x; \mathscr{T} \rangle \in \mathscr{K}$ only for $x = (1,0,0)$.

13.2.7 Example. Let v be the four-person simple voting game with symbol $[3; 2, 1, 1, 1]$. Here, for $\mathscr{T} = \{\{1,2\},\{3\},\{4\}\}$, we find $\langle x; \mathscr{T} \rangle \in \mathscr{K}$ only for $x = \left(\frac{1}{2},\frac{1}{2},0,0\right)$. On the other hand, for $\mathscr{T} = \{N\}$, the only point in the kernel is given by $x = \left(\frac{2}{5},\frac{1}{5},\frac{1}{5},\frac{1}{5}\right)$.

In the examples above, it may be verified directly that all the points in the kernel also belong to the bargaining set $\mathscr{M}_1^{(i)}$. In fact, this is generally true, as the following theorem shows:

13.2.8 Theorem. For any game $\mathscr{K} \subset \mathscr{M}_1^{(i)}$.

Proof. We shall show that, if $\langle \mathbf{x}; \mathscr{T} \rangle \notin \mathscr{M}_1^{(i)}$, then $\langle x; \mathscr{T} \rangle \notin \mathscr{K}$.

Suppose, then, $\langle x; \mathscr{T} \rangle$ is an i.r.p.c. not in $\mathscr{M}_1^{(i)}$. This means there exist $i, j \in T_k \in \mathscr{T}$ such that i has an objection $\langle y; \mathscr{U} \rangle$ against j and j has no counterobjection. We shall show $i \succ\succ j$.

Since j has no counterobjection, it is clear that $x_j > v(\{j\})$, for, otherwise, j could always counterobject with some $\langle z, \mathscr{V} \rangle$ such that $z_j = v(\{j\})$ and $\{j\} \in \mathscr{V}$.

Let $S = P(\{i\}, \mathscr{U})$; it is clear we must have $j \notin S$, and so

$$s_{ij}(x) \geq e(S, x)$$

Suppose, now, that $s_{ji}(x) \geq s_{ij}(x)$. Then there is some S', with $j \in S'$, $i \notin S'$, such that

$$e(S', x) = s_{ji}(x) \geq s_{ij}(x)$$

We claim, then, that j has a counterobjection, using the coalition S' as his partners. In fact, it suffices to show that S' can afford to give at least x_l to its members, at least y_l to its members $l \in S' \cap S$. We have, in fact,

$$\sum_{l \in S' \cap S} y_l = \sum_{l \in S} y_l - \sum_{S-S'} y_l \leq v(S) - \sum_{S-S'} y_l \leq v(S) - \sum_{S-S'} x_l$$

and so

(13-22)

1000 am

$$\sum_{l \in S' \cap S} (y_l - x_l) \le v(S) - \sum_S x_l \le s_{ij}(x)$$

Thus

$$\sum_{l \in S' \cap S} y_l + \sum_{l \in S' - S} x_l \le s_{ij}(x) + \sum_{l \in S'} x_l = s_{ij}(x) + v(S') - e(S', x) = s_{ij}(x) + v(S') - s_{ji}(x)$$

and, assuming $s_{ji}(x) \ge s_{ij}(x)$, this gives us

$$\sum_{l \in S' \cap S} y_l + \sum_{l \in S' \cap S} x_l \le v(S')$$

Since $i \notin S'$, j has a counterobjection to i, in which his partners are S'. But we had assumed j had no counterobjection. This contradiction proves that

$$s_{ji}(x) < s_{ij}(x)$$

and so $i \succ\succ j$. Thus $\langle x; \mathcal{T} \rangle \notin \mathcal{K}$. We conclude that $\mathcal{K} \subset \mathcal{M}_1^{(i)}$.

Thus, every i.r.p.c. in the kernel belongs to $\mathcal{M}_1^{(i)}$. The converse is not, however, true; frequently $\mathcal{K}$ is a proper subset of $\mathcal{M}_1^{(i)}$. As an example, consider the four-person game of Example 13.2.7.: when $\mathcal{T} = \{\{1,2\},\{3\},\{4\}\}$, we find the $\langle x; \mathcal{T} \rangle \in \mathcal{K}$ only if $x = (\frac{1}{2}, \frac{1}{2}, 0, 0)$; on the other hand, $\langle x; \mathcal{T} \rangle \in \mathcal{M}_1^{(i)}$ for all x of the form $(t, 1 - t, 0, 0)$, with $\frac{1}{2} \le t \le \frac{2}{3}$.

We know, of course, that $\mathcal{M}_1^{(i)}$ is nonempty; more exactly, that given any $\mathcal{T}$ there is some x with $\langle x; \mathcal{T} \rangle \in \mathcal{M}_1^{(i)}$. The obvious question is whether the same is true of $\mathcal{K}$. In fact it is, and the reader may enjoy modifying the proof of Theorem 13.1.9 from $\mathcal{M}_1^{(i)}$ to $\mathcal{K}$. In the next section, we give a more interesting proof of this statement.

13.3. The Nucleolus

Still another solution concept related to the bargaining sets is the *nucleolus*.

For a given *n*-person game v and payoff $x = (x_1, \ldots, x_n)$, we define the 2^n-vector $\theta(x)$ as the vector whose components are the excesses of the 2^n subsets $S \subset N$, arranged in decreasing order, i.e.

$$\theta_k(x) = e(S_k, x) \tag{13.16}$$

where $S_1, S_2, \ldots, S_{2^n}$ are the subsets of N, arranged by

$$e(S_k, x) \ge e(S_{k+1}, x) \tag{13.17}$$

As an example, in the three-person simple game v where $v(S) = 1$ if S has two or three players and $v(S) = 0$ otherwise, the payoff vector $(.3, .5, .2)$ gives us the following excesses:

S	$e(S, x)$
$\{1, 3\}$	.5
$\{2, 3\}$	.3
$\{1, 2\}$	.2
N	0
$\emptyset$	0
$\{3\}$	$-.2$
$\{1\}$	$-.3$
$\{2\}$	$-.5$

and so $\theta(x) = (.5, .3, .2, 0, 0, -.2, -.3, -.5)$.

Similarly, if we set $y = (.1, .5, .4)$, a similar analysis will give us

$$\theta(y) = (.5, .4, .1, 0, 0, -.1, -.4, -.5)$$

We will now order the several vectors $\theta(x)$ by the lexicographic order. Generally speaking, if we are given two vectors $\alpha = (\alpha_1, \dots, \alpha_q)$ and $\beta = (\beta_1, \dots, \beta_q)$, we say that α is lexicographically smaller than β if there is some integer k, $1 \leq k \leq q$, such that

$$\alpha_l = \beta_l \quad \text{for} \quad 1 \leq l < k$$
$$\alpha_k < \beta_k$$

We shall write $\alpha <_L \beta$ for this relation, and $\alpha \leq_L \beta$ if either $\alpha <_L \beta$ or $\alpha = \beta$.

In the example above, we find that $\theta(x) <_L \theta(y)$ because $\theta_1(x) = \theta_1(y)$, but $\theta_2(x) < \theta_2(y)$.

Now, the lexicographic ordering on $\theta(x)$ can be used to induce an order on the payoff vectors x themselves, i.e., we shall write

$$x \precsim y$$

if and only if $\theta(x) \leq_L \theta(y)$ and

$$x \prec y$$

if and only if $\theta(x) <_L \theta(y)$.

With this in mind, we can now define the nucleolus of a game:

13.3.1 Definition. Let v be an n-person game, and let X be a set of n-vectors (payoff vectors). Then, the nucleolus of v over the set X is the set $v(x)$ defined by

$$v(X) = \left\{ x \,\middle|\, \begin{array}{l} x \in X \\[4pt] \text{If } y \in X, \text{then } x \precsim y \end{array} \right\}$$

Thus the nucleolus consists of those points x, which minimize the function $\theta(x)$ (in the lexicographic order) over the set X. Thus v depends, not only on the game v but also on the particular set X, which has been chosen. Usually, X is chosen to be the set of all imputations; more generally, for a given coalition structure $\mathscr{T}$, X may be $X(\mathscr{T})$ the set of all x satisfying conditions (13.1) and (13.2), so that $\langle x, \mathscr{T} \rangle$ is an i.r.p.c. Other choices may be possible for X.

The question naturally arises as to whether v is empty; assuming v is nonempty, we would like to know something about its structure. In fact, we shall see that, under very reasonable conditions, v is nonempty and reduces to a single point.

13.3.2 Theorem. If X is a nonempty compact set, then $v(X)$ is also nonempty and compact.

Proof. Let us define a set X' by

$$X' = \left\{ x \,\middle|\, \begin{array}{c} x \in X \\ \text{If } y \in S, \text{then } \theta_1(x) \leq \theta_1(y) \end{array} \right\}$$

and generally, a sequence $X', X'', \ldots, X^{(2^n)}$ by the inductive definition

$$X^{(k)} = \left\{ x \,\middle|\, \begin{array}{c} x \in X^{(k-1)} \\ \text{If } y \in X^{(k-1)}, \text{then } \theta_k(x) \leq \theta_k(y) \end{array} \right\}$$

for $k = 1, 2, \ldots, 2^n$.

It is not too difficult to see that, for each k, $\theta_k(x)$ is a continuous function of x. Thus, if we assume $X^{(k-1)}$ is compact and nonempty, it is clear that $X^{(k)}$, which minimizes θ_k over $X^{(k-1)}$ is also compact and nonempty. By induction, we conclude that $X^{(2^n)}$ is also compact and nonempty. We now claim that

$$v(X) = X^{(2^n)}$$

In fact, suppose $v(X) \neq X^{(2^n)}$. This means there is some $y \in X - X^{(2^n)}$, $x \in X^{(2^n)}$, such that $\theta(y) \leq_L \theta(x)$. Let k be the smallest integer such that $y \notin X^{(k)}$. Then $\theta_k(y) > \theta_k(x)$. But $y \in X^{(l)}$ for all $l \leq k-1$, and so $\theta_l(y) = \theta_l(x)$ for $l \leq k-1$. This means, however, that

$$\theta(x) <_L \theta(y)$$

The contradiction proves that $v(X) = X^{(2^n)}$.

Thus, the nucleolus will be a nonempty compact set if X is also nonempty and compact. In fact, it is not necessary that X be compact for, if S is merely closed and bounded above, then X' is easily seen to be nonempty and compact. In particular, if

X is not the set of imputations, but rather the set of vectors whose components sum to $v(N)$ (without the individual rationality requirement), X' will be nonempty and compact. From this, the induction steps will guarantee that $X^{(2^n)}$ is also nonempty and compact.

We prove next, that in the most interesting cases, v reduces to a single point. The following lemma is important.

13.3.3 Lemma. Let $\theta(x) = \theta(y)$, $x \neq y$. Then, for $0 < r < 1$

$$\theta(rx + (1 - r)y) <_L \theta(x)$$

In other words, θ is *strictly quasi-convex* in the lexicographic ordering.

Proof. Let us set $z = rx + (1 - r)y$. We have

$$\theta(x) = (e(S_1, x), e(S_2, x), \ldots, e(S_{2^n}, x))$$
$$\theta(y) = (e(T_1, y), e(T_2, y), \ldots, e(T_{2^n}, y))$$

where the $S_1, \ldots, S_{2^n}$, $T_1, \ldots, T_{2^n}$ are two permutations of the subsets of N, ordered as follows:

(a) $e(S_k, x) \geq e(S_{k+1}, x)$
(b) If $e(S_k, x) = e(S_{k+1}, x)$, then $e(S_k, y) \geq e(S_{k+1}, y)$
(c) $e(T_k, y) \geq e(T_{k+1}, y)$

Now let c be the first index such that

$$e(S_c, x) \neq e(S_c, y)$$

and let

$$q = e(S_c, x)$$

For any S, we have

$$e(S, z) = v(S) - \sum_{i \in S}(rx_i + (1 - r)y_i)$$
$$= v(S) - r\sum_{i \in S}x_i - (1 - r)\sum_{i \in S}y_i$$
$$= r\left[v(S) - \sum_{i \in S}x_i\right] + (1 - r)\left[v(S) - \sum_{i \in S}y_i\right]$$

so that

$$e(S, z) = re(S, x) + (1 - r)e(S, y)$$

Now, for $k < c$, we have $e(S_k, x) = e(S_k, y)$ and so

$$e(S_k, z) = e(S_k, x)$$

For $k \geq c$, we see that we cannot have

$$e(S_k, y) > e(S_c, x) = q$$

for if we did, then every coalition S such that $e(S, x) > q$ would also have $e(S, y) > q$; moreover, at least one coalition (namely, S_k) would have $e(S_k, y) > q$ while $e(S_k, x) \leq q$. Thus it is not possible that $\theta(x) = \theta(y)$ as we have assumed. Thus we must have

$$e(S_k, y) \leq e(S_c, x)$$

for all $k \geq c$. In particular, we must have

$$e(S_c, y) < e(S_c, x)$$

since equality here is not possible; moreover, if, for any $k > c$, we have

$$e(S_k, x) = e(S_c, x)$$

then by our condition (b) above, we have

$$e(S_k, y) \leq e(S_c, y) < q$$

We see then, that, for $k \geq c$, we have either

$$e(S_k, x) < q, \quad e(S_k, y) \leq q$$

or

$$e(S_k, x) \leq q, \quad e(S_k, y) < q$$

and so

$$e(S_k, z) = re(S_k, x) + (1 - r)e(S_k, y) < q$$

We see, thus, that, for $k < c$, we will have $e(S_k, z) = e(S_k, x)$. For $k \geq c$, we have $e(S_k, z) < q$. This will mean that

$$\theta_k(z) = \theta_k(x), \quad 1 \leq k \leq c - 1$$

$$\theta_c(z) < \theta_c(x)$$

and so $\theta(z) <_L \theta(x)$, as desired.

This lemma allows us to prove the following:

13.3.4 Theorem. If X is nonempty, compact, and convex, then $v(X)$ consists of a single point.

Proof. We have already proved that $v(X)$ is nonempty. Suppose, then, that x and y are two distinct points of $v(X)$. By convexity of X we see that, if $0 < r < 1$, then

$$z = rx + (1 - r)y$$

also is in X. But then by Lemma 13.3.4, $\theta(z) <_L \theta(x)$, and thus $x \notin \nu(X)$. This contradiction proves $\nu(X)$ has a unique element.

Since, in the compact convex case, ν reduces to a single point, we can, by a certain abuse of language, identify the set with its element, and say that this point $\tilde{x}$ is the nucleolus.

13.3.5 Example. Let ν be the three-person simple majority game. If we let X be the set of all imputations, then

$$\tilde{x} = (\tfrac{1}{3}, \tfrac{1}{3}, \tfrac{1}{3})$$

is the nucleolus. If, instead, $X = X(\mathcal{T})$, where $\mathcal{T} = \{\{1,2\},\{3\}\}$, then

$$\tilde{x} = (\tfrac{1}{2}, \tfrac{1}{2}, 0)$$

is the nucleolus.

13.3.6 Example. Let ν be the four-person simple game $[3; 2, 1, 1, 1]$. If we let X be the set of all imputations then

$$\tilde{x} = (\tfrac{2}{5}, \tfrac{1}{5}, \tfrac{1}{5}, \tfrac{1}{5})$$

If on the other hand, we set $X = X(\mathcal{T})$, where $\mathcal{T} = \{\{1,2\},\{3\},\{4\}\}$, then

$$\tilde{x} = (\tfrac{1}{2}, \tfrac{1}{2}, 0, 0)$$

From the examples above, it may be seen that if $\tilde{x}$ is the nucleolus for $X(\mathcal{T})$ then $\langle \tilde{x}; \mathcal{T} \rangle$ belongs to the kernel. The obvious question is whether this is true in general: the answer, given by the following theorem, is affirmative, and provides us with an alternative proof of the nonemptiness of the kernel.

13.3.7 Theorem. For a given ν and a coalition structure $\mathcal{T}$, let $X(\mathcal{T})$ be the set of all n-vectors satisfying (13.1)–(13.2), and let $\tilde{x}$ be the nucleolus of ν over $X(\mathcal{T})$. Then

$$\langle \tilde{x}, \mathcal{T} \rangle \in \mathcal{K}$$

Proof. Let us suppose $\langle \tilde{x}; \mathcal{T} \rangle \notin \mathcal{K}$. This means there exist $i, j \in T_k \in \mathcal{T}$ that $i \succ\succ j$. Then

$$s_{ij}(\tilde{x}) > s_{ji}(\tilde{x})$$

$$\tilde{x}_j > \nu(\{j\})$$

Let

$$\delta = \min\{\tfrac{1}{2}(s_{ij} - s_{ji}), \tilde{x}_j - \nu(\{j\})\}$$

Then $\delta > 0$. Let y be given by

$$y_i = \tilde{x}_i + \delta$$
$$y_j = \tilde{x}_j - \delta$$
$$y_l = \tilde{x}_l \quad \text{for} \quad i \neq l \neq j$$

We have, as before,

$$\theta(\tilde{x}) = (e(S_1, \tilde{x}), e(S_2, \tilde{x}), \dots, e(S_{2^n}, \tilde{x}))$$

where, as always, the S_k are the 2^n subsets of N, in order of decreasing excess. We let S_c be the first of these containing i but not j. Thus

$$e(S_c, \tilde{x}) = s_{ij}$$

and for all $k < c$, we have either i, j both in S_k or i, j both in $N - S_k$. There is no S_k such that $k \leq c, j \in S_k, i \notin S_k$, as all such S_k have lower excess than S_c. We shall make the further assumption that, if there is some S_k with

$$e(S_k, \tilde{x}) = e(S_c, \tilde{x})$$

such that either $i, j \in S_k$ or $i, j \in N - S_k$, then $k < c$.

We see, now, that $\theta(y) <_{\mathrm{L}} \theta(\tilde{x})$. In fact, we have, for $k < c$

$$\sum_{l \in S_k} y_l = \sum_{l \in S_k} \tilde{x}_l$$

since, if $i, j \in S_k$, then one term in the left-hand sum has increased by δ, while another has decreased by δ; the rest are all unchanged. If $i, j \in N - S_k$, then the terms in the two sums coincide. Thus for $k < c$

$$e(S_k, y) = e(S_k, \tilde{x})$$

Suppose, now, $k \geq c$. Then there are three possibilities:

(a) $i \in S_k, i \notin S_k$. In this case

$$e(S_k, \tilde{x}) \leq s_{ij}$$

and so

$$e(S_k, y) \leq s_{ij} - \delta < s_{ij}$$

(b) $i \notin S_k, j \in S_k$. In this case

$$e(S_k, \tilde{x}) \leq s_{ji} + \delta < s_{ij}$$

(c) Either $i, j \in S_k$ or $i, j \in N - S_k$. Then

$$e(S_k, \tilde{x}) < s_{ij}$$

and so

$$e(S_k, y) < s_{ij}$$

We see, then, that $e(S_k, y) = \theta_k(\tilde{x})$ for $k < c$ and $e(S_k, y) < \theta_c(\tilde{x})$ for $k \geq c$. We conclude that we have

$$\theta(y) <_{\mathrm{L}} \theta(\tilde{x})$$

Moreover, it is easily seen that $y \in S(\mathcal{T})$, which contradicts the hypothesis that $\tilde{x}$ is the nucleolus. The contradiction proves the theorem.

The following theorem, due to E. Kohlberg, elegantly characterizes the nucleolus in the special case where X is the set of all imputations. For a given $x \in X$, we define a sequence of collections $\mathcal{B}_1, \mathcal{B}_2, \ldots, \mathcal{B}_q$, such that $\mathcal{B}_1$ is the collection of coalitions with maximal excess. $\mathcal{B}_2$ that of coalitions with the next largest excess, etc. Mathematically, there are q numbers α_k

$$\alpha_1 > \alpha_2 > \cdots > \alpha_q$$

such that, for all $S \in \mathcal{B}_k$

$$e(S, x) = \alpha_k$$

and every $S \subset N$ belongs to one of the $\mathcal{B}_k$, i.e.

$$\bigcup_{k=1}^{q} \mathcal{B}_k = 2^N$$

We define, moreover

$$\mathcal{B}_0 = \{\{i\} | x_i = v(\{i\})\}$$

and, for $k = 1, \ldots, q$

$$\mathcal{C}_k = \bigcup_{j=1}^{k} \mathcal{B}_j$$

Finally, we shall say that $\mathcal{B}_0, \mathcal{B}_1, \ldots, \mathcal{B}_q$ is *the array corresponding to x*. Then

13.3.8 Theorem. A necessary and sufficient condition for $x \in X$ to be $\nu(X)$ is that, for every $k = 1, \ldots, q$, there exists a balanced collection $\mathcal{A}_k$ satisfying

$$\mathcal{C}_k \subset \mathcal{A}_k \subset \mathcal{C}_k \cup \mathcal{B}_k$$

Proof. In effect, the fact that some $\mathcal{A}_k$ is balanced means that there exist balancing coefficients λ_s, satisfying

$$\sum_{\substack{S \in \mathcal{C}_k \cup \mathcal{B}_0 \\ i \in S}} \lambda_s = 1 \quad \text{for all} \quad i \in N \tag{13.18}$$

$$\lambda_s > 0 \quad \text{for} \quad S \in \mathcal{C}_k \tag{13.19}$$

$$\lambda_s \geq 0 \quad \text{for} \quad S \in \mathcal{B}_0 \tag{13.20}$$

If we let w be the reciprocal of the smallest positive λ_s, this can be rewritten as

$$\sum_{\substack{S \in \mathcal{C}_k \cup \mathcal{B}_0 \\ i \in S}} \lambda'_s - w = 0 \quad \text{for all} \quad i \in N \tag{13.21}$$

$$\lambda'_s \geq 1, \quad S \in \mathcal{C}_k \tag{13.22}$$

$$\lambda'_s \geq 0 \quad S \in \mathcal{B}_0 \tag{13.23}$$

We think of this as a linear program with the trivial objective function 0 to be minimized. Its dual program then has the form

$$\text{Maximize} \quad \sum_{S \in \mathcal{C}_k} z_S \tag{13.24}$$

subject to

$$\sum_{i \in S} u_i - z_S = 0, \quad S \in \mathcal{C}_k \tag{13.25}$$

$$u_i \geq 0, \quad \{i\} \in \mathcal{B}_0 \tag{13.26}$$

$$\sum_{i \in N} u_i = 0 \tag{13.27}$$

$$z_S \geq 0, \quad S \in \mathcal{C}_k \tag{13.28}$$

The dual program, (13.24)–(13.28), is obviously feasible, as $z_S = 0$, $u_i = 0$ satisfies the constraints. Thus its value is at least 0. On the other hand, the primal program (13.21)–(13.23), if feasible, has value 0. Applying Theorem 3.2.6, we see that the balancedness of $\mathcal{A}_k$ is equivalent to whether program (13.24)–(13.28) has value 0. As the z_S are non-negative, we see that (13.24)–(13.28) has positive value if and only if there is some feasible vector (u_i, z_S) for which at least one $z_S > 0$. Thus, $\mathcal{A}_k$ is balanced if and only if, for any vector u_i satisfying

$$\sum_{i \in S} u_i \geq 0, \quad S \in \mathcal{C}_k \cup \mathcal{B}_0 \tag{13.29}$$

$$\sum_{i \in N} u_i = 0 \tag{13.30}$$

we also have

$$\sum_{i \in S} u_i = 0 \quad \text{for all} \quad S \in \mathcal{C}_k \tag{13.31}$$

We therefore restate Theorem 13.3.8 in the form of a lemma.

13.3.9 Lemma. A necessary and sufficient condition for $x \in X$ to be $v(X)$ is that, for every k, any vector u_i satisfying (13.29) and (13.30) also satisfies (13.31).

Proof. Suppose first that x is not the nucleolus $v(X)$. Then there is some $y \in X$ such that $\theta(y) <_L \theta(x)$. Let, then,

$$u = y - x$$

Since $y \in X$, we will have $u_i \geq 0$ for $\{i\} \in \mathcal{B}_0$, and also (13.30). Now, let S_c be the first coalition such that

$$e(S_c, y) < e(S_c, x)$$

and assume $S_c \in \mathscr{B}_k$.

Clearly, we have $e(S_l, y) \geq e(S_l, x)$ for $l < c$, and, moreover

$$e(S_l, y) \leq e(S_l, x)$$

for all $S_l \in \mathscr{B}_k$. This will mean

$$\sum_{S_l} u_i = 0 \quad \text{for} \quad S_l \in \mathscr{C}_{k-1}$$

$$\sum_{S_l} u_i \geq 0 \quad \text{for} \quad S_l \in \mathscr{B}_k$$

$$\sum_{S_c} u_i > 0$$

and so the condition does not hold.

Suppose, next, that for some k the condition does not hold: there exists u satisfying (13.29) and (13.30), and such that, for at least one $S \in \mathscr{C}_k$

$$\sum_{i \in S} u_i > 0$$

Set, then

hence, for small $t > 0$, we find that $\quad y = x + tu$

$$\sum_N y_i = \sum_N x_i + t \sum_N u_i = v(N)$$

and

$$y_i = x_i + tu_i \geq v(\{i\})$$

so that $y \in X$.

We find, then, that, for $S \in \mathscr{C}_k$

$$e(S, y) \leq e(S, x)$$

with inequality holding in at least one case. Moreover, for $S \notin \mathscr{C}_k$, we will have

$$e(S, y) = e(S, x) - t \sum_S u_i \leq \alpha_{k+1} + t \sum |u_i|$$

and, for sufficiently small t, this is

$$e(S, y) < \alpha_k$$

Thus $\theta(y) <_L \theta(x)$ and $x \neq v(X)$.

This proves the lemma. Since the condition in the lemma is equivalent to that in Theorem 13.3.8, this completes a proof of the theorem.

We have effectively characterized the nucleolus, at least over the set of all imputations. It remains to compute it. Generally speaking, we may assume that the set X is a convex polyhedral set, determined by a system of linear equations and inequalities. Consider, then, the linear program

Minimize α (13.32)

subject to

$$\sum_{i \in S} x_i + \alpha \geq v(S), \quad S \subset N \tag{13.33}$$

$$x \in X \tag{13.34}$$

Let α_1 be the minimum of this program. If this is attained at a unique point $\tilde{x}$, then $\tilde{x}$ is $v(X)$. Generally, however, this minimum is attained over a certain set X^1. If so, there will generally be a certain collection $\mathscr{B}_1$ of sets such that, for all $S \in \mathscr{B}_1$ and $x \in X^1$

$$e(S, x) = \alpha_1$$

We then solve the program

Minimize α

subject to

$$\sum_{i \in S} x_i + \alpha \geq v(S) \quad \text{for} \quad S \in 2^N - \mathscr{B}_1$$

$$x \in X^1$$

The solution of this will give us the "second-largest" excess α_2, and the collection $\mathscr{B}_2$, as in (13.23). Continuing in this way, we will eventually obtain a unique point that is the solution of a sequence of linear programs; this is $v(X)$.

13.3.10 Example. Consider the four-person game v such that

$$v(N) = 100, \quad v(\{1, 2, 3\}) = 95, \quad v(\{1, 2, 4\}) = 85$$

$$v(\{1, 3, 4\}) = 80, \quad v(\{2, 3, 4\}) = 55$$
$$v(\{i, j\}) = 50 \qquad \text{for all} \quad i \neq j$$
$$v(\{i\}) = 0 \qquad \text{for all} \quad i$$

Let X be the set of all imputations. We can disregard the two sets N and $\emptyset$, as their excesses are always 0 for $x \in X$. Then we have the LP

Minimize α

subject to

$$x_1 + x_2 + x_3 + \alpha \geq 95$$
$$x_1 + x_2 + x_4 + \alpha \geq 85$$
$$x_1 + x_3 + x_4 + \alpha \geq 80$$
$$x_2 + x_3 + x_4 + \alpha \geq 55$$
$$x_i + x_j + \alpha \geq 50$$
$$x_i + \alpha \geq 0$$
$$x_1 + x_2 + x_3 + x_4 = 100$$
$$x_i \geq 0$$

Solution of this gives us

$$\alpha_1 = 10$$

obtained over the set X^1 given by

$$x_1 + x_2 = 60$$
$$x_1 \geq 30, \quad x_2 \geq 25$$
$$x_3 = 25, x_4 = 15$$

and it may be seen that

$$\mathscr{B}_1 = \{\{1,2,3\}, \{1,2,4\}, \{3,4\}\}$$

as $e(S,x) = 10$ for all $S \in \mathscr{B}_1$ and $x \in X^1$. As was to be expected, $\mathscr{B}_1$ is balanced. The second LP, now, is

Minimize α

subject to

$$x_1 + x_3 + x_4 + \alpha \geq 80$$
$$x_2 + x_3 + x_4 + \alpha \geq 55$$
$$x_1 + x_3 + \alpha \geq 50$$
$$x_1 + x_4 + \alpha \geq 50$$
$$x_2 + x_3 + \alpha \geq 50$$
$$x_2 + x_4 + \alpha \geq 50$$
$$x_i + \alpha \geq 0$$
$$x \in X^1$$

Some simplification (using the definition of X^1) reduces this to

Minimize α

subject to

$$x_1 + \alpha \geq 40$$
$$x_2 + \alpha \geq 15$$
$$x_1 + \alpha \geq 25$$
$$x_1 + \alpha \geq 35$$
$$x_2 + \alpha \geq 25$$
$$x_2 + \alpha \geq 35$$
$$x_1 + x_2 = 60$$
$$x_1 \geq 30$$
$$x_2 \geq 25$$

or more simply

Minimize α

subject to

$$x_1 + \alpha \geq 40$$
$$x_2 + \alpha \geq 35$$
$$x_1 + x_2 = 60$$
$$x_1 \geq 30$$
$$x_2 \geq 25$$

and this gives us the solution $\alpha_2 = 7.5$ with $x_1 = 32.5$, $x_2 = 27.5$. Thus we have obtained $v(X)$: it is

$$\tilde{x} = (32.5, 27.5, 25, 15)$$

We have, moreover,

$$\mathscr{B}_1 = \{\{1, 2, 3\}, \{1, 2, 4\}, \{3, 4\}\}$$
$$\mathscr{B}_2 = \{\{1, 3, 4\}, \{2, 3, 4\}\}$$

It may be seen that both $\mathscr{B}_1$ and $\mathscr{C}_2 = \mathscr{B}_1 \cup \mathscr{B}_2$ are balanced. Moreover, any collection that contains all the three-person subsets of N is balanced. Thus, all the sets $\mathscr{C}_3, \mathscr{C}_4, \ldots, \mathscr{C}_q$ are also balanced, and the Kohlberg criterion verifies that this is indeed the nucleolus.

13.4.   The Airport Game

We consider here an application of game-theoretic analysis to the determination of airport landing fees. Essentially, our desire is to develop a "fair" schedule of fees. We discuss here an airport, and indeed the particular cost structure yields an especially elegant result; nevertheless, such analysis should be applicable to the assessment of charges for common facilities.

Generally speaking, expenses at an airport consist of two parts: there is a variable expense, which is proportional to the number of planes using the airport (though to be exact this expense will generally be greater for the larger planes) and a fixed capital cost, which must be amortized over a certain period. There is generally no problem in allocation of the variable costs as these are directly incurred by the several planes; the problem lies in allocation of the capital costs (e.g., runway and terminal construction) to the planes.

Generally, it turns out that the capital costs of the airport depend essentially on the "largest" plane, which is to land there; this is due to the runway. Once the runway has been built, no further capital expense is necessary—until a new larger plane (i.e., one needing a longer runway) is introduced.

We can now divide the planes into m types. The cost of a runway adequate for planes of type j $(j = 1, \ldots, m)$ is C_j, where we assume

$$0 = C_0 < C_1 < C_2 < \cdots < C_{m-1} < C_m$$

Let N_j be the set of landings by planes of type j and assume there are n_j landings by planes of this type. Then

$$N = \bigcup_{j=1}^{m} N_j$$

is the set of all landings, and, for $S \subset N$, we shall write

$$j(S) = \max\{j \mid S \cap N_j \neq \varnothing\}$$

It may be seen that the cost of a runway adequate to receive all landings in set S is

$$c(S) = C_{j(S)}$$

while

$$c(\varnothing) = 0$$

To obtain a set of "fair" landing fees, we shall treat this cost structure as a game with player set N and characteristic function

$$v(S) = -c(S)$$

The reader may prove that v is superadditive, so that it is a proper game. An imputation for v is a vector x satisfying

$$x_i \geq -c(\{i\}) = -C_j \quad \text{if} \quad i \in N_j$$

$$\sum_{i \in N} x_i = -c(N) = -C_m$$

Thus $-x = (-x_1, \ldots, -x_m)$ is a possible fee schedule, which will be just enough to recapture the capital costs of the airport.

As a "fair" schedule, two imputations seem possible candidates: these are the Shapley value and the nucleolus. We consider both of these in turn.

To compute the Shapley value, let us set

$$R_k = \bigcup_{j=k}^{m} N_j, \qquad r_k = \sum_{j=k}^{m} n_j$$

and consider the m characteristic functions $v_1, v_2, \ldots, v_m$ defined by

$$v_k(S) = \begin{cases} 0 & \text{if} \quad S \cap R_k = \emptyset \\ C_{k-1} - C_k & \text{if} \quad S \cap R_k \neq \emptyset \end{cases}$$

It is not too difficult to see that, if $k \leq j(S)$, then $S \cap R_k \neq \emptyset$, while, if $k > j(S)$, then $S \cap R_k = \emptyset$. Thus

$$\sum_{k=1}^{m} v_k(S) = \sum_{k=1}^{j(S)}(C_{k-1} - C_k) = C_0 - C_{j(S)} = -c(S)$$

and so

$$v = \sum_{k=1}^{m} v_k$$

Thus, by the additivity property

$$\varphi[v] = \sum_{k=1}^{m} \varphi[v_k]$$

Now it is obvious that v_k is a game with carrier R_k, symmetric over that carrier. Thus

$$\varphi_i[v_k] = \begin{cases} 0, & i \notin R_k \\ \dfrac{C_{k-1} - C_k}{r_k}, & i \in R_k \end{cases}$$

Now, if $i \in N_j$ then $i \in R_k$ for all $k \leq j$ (and no other k) thus

$$\varphi_i[v] = \sum_{k=1}^{j} \frac{C_{k-1} - C_k}{r_k} \quad \text{if} \quad i \in N_j$$

and the Shapley value cost schedule is merely the negative of this, or

$$\varphi_i[c] = \sum_{k=1}^{j} \frac{C_k - C_{k-1}}{r_k}, \quad i \in N_j$$

Alternatively, we can use the nucleolus as a fair schedule of landing fees. To compute the nucleolus (over X, the set of all imputations), we note first of all that the game has a nonempty core. The imputation given by $x_i = -C_m$ for one player

$i \in N_m$ and $x_l = 0$, for all other players certainly belongs to the core. Thus the maximal excess will be 0, corresponding to N and $\emptyset$, and all other excesses will be nonpositive (and, in fact, usually negative). The computation of the nucleolus then starts by considering the LP

$$\text{Minimize} \quad \alpha \tag{13.35}$$

subject to

$$\sum_{i \in N} x_i = -C_m \tag{13.36}$$

$$\alpha + \sum_{i \in S} x_i \geq -C_{j(S)} \quad \text{for} \quad S \subset N, \; S \neq N, \; \emptyset \tag{13.37}$$

$$x_i \geq -C_j \quad \text{for} \quad i \in N_j, \; j \in M \tag{13.38}$$

By symmetry, it is not difficult to see that the optimum of this program is obtained by letting x_i be equal for all i in a given N_j; we therefore set

$$x_i = -f_j, \quad i \in N_j$$

so that f_j is the fee for planes of type j. Letting s_j be the number of players of type j in a given set S, we have then

$$\sum n_j f_j = C_m \tag{13.39}$$

$$-\alpha + \sum s_j f_j \leq C_{j(S)} \quad \text{for} \quad S \subset N, S \neq N, \emptyset \tag{13.40}$$

$$f_j \leq C_j, \quad j = 1, \ldots, m \tag{13.41}$$

If, in particular, we set $S = N - \{l\}$, where $l \in N_k$, this gives us

$$-\alpha + \sum n_j f_j - f_k \leq C_{j(S)} \leq C_m \tag{13.42}$$

and, subtracting from (13.40), we have

$$\alpha + f_k \geq 0 \tag{13.43}$$

for $k = 1, \ldots, m$. Since the minimum of α is nonpositive, this will mean all f_k are nonnegative.

Since all f_k are non-negative, we see that the constraints (13.40) are strongest when S is of the form

$$S = N_1 \cup N_2 \cup \cdots \cup N_k, k = 1, 2, \ldots, m - 1$$
$$= N - \{l\}$$

since every T other than N or $\varnothing$ is contained in some S of this form, with $v(S) = v(T)$. Thus we are finally left with the program

Minimize α $\qquad\qquad\qquad\qquad\qquad\qquad\qquad\qquad\qquad\qquad$ (13.44)

subject to

$$- \alpha + \sum_{j=1}^{k} n_j f_j \le C_k, \quad k = 1, \ldots, m-1 \qquad\qquad (13.45)$$

$$\alpha + f_k \ge 0, \quad k = 1, \ldots, m \qquad\qquad\qquad (13.46)$$

$$\sum_{j=1}^{m} n_j f_j = C_m \qquad\qquad\qquad\qquad (13.47)$$

Since $f_k \ge -\alpha$ for each k, this give us

$$- \alpha(t_k + 1) \le C_k \quad 1 \le k \le m-1$$
$$- \alpha t_m \le C_m$$

where

$$t_k = \sum_{j=1}^{k} n_j, \quad 1 \le k \le m$$

and so

$$- \alpha \le \min\left\{ \min_{1 \le k \le m-1} \left\{ \frac{C_k}{t_k + 1} \right\}, \frac{C_m}{t_m} \right\}$$

Let k_1 be the largest value of k for which this minimum is attained, and set

$$\alpha_1 = \begin{cases} - C_{k_1}/(t_{k_1} + 1) & \text{if} \quad k_1 \le m-1 \\ - C_m/t_m & \text{if} \quad k_1 = m \end{cases}$$

Then we claim that α_1 is indeed the value of LP (13.44)–(13.47). In fact, it is clear that α can be made no smaller given the constraints. If, moreover, we set

$$f_k = -\alpha_1, \quad 1 \le k \le m-1$$
$$f_m = C_m + \alpha_1 t_{m-1}/n_m$$

then it is not difficult to see that these f_k, together with α_1, satisfy the constraints. Since a_1 can be made no smaller, this is certainly a solution of the LP. Moreover, it is easy to see that the constraints

$$- \alpha + \sum_{j=1}^{k_1} n_j f_j \le C_{k_1}, \quad \alpha_1 + f_j \ge 0$$

mean that we must have $f_j = -\alpha_1$ for $1 \le j \le k_1$ as a condition for x to be the nucleolus.

If $k_1 = m$ or $m-1$, this, of course, determines the nucleolus exactly. If $k_1 \leq m-2$, we must still determine f_j for $k_1 + 1 \leq j \leq m$. This will give us a new LP

$$\text{Minimize} \quad \alpha \tag{13.48}$$

subject to

$$-\alpha + \sum_{j=k_1+1}^{k} n_j f_j \leq C_k + t_k \alpha_1, \quad k_1 + 1 \leq k \leq m-1 \tag{13.49}$$

$$\alpha + f_k \geq 0, \quad k_1 + 1 \leq k \leq m \tag{13.50}$$

$$\sum_{j=k_1+1}^{m} n_j f_j = C_m + \alpha_1 t_{k_1} \tag{13.51}$$

It may be seen that this is of the same form as (13.44)–(13.47), but with k_1 fewer constraints and variables. Its solution will therefore have the form

$$-\alpha_2 = \min\left\{ \min_{k_1+1 \leq k \leq m-1}\left\{ \frac{C_k + t_{k_1}\alpha_1}{t'_k + 1} \right\}, \frac{C_m + t_{k_1}\alpha_1}{t'_m} \right\}$$

where

$$t'_k = \sum_{j=k_1+1}^{k} n_j, \quad k_1 + 1 \leq k \leq m$$

Thus we will have

$$f_j = -\alpha_2 \quad \text{for} \quad k_1 + 1 \leq j \leq k_2$$

where k_2 is the value of k, which gives us the minimum defining α_2. In this fashion, we eventually obtain a sequence of numbers α_q

$$0 > \alpha_1 > \alpha_2 > \cdots > \alpha_p$$

and a corresponding sequence of integers k_q

$$1 \leq k_1 < k_2 < \cdots < k_p = m$$

given inductively by

$$\alpha_{q+1} = -\min\left\{ \min_{k_q+1 \leq k \leq m-1}\left\{ \frac{C_k + \sum_{r=1}^{q} t_{k_r}^{(r)}\alpha r}{t_k^{(r+1)} + 1} \right\}, \frac{C_m + \sum t_{k_r}^{(r)}\alpha r}{t_m^{(r+1)}} \right\}$$

where

$$t_k^{(r)} + \sum_{j=k_{r-1}+1}^{k} n_j$$

and k_{q+1} is the value of k, which gives the minimum defining α_{q+1}. The (negative) nucleolus is then given by the fee schedule

$$f_j = -\alpha_q, \quad k_{q-1} + 1 \leq j \leq k_q$$

Table 13.1: Aircraft movements, runway costs, and movement fees at Birmingham Airport, 1968–1969.

Aircraft type	Subscript	No. of aircraft movements	Operating cost per movement	Annual capital cost	Nucleolus	Nucleolus based fee	Shapley value	Shapley value-based fee	Actual landing fee
	i	m_i	a_i	C_i	v_i	$v_i + a_i$	φ_i	$\varphi_i + a_i$	
Fokker Friendship 27	1	42	£5.23	£65,899	£7.89	£13.12	£4.86	£10.09	£5.80
Viscount 800	2	9,555	6.09	76,725	7.89	13.98	5.66	11.75	11.40
Hawker Siddeley Trident	3	288	7.55	95,200	7.89	15.44	1.030	17.85	21.70
Britannia 100	4	303	7.71	97,200	7.89	15.60	10.85	18.56	29.80
Casavelle (500)	5	151	7.73	97,436	7.89	15.62	10.92	18.65	20.30
BAC 111(500)	6	1,315	7.79	98,142	7.89	15.68	11.13	18.92	16.70
Vanguard 953	7	505	8.13	102,496	7.89	16.02	13.40	21.53	26.40
Comet 4B	8	1,128	8.32	104,849	7.89	16.21	15.07	23.39	29.40
Britannia 300	9	151	8.99	113,322	40.16	49.13	44.80	53.79	34.70
Convair Corronado	10	112	9.16	115,440	40.16	49.30	6.061	69.77	48.30
Boeing 707	11	22	9.34	117,676	103.46	112.80	162.24	171.58	66.70

Source: From Littlechild, S. C., & Owen, G. (1974). *Management Science, 20,* 371 and Littlechild, S. C. (1974). *International Journal of Game Theory, 3,* 27.

As an example, we give in Table 13.1 the total landings at Birmingham airport (Birmingham, UK) during the year 1972. Costs include a direct cost plus a general capital cost. The plane's share of the capital cost has been computed according to both the Shapley value and the nucleolus, and the direct cost added to this to obtain two possible "fair" schedules of fees for the airport. These may be compared with the actual landing fees charged.

Problems

1. For the weighted majority game [5; 3, 2, 2, 1], find the vectors x such that $\langle x; \mathcal{T} \rangle \in \mathcal{M}_1^{(i)}$, for each of the following choices of $\mathcal{T}$:

   (a) $\{\{1, 2\}, \{3\}, \{4\}\}$
   (b) $\{\{1\}, \{2, 3, 4\}\}$
   (c) $\{\{1, 2, 4\}, \{3\}\}$
   (d) $\{N\}$

2. For the game of Problem 1, find all x such that $\langle x; \mathcal{T} \rangle \in \mathcal{K}$ for each choice $\mathcal{T}$ given there.

3. Let v be the seven-person "projective game," a simple game with minimal winning coalitions $\{1, 2, 4\}, \{2, 3, 5\}, \{3, 4, 6\}, \{4, 5, 7\}, \{1, 5, 6\}, \{2, 6, 7\}, \{1, 3, 7\}$.

   (a) Find the nucleolus of v over X, where X is the set of all imputations.
   (b) Find the set of all x such that $\langle x; \{N\} \rangle \in \mathcal{K}$.

4. For a game v, define the prekernel $\mathcal{K}^*$ as the set of all *preimputations* x (see Problem 10.5) such that
$$s_{ij}(x) = s_{ji}(x)$$
   for each pair i, j. Show that, if v satisfies
$$v(S \cup \{i\}) \geq v(S) + v(\{i\})$$
   whenever $i \notin S$, then the kernel and prekernel coincide, in the sense that $x \in \mathcal{K}^*$ if and only if $\langle x; \{N\} \rangle \in \mathcal{K}$.

5. Prove that the nucleolus $\tilde{x}$ [taken over the set $A(v)$ of all imputations] always belongs to $LC(v)$.

6. For a given v and a payoff vector x, let
$$\theta(x) = (\theta_1(x), \theta_2(x), \dots, \theta_{2^n}(x))$$
   be the vector of excesses listed in decreasing order. For real α, let
$$F(x, \alpha) = \sum_{k=1}^{2^n} \alpha^k \theta_k(x)$$

Prove that for all $\alpha > 0$ sufficiently small, the nucleolus $\tilde{x}$ minimizes the function $F(\cdot, \alpha)$ over the set of all imputations.

7. For a game v and any pair i, $j(i \neq j)$ of players, define a *demand function* as a continuous function $d_{ij}(x)$ such that

$$d_{ij} = 0 \quad \text{if} \quad s_{ij}(x) \leq s_{ji}(x)$$

$$0 \leq d_{ij}(x) \leq \min\{\tfrac{1}{2}(s_{ij}(x) - s_{ji}(x)), x_j - v(\{j\})\}$$
$$\text{otherwise}$$

suppose that the vector $x = x(t)$ satisfies the differential equations
$$\tfrac{dx_i}{dt} = \sum_{j \neq i}(d_{ij}(x) - d_{ji}(x)), \quad i = 1, \dots, n$$

(a) Prove that the sum $\sum x_i(t)$ is constant with respect to time.
(b) Show that if $x(0)$ is an imputation, then $x(t)$ will also be an imputation for all $t \geq 0$.
(c) Show that the function

$$F\left(x, \tfrac{1}{2}\right) = \sum_{k=1}^{2^n} 2^{-k}\theta_k(x)$$

is monotonic nondecreasing as a function of t.

8. For $x \in C(v)$, define $\delta_{ij}(x)$ as the largest δ such that the point x', where

$$x'_j = x_i + \delta$$
$$x'_j = x_j - \delta$$

$$x'_k = x_k \quad \text{if} \quad i \neq k \neq j$$

also lies in $C(v)$.
Let $R_{ij}(x)$ be the line segment whose endpoints are x', x'', given by

$$x'_j = x_i + \delta_{ij}, \qquad x''_i = x_i - \delta_{ij},$$
$$x'_j = x_j - \delta_{ij}, \qquad x''_j = x_j + \delta_{ji},$$
$$x'_k = x_k \text{ if } i \neq k \neq j, \quad x''_k = x_k \text{ if } i \neq k \neq j$$

Assuming that $x \in C(v)$, prove that $\langle x, \{N\} \rangle \in \mathcal{K}$ if and only if x bisects the segment $R_{ij}(x)$ for every pair $i \neq j$. (Thus, the set $\mathcal{K} \cap C$ depends, in some sense, only on the geometric shape of the set C.)

Chapter 14

Nonatomic Games

14.1. Games with a Continuum of Players

In the attempt to apply n-person game theory to economic analysis, it becomes quite clear that small games (i.e., games with a small number of players) are hardly adequate to represent free-market situations. We want, for this purpose, games with such a large number of players that any single player will have a negligible effect on the payoffs to the other players. How large should this number be? As large as the number of points on a line. (Generally, this line will be the unit interval, [0,1].) This can best be defined as follows.

14.1.1 Definition. By a *game for a continuum of players* we shall mean a σ-algebra, ϑ of subsets of [0,1], together with a real-valued function v, defined on ϑ, satisfying

(i) $v(\varnothing) = 0$

(ii) $v(A \cup B) \geq v(A) + v(B)$ if $A \cap B = \varnothing$

As before, we point out that the elements of ϑ are called coalitions, while the elements of [0,1] are called players.

One must tread more carefully here than in the case of finite games. For example, the definition of (0,1) normalization must be altered: it is not enough now to demand that $v(\{x\}) = 0$ for any point $x \in [0, 1]$, and $v([0, 1]) = 1$; this would be satisfied by letting v be equal, say, to Lebesgue measure. But a measure is additive, and such a game would therefore be inessential. Instead, we redefined (0,1) normalization by these conditions

$$v(0, 1) = 1 \tag{14.1}$$

$$v(S) \geq 0 \quad \text{if} \quad S \in \vartheta \tag{14.2}$$

$$\text{If } \alpha \text{ is a measure on } \vartheta \text{ and } \alpha \leq v, \text{ then } \alpha \equiv 0 \tag{14.3}$$

A restatement of (14.3) is given by observing that, if $v \geq 0$, we can define the set function

$$\alpha(S) = \inf \sum v(S_i) \tag{14.4}$$

where the infimum is taken over all sequences of sets S_i from ϑ such that $\cup S_i \supset S$. This implies that α is an outer measure. But the superadditivity of v means that this infimum is obtained by taking "fine" partitions of the set S, and this means that all elements of ϑ are measurable under α. Thus α is a measure on ϑ. Now, it is easily seen that α is the largest measure such that $\alpha \leq v$. Thus (14.3) can be replaced by

$$\text{If } \alpha \text{ is as defined by (14.4), then } \alpha \equiv 0 \qquad (14.5)$$

It should be pointed out that, even if v is not non-negative, a signed measure α can be defined in a somewhat modified manner.

In general, this measure, or signed measure, α will replace the numbers $v(\{i\})$. Thus, an imputation for the game v is defined to be any (signed) measure σ such that

$$\sigma([0, 1]) = v([0, 1]) \qquad (14.6)$$

$$\sigma(S) \geq \alpha(S) \quad \text{for} \quad S \in \vartheta \qquad (14.7)$$

If the game is in $(0,1)$ normalization, of course, condition (14.7) is replaced by saying simply that σ must be a measure rather than a signed measure.

The domination relation must also be redefined. In fact, to have $\sigma \vdash \tau$ through a set S, the relation $\sigma(S) \leq v(S)$ must certainly hold; it is not possible, however, to demand that $\sigma(A) > \tau(A)$ for all $A \subset S$, as this can certainly not hold if S is nondenumerable and all one-point sets belong to ϑ. On the other hand, it is certainly not sufficient to demand merely $\sigma(A) \geq \tau(A)$ for $A \subset S$, as this might lead to the relation $\sigma \vdash \sigma$. We compromise by demand that the inequality $\sigma(A) > \tau(A)$ must hold for all subsets A of ϑ, which are such that

$$\sup_{B \in \vartheta} \{v(A \cup B) - v(B)\} > 0$$

This effectively allows us to dispense with "small" (e.g., finite) sets.

With these definitions to imputation and domination, stable sets can be defined exactly as for finite games. In general, not much is known about stable sets for such games — just as little is known about stable sets for, n-person games, when n is large — except for special types. In particular, stable sets will exist for simple games: if S is a minimal winning coalition, the set of all imputations σ such that $\sigma \equiv 0$ on $[0, 1] - S$ will be stable. (We assume $(0,1)$ normalization here.)

Although little work has been done on stable sets, much work has been done on the Shapley value of games for a continuum of players.

Generally, the Axioms S1–S3 given in the definition of the Shapley value in Chapter 12 can be adapted directly, *mutatis mutandis*. A certain difficulty arises, however, with S2 (the symmetry axiom); the permutations of N, which were treated there, must be replaced by one–one measurable transformations of $[0,1]$ into itself. While this might not, at first sight, present any problems, we shall see that it leads to a contradiction in certain degenerate cases.

The actual determination of the value for a particular game, of course, presents considerable difficulty, inasmuch as the formula is not adaptable to the infinite case.

Generally, the method used to determine the value for an arbitrary game is to partition the interval [0,1] into subsets (not necessarily subintervals). These subsets are then treated as single players and the Shapley value for the corresponding finite game is calculated. Finer and finer partitions of [0,1] are taken, and the Shapley value $\varphi[v]$ (an additive set function rather than a vector) is then defined as the limit of the values for these finite games (if this limit exists).

Let us suppose, now, that there is a measure μ defined on ϑ such that $v(S)$ depends merely on $\mu(S)$. For example, we might have $v(S) = [\mu(S)]^2$. Then, by the symmetry axiom, if $\mu(S) = \mu(T)$, we would have $\varphi[v](S) = \varphi[v](T)$. But as $\varphi[v]$ must be an additive set function, it follows that $\varphi[v]$ must be equal to the measure μ, divided by a constant factor to ensure that $\varphi[v]([0, 1]) = v([0, 1])$.

14.1.2 Example. Let $\lambda(S)$ be Lebesgue measure, and let $\mu(S)$ be given by

$$\mu(S) = 2 \int_S x \ dx$$

Now, if v is a function of λ alone, then $\varphi[v] = \lambda$; similarly, if v is a function of μ alone, then $\varphi[v] = \mu$. Suppose, now, that

$$v(S) = \mu(S)\lambda(S) \tag{14.8}$$

for all $S \in \vartheta$. If we write $v_1 = \frac{1}{2}\mu^2$, $v_2 = \frac{1}{2}\lambda^2$, then

$$v_3 = v + v_1 + v_2 = \frac{1}{2}(\lambda + \mu)^2$$

Now, by additivity, we must have

$$\varphi[v] = \varphi[v_3] - \varphi[v_1] - \varphi[v_2]$$

As λ, μ and $\lambda + \mu$ are measures, the right side of this equation is easy to evaluate. In fact, v_3 depends only on $\lambda + \mu$; as $v_3[0, 1] = 2$ $[\lambda + \mu]([0, 1]) = 2$ also, then

$$\varphi[v_3] = \lambda + \mu \tag{14.9}$$

Next, v_1 depends only on μ. We have $v_1([0, 1]) = \frac{1}{2}$; hence

$$\varphi[v_1] = \frac{1}{2}\mu \tag{14.10}$$

and similarly

$$\varphi[v_2] = \frac{1}{2}\lambda \tag{14.11}$$

Putting all these together, we obtain

$$\varphi[v] = \frac{1}{2}(\lambda + \mu) \tag{14.12}$$

as the Shapley value of v defined by (14.8). In a similar way, we can find the Shapley value of any polynomial function of μ and λ, and for that matter, of any finite set of measures.

Let us suppose now, however, that v is given by

$$v(S) = \begin{cases} 0 & \text{if} \quad \lambda(S) < 1 \\ 1 & \text{if} \quad \lambda(S) = 1 \end{cases}$$

Clearly, here, v is a function of λ, and so we must have $\varphi[v] = \lambda$. It is easy to see, however, that $\mu(S) = 1$ if and only if $\lambda(S) = 1$. Thus we will also have

$$v(S) = \begin{cases} 0 & \text{if} \quad \mu(S) < 1 \\ 1 & \text{if} \quad \mu(S) = 1 \end{cases}$$

and so we must have $\varphi[v] = \mu$. This contradiction shows that the Shapley value $\varphi[v]$ cannot exist for this game. What has happened is that the infinite process described above does not have a limit.

14.2. Values of Nonatomic Games

Aumann and Shapley [14.2] have developed an important generalization of the Shapley value to games with a continuum of players.

It is assumed that the set of players is $(I, \mathscr{B})$, where I is the interval $[0,1]$ and $\mathscr{B}$ is the σ-field of Borel sets.

14.2.1 Definition. For any game v, a *null set* is the complement of any carrier. An *atom* is any measurable non-null set S such that, if

$$S = T \cup T', \quad T \cap T' = \varnothing$$

with T, T' measurable, then either T or T' is a null set.

14.2.2 Definition. A game v is *nonatomic* if it has no atoms.

In what follows, we shall deal exclusively with nonatomic games.

Given a game v, the value of v is a finitely additive set function $\varphi[v]$ (a signed measure) defined on the Borel subsets of $[0,1]$. If θ is an *automorphism* of $[0,1]$ (i.e., a one–one, measurable mapping of $[0,1]$ onto itself), then, for any v, we define the game $\theta*v$ by

$$\theta*v(S) = v(\theta(X))$$

for all $S \in \mathscr{B}$. The following axioms seem reasonable.

A1. For any v

$$\varphi[v](I) = v(I)$$

A2. For any game v and automorphism θ

$$\varphi[\theta*v] = \theta*\varphi[v]$$

A3. For games v, w and scalars α, β

$$\varphi[av + \beta w] = \alpha\varphi[v] + \mathscr{B}\varphi[w]$$

A4. If v is monotone (i.e., $v(S) \geq v(T)$ whenever $T \subset S$), then $\varphi[v]$ is also monotone.

Essentially, the first three axioms are very similar to Shapley's three axioms (12.1.3), though A1 is actually weaker than SI. On the other hand, A4 (the *monotonicity* axiom) is new. It seems, however, quite reasonable. After all, if v is monotone, we see that any coalition S always makes non-negative contributions to any set that it may join; it seems only reasonable, then, that it should receive a non-negative value (which is all that A4 requires).

Given these four axioms, we look for some value system φ, which will satisfy them. Unfortunately, we find immediately that such a value cannot be defined for all games. For, letting λ be Lebesgue measure, consider the game v defined by

$$v(S) = \begin{cases} 0 & \text{if} \quad \lambda(S) < 1 \\ 1 & \text{if} \quad \lambda(S) = 1 \end{cases}$$

which we can call the "almost-unanimity" game. Given any two Borel sets S and T, with

$$\lambda(S) = \lambda(T) > 0$$

there is a Lebesgue measure-preserving automorphism θ such that

$$\theta(S) = T$$

It is easy to see, now, that $\theta_* v = v$; thus, we must have, by A2

$$\varphi[v](S) = \varphi[v](T)$$

and, as this holds for any two sets of equal Lebesgue measure, we must have

$$\varphi[v](S) = f(\lambda(S))$$

where f is a monotone function satisfying $f(1) = 1$. Since, moreover, $\varphi[v]$ is an additive set function, we must have

$$f(a + b) = f(a) + f(b)$$

Thus $f \circ \lambda$ is λ itself, and so

$$\varphi[v](S) = \lambda(S)$$

Thus the value $\varphi[v]$ must coincide with Lebesgue measure.

Suppose, however, that μ is some measure equivalent to Lebesgue measure, e.g.,

$$\mu(S) = \int_S 2x\, d\lambda(x)$$

for any $S \in \mathscr{B}$. The two measures μ and λ are equivalent in the sense that $\mu(S) = 0$ if and only if $\lambda(S) = 0$. We see then that v can also be expressed in the form

$$v(S) = \begin{cases} 0 & \text{if} \quad \mu(S) < 1 \\ 1 & \text{if} \quad \mu(S) = 1 \end{cases}$$

By precisely similar reasoning, we conclude that $\varphi[v]$ must coincide with μ. But $\mu \neq \lambda$, and we have reached a contradiction.

There is an evident lesson to be learned from this example, namely, that we cannot hope to define a value φ over the entire space of all nonatomic games — not, i.e., if we wish it to satisfy our axioms. Since the axioms seem reasonable, we conclude that we should restrict ourselves to smaller spaces of games.

Let

$$\Omega = \{S_0, S_1, \ldots, S_m\}$$

be a sequence of sets in $\mathscr{B}$, satisfying

$$\varnothing = S_0 \subset S_1 \subset \cdots \subset S_m = I$$

We call such a sequence a *chain* of sets. For such Ω and a set function v, we write

$$\|v\|_\Omega = \sum_{i=1}^m |v(S_i) - v(S_{i-1})| \tag{14.13}$$

Setting now

$$\|v\| = \sup \|v\|_\Omega \tag{14.14}$$

where the supremum is taken over all possible chains Ω, we call $\|v\|$ the *total variation* of the function v. Let, now BV be the space of all functions of bounded variation (i.e., functions v such that $\|v\|$ is finite).

14.2.3 Theorem. The space BV is a Banach space with norm $\|\bullet\|$.

Proof. We must show that $\|\bullet\|$ is a norm, and that BV is complete in this norm. In fact, it is easily seen that, for any function v, and scalar α

$$\|\alpha v\| = |\alpha| \|v\|$$

as, clearly, $\|\alpha v\|_\Omega = |\alpha| \|v\|_\Omega$ for all Ω. Also, for functions v and w

$$\|v + w\| \leq \|v\| + \|w\|$$

as again $\|v + w\|_\Omega \leq \|v\|_\Omega + \|w\|_\Omega$. Finally, it is clear that, unless v is identically zero

$$\|v\| > 0$$

Thus, the total variation serves as a norm. We must next show completeness. In fact, assume the sequence $v_1, v_2, \ldots, v_n, \ldots$ is Cauchy in the variation norm. Then, it is easy to see that, for each $S \in \mathscr{B}$, the sequence of real numbers $v_1(S), \ldots, v_n(S), \ldots$ is a Cauchy sequence, and so has a limit $v(s)$. We must show $v \in$ BV. In fact, choose N large enough that $\|v_n - v_N\| \le 1$ for all $n \ge N$. Then, for any Ω

$$\|v_n\|_\Omega \le \|v_n - v_N\|_\Omega + \|v_N\|_\Omega \le 1 + \|v_N\|_\Omega$$

and, taking limits as $n \to \infty$, we have

$$\|v\|_\Omega \le 1 + \|v_N\|_\Omega$$

for every Ω. Thus

$$\|v\| \le 1 + \|v_N\|$$

and $v \in$ BV. Finally, it is easy to check that $\|v_n - v\| \to 0$, and thus v is the limit of the sequence, as desired. Thus BV is complete and forms a Banach space.

Given that BV is a Banach space, but that it is not possible to define a value satisfying axioms A1–A4 on that space, we look for closed subspaces of BV on which this can be done.

Let NA^+ be the space of nonatomic measures μ satisfying $\mu(I) = 1$. It is easily seen that any power of such a measure has finite total variation. We shall let pNA denote the closed subspace of BV, which is spanned by powers of NA^+ measures.

Let us suppose, next, that the set function u is monotone. In that case, for any Ω, it is easy to see that

$$\|u\|_\Omega = \sum(u(S_i) - u(S_{i-1})) = u(I)$$

and so

$$\|u\| = u(I)$$

If a function v can be expressed as the difference of two monotone functions $u - w$, then we have

$$\|v\| = \|u - w\| \le \|u\| + \|w\| = u(I) + w(I)$$

and thus every such function has bounded variation. Conversely, we have the following theorem:

14.2.4 Theorem. Every function of bounded variation v can be expressed as the difference of two monotone functions u and w and, moreover, they can be chosen so that

$$\|v\| = \|u\| + \|w\| = u(I) + w(I)$$

Proof. We shall construct the two functions u and w, which we shall call the *upper variation* and *lower variation*, respectively, of v.

Let $S \in \mathscr{B}$, and let Λ be any sequence of sets $S_0, \ldots, S_n$ satisfying

$$\varnothing = S_0 \subset S_1 \subset \cdots \subset S_{n-1} \subset S_n = S$$

Define then

$$u_\Lambda(S) = \sum_{i=1}^{n} \max\{v(S_i) - v(S_{i-1}), 0\} \qquad (14.15)$$

and set

$$u(S) = \sup_\Lambda u_\Lambda(S)$$

where the supremum is taken over all such sequences, Λ. Define, similarly

$$w_\Lambda(S) = \sum_{i=1}^{n} \max\{v(S_{i-1}) - v(S_i), 0\}$$

and

$$w(S) = \sup_\Lambda w_\Lambda(S)$$

It is not difficult to see that, for any Λ

$$u_\Lambda(S) = v(S) + w_\Lambda(S)$$

and therefore

$$u(S) = v(S) + w(S)$$

or, equivalently

$$v(S) = u(S) - w(S)$$

The two functions u and w are the *upper* and *lower variations* of v, respectively, and are easily seen to be monotone. Thus

$$\|u\| = u(I), \quad \|w\| = w(I)$$

Moreover, for any chain Ω

$$\|v\|_\Omega = u_\Omega(I) + w_\Omega(I)$$

and so

$$\|v\| = u(I) + w(I)$$

so that

$$\|v\| = \|u\| + \|w\|$$

as desired.

14.2.5 Corollary. For any $v \in BV$

$$\|v\| = \min(u(I) + w(I))$$

where the minimum is taken over all monotone functions u, w satisfying $u(\varnothing) = 0$, $w(\varnothing) = 0$, $v = u - w$.

Inasmuch as the value mapping φ is to be thought of as a linear mapping from some Banach space into itself, it becomes natural to discuss the norm of φ, $\|\varphi\|$, defined as

$$\|\varphi\| = \max \frac{\|\varphi[v]\|}{\|v\|} \tag{14.16}$$

where the maximum is taken over all functions $v \neq 0$, or even over all v with $\|v\| = 1$. The following theorem is of interest:

14.2.6 Theorem. If φ is a value over BV satisfying A1, A3, and A4, then $\|\varphi\| = 1$.

Proof. Suppose that u is monotone. Then by A4, $\varphi[u]$ is also monotone, and so

$$\|\varphi[u]\| = \varphi[u](I), \quad \|u\| = u(I)$$

By A1, however, $u(I) = \varphi[u](I)$, and so

$$\|\varphi[u]\| = \|u\|$$

Let, now, $v \in \mathrm{BV}$. By Theorem 14.2.4. we can write

$$v = u - w, \quad \|v\| = \|u\| + \|w\|$$

where u and w are monotone. Then we have, by A3

$$\varphi[v] = \varphi[u] - \varphi[w]$$

and, since $\varphi[u]$ and $\varphi[w]$ are monotone

$$\|\varphi[v]\| \leq \|\varphi[u]\| + \varphi[w]\|$$

or

$$\|\varphi[v]\| \leq \|u\| + \|w\|$$

Thus

$$\|\varphi[v]\| \leq \|v\|$$

and we conclude that

$$\max \frac{\|\varphi[v]\|}{\|v\|} = 1$$

or

$$\|\varphi\| = 1$$

Since φ has finite norm, it will follow that this is continuous. Unfortunately, a certain difficulty arises. The problem is that, as we have seen, φ cannot be defined over the entire space BV. We would therefore like to define φ over some subspace $Q \subset$ BV. Unfortunately, we may find that, if we restrict ourselves to elements of Q, it is no longer true that we can write

$$\|v\| = \|u\| + \|w\|$$

where u and w are monotone elements of Q, with $v = u - w$. In that case, the proof given above of Theorem 14.2.6 will no longer be valid. We, therefore, introduce the following definition.

14.2.7 Definition. A subspace Q of BV is said to be *internal* if, for every $v \in Q$, we have

$$\|v\| = \inf\{\|u\| + \|w\|\}$$

where the infimum is taken over all monotone $u, w \in Q$ such that $v = u - w$.

14.2.8 Theorem. If Q is an internal subspace of BV and φ is a value on Q satisfying A1, A3, and A4, then we must have

$$\|\varphi\| = 1$$

Proof. Essentially as for (14.18).

We are now in a position to consider certain subspaces of BV. Perhaps most obvious is the space P, consisting of functions of the form

$$v = p(\mu_1, \mu_2, \ldots, \mu_n)$$

where p is a polynomial, and $\mu_1, \ldots, \mu_n$ are measures in NA$^+$. Unfortunately, P is not closed, because the limit of polynomials need not be a polynomial. Nevertheless, a value φ satisfying the axioms A1–A4 will be uniquely defined on P.

To prove this, we use the following sequence of lemmas:

14.2.9 Lemma. If μ and v are measures, then any homogeneous polynomial of degree n in μ and v can be written as a linear combination of nth powers of measures.

Proof. Let us write

$$p(\mu, v) = \sum_{j=0}^{n} \binom{n}{j} b_j \mu^j v^{n-j}$$

where the $\binom{n}{j}$ are the usual factorial coefficients. Choose $n + 1$ distinct positive numbers $\alpha_0, \alpha_1, \ldots, \alpha_n$, and set

$$\lambda_i = \alpha_i \mu + v$$

which is also a measure. We wish to write

$$p(\mu, v) = \sum_{i=0}^{n} s_i \lambda_i^n$$

or

$$\sum_{j=0}^{n} \binom{n}{j} b_j \mu^j v^{n-j} = \sum_{i=0}^{n} s_i (\alpha_i \mu + v)^n$$

This can be done by equating coefficients of $\mu^j v^{n-j}$ for each j; thus we must have

$$\binom{n}{j} b_j = \sum_{i=0}^{n} \binom{n}{j} s_i \alpha_i^j$$

or equivalently

$$\sum_{i=0}^{n} \alpha_i^j s_i = b_j, \quad j = 0, 1, \ldots, n$$

This is a system of $n+1$ equations in $n+1$ unknowns, and hence will have a solution if the matrix of coefficients is nonsingular. In fact, we have

$$A = \begin{bmatrix} 1 & \alpha_0 & \alpha_0^2 & \cdots & \alpha_0^n \\ 1 & \alpha_1 & \alpha_1^2 & \cdots & \alpha_1^n \\ \vdots & \vdots & \vdots & & \vdots \\ 1 & \alpha_n & \alpha_n^2 & \cdots & \alpha_n^n \end{bmatrix}$$

and the determinant of this matrix is the familiar van der Monde determinant, which is not zero if all the α_i are distinct. Thus, in fact

$$p = \sum_{i=1}^{n} s_i \lambda_i^n$$

as desired. Note, finally, that, if both μ and v are in NA^+, we can also choose λ_i', in NA^+: simply set

$$s_i' = s_i (1 = \alpha_i)^n \tag{14.17}$$

$$\lambda_i' = \frac{\lambda_i}{1 + \alpha_i} \tag{14.18}$$

Then $\lambda_i'(I) = 1$ for each i, and we have

$$p = \sum s_i' \lambda_i'^n$$

as desired.

14.2.10 Lemma. Any homogeneous polynomial of degree n in the measures $\mu_1, \mu_2, \ldots, \mu_k$, in NA^+, can be expressed as a linear combination of nth powers of measures in NA^+.

Proof. We prove by induction on k. In fact, we have already proved this for $k = 2$. Assume this, then, for all numbers up to $k - 1$, and let

$$p(\mu_1, \mu_2, \ldots, \mu_k)$$

be homogeneous of degree n. We can then write

$$p = \sum_{j=0}^{n} q_j(\mu_1, \ldots, \mu_{k-1}) \mu_k^{n-j}$$

where q_j is a polynomial of degree j in $\mu_1, \ldots, \mu_{k-1}$. By the induction hypothesis, we have

$$q_j = \sum_{i=1}^{m} s_{ij} \lambda_{ij}^{j}$$

where each λ_{ij} is a measure in NA^+. Thus

$$p = \sum_{j=0}^{n} \sum_{i=1}^{m} s_{ij} \lambda_{ij}^{j} \mu_k^{n-j}$$

and this is a linear combination of terms

$$\lambda_{ij}^{j} \mu_k^{n-j}$$

each of which is a homogeneous polynomial of degree n in λ_{ij} and μ_k. By the previous lemma, each of these can be expressed as a linear combination of nth products of measures:

$$\lambda_{ij}^{j} \mu_k^{n-j} = \sum t_{ijl} \lambda_{ijl}^{n}$$

and so

$$p = \sum_i \sum_j \sum_l s_{ij} t_{ijl} \lambda_{ijl}^{n} \tag{14.19}$$

as desired.

14.2.11 Theorem. Any polynomial of degree n in measures $\mu_1, \ldots, \mu_k$ in NA^+ can be written as a linear combination of nth, or lower, powers of measures in NA^+.

Proof. This follows directly from (14.21) and the fact that each such polynomial can be written as the sum of $n + 1$ homogeneous polynomials, each of degree n or lower.

We can now state the following:

14.2.12 Theorem. There exists a unique value φ defined on the space P of polynomials in NA^+ measures.

Proof. (Uniqueness). Let μ be a measure in NA$^+$, and suppose

$$v = \mu^n$$

If S and T are Borel sets with

$$\mu(S) = \mu(T)$$

then there exists a measurable transformation θ of I, which transforms S into T and preserves μ-measure. Since $v = \theta_* v$, by A2 we conclude that we must have

$$\varphi[v](S) = \varphi[v](T)$$

and we see that $\varphi[v]$ must depend only on the measure μ. Now, $\varphi[v]$ must be an additive set function, and so we must have

$$\varphi[v] = c\mu$$

for some constant c. However

$$v(I) = \mu^n(I) = 1$$

and so, by A1, we must have

$$\varphi[v](I) = 1$$

We conclude $c = 1$, and so

$$\varphi[\mu^n] = \mu$$

for any $\mu \in$ NA$^+$, and any integer n.

Let, now, v be any polynomial in measures $\mu_i \in$ NA$^+$. As we saw, this can be rewritten in the form

$$v = \sum_i \sum_j s_{ij} \lambda_{ij}^j$$

where the λ_{ij} are in NA$^+$, and the s_{ij} are scalars. Then, by A3, we must have

$$\varphi[v] = \sum \sum s_{ij} \lambda_{ij}$$

We see, thus, that there can be at most one value φ satisfying A1–A3 over the space P. The question now is as to existence, i.e., it is possible that we might write

$$v = \sum \sum s_{ij} \lambda_{ij}^j = \sum \sum t_{ij} v_{ij}^j$$

and this might give two different values for $\varphi[v]$.

To prove that this cannot happen, we use Lyapunov's theorem, which states that, if $\mu_1, \ldots, \mu_k$ are nonatomic measures on $[0,1]$, then for any $S \in \mathcal{B}$ and any t, $0 < t < 1$, there exists a set $tS \subset S$ such that

$$\mu_i(tS) = t\mu_i(S), \quad i = 1, \ldots, k$$

Assume, then, that

$$v = \sum_{i=1}^{k} s_i f_i \circ \mu_i$$

where μ_i, is a measure in NA^+, each f_i is continuously differentiable over $[0,1]$, with $f_i(0) = 0, f_i(1) = 1$, and the s_i are real numbers. For a given S and t, we define sets $tS \subset S$ and $t(I - S) \subset I - S$ such that

$$\mu_i(tS) = t\mu_i(S), \quad i = 1,\ldots,k$$

$$\mu_i(t(I - S)) = t - t\mu_i(S)$$

For any $\tau, 0 < \tau < 1 - t$, we can define a set $\tau S \subset S$ such that

$$\tau S \cap tS = \varnothing$$

$$\mu_i(\tau S) = \tau\mu_i(S), \quad i = 1,\ldots,k$$

Set, now

$$tI = tS \cup t(I - S)$$

so that

$$\mu_i(tI) = t, \quad i = 1,\ldots,k$$

and we see that

$$\lim_{\tau \to 0} \frac{v(tI \cup \tau S) - v(tI)}{\tau} = \sum_{i=1}^{k} s_i f_i'(t)\mu_i(S) \tag{14.20}$$

Integrating with respect to t gives us

$$\int_0^1 \lim \frac{v(tI \cup \tau S) - v(tI)}{\tau} dt = \sum_{i=1}^{k} s_i \mu_i(S) \tag{14.21}$$

To see, now, that any two representations of the same v lead to the same $\varphi[v]$, it suffices to prove that, for v identically 0, we necessarily obtain $\varphi[v] = 0$. But (14.21), in the form

$$\varphi[v](S) = \int_0^1 \lim \frac{v(tI \cup \tau S) - v(tI)}{\tau} dt \tag{14.22}$$

clearly shows this, as the integrand obviously vanishes if $v = 0$.

The above shows that the expression (14.20), as a value $\varphi[v]$, is well defined. It remains, of course, to show that it satisfies the four axioms, A1–A4, given for values. In fact, this is not difficult to prove, especially if we use (14.22). This will be left as an exercise to the reader.

We see, thus, that our axioms lead us to define a unique value $\varphi[v]$, given by either expression (14.20) or (14.22), for all games $v \in P$ (polynomials in NA^+ measures).

We would like, now, to extend this value to larger spaces. One obvious extension would be to the space of polynomials in signed measures (i.e., additive functions, which take negative values). This is, however, no extension at all, since, by a well-known theorem of measure theory, any signed measure can be expressed as the difference of two measures (the Hahn decomposition). Thus any polynomial in signed measures is actually a polynomial in measures.

Since P is not a closed subspace (i.e., the limit of polynomials is not necessarily a polynomial), we might consider its closure, which is designated pNA. This space will contain all functions of the form

$$v = f \circ (\mu_1, \mu_2, \ldots, \mu_n) \tag{14.23}$$

where the $\mu_i \in NA^+$ and f are functions of n variables, continuously differentiate over the cube $[0, 1]^n$ and satisfying $f(0) = 0$. (In fact, it is not even necessary that f be defined over the entire unit cube; it suffices that it be continuously differentiate over R, where R is the range of the mapping

$$\mu : \mathscr{B} \to [0, 1]^n$$

and $\mu = (\mu_1, \ldots, \mu_n)$ is the n-vector of measures.)

It can be proved that the space pNA is internal. Thus any value on pNA, satisfying the axioms A1–A4, must be continuous. (In fact, by Theorem 14.2.12, its norm must be 1.) We have seen that there is a unique value φ defined on the space P; since pNA is the closure of P, it continuity tells us that there can be at most one extension of this from P to pNA.

Suppose, then, that v is as in (14.23). Assuming that the function f is defined over the entire cube $[0, 1]^n$, let

$$\alpha_i = \int_0^1 f_i(t, t, \ldots, t) dt \tag{14.24}$$

Then

$$\varphi[v] = \sum_{i=1}^n \alpha_i \mu_i \tag{14.25}$$

will be the desired value. Much as in the case of the space P, it may be seen that this also gives us

$$\varphi[v](S) = \int_0^1 \lim_{\tau \to 0} \frac{v(tI \cup \tau S) - v(tI)}{\tau} dt$$

14.2.13 Example. Let μ_1, μ_2, μ_3 be three distinct measures on $[0,1]$, with $\mu_i(I) = 1$, and suppose

$$v = \mu_1 \mu_2 + \mu_2 \mu_3^2$$

We have two possibilities here: we might write

$$v = \tfrac{1}{2}(\mu_1 + \mu_2)^2 - \tfrac{1}{2}\mu_1^2 - \tfrac{1}{2}\mu_2^2 - \tfrac{1}{3}(\mu_2 + \mu_3)^3 + \tfrac{1}{6}(\mu_2 + 2\mu_3)^3 + \tfrac{1}{6}\mu_2^3 - \mu_3^3$$

or

$$v = 2\left(\frac{\mu_1 + \mu_2}{2}\right)^2 - \frac{1}{2}\mu_1^2 - \frac{1}{2}\mu_2^2 - \frac{8}{3}\left(\frac{\mu_2 + \mu_3}{2}\right)^3 + \frac{9}{2}\left(\frac{\mu_2 + 2\mu_3}{3}\right)^3 + \frac{1}{6}\mu_2^3 - \mu_3^3$$

so that v is a linear combination of powers of measures in NA^+. Then by (14.20)

$$\varphi[v] = 2\left(\frac{\mu_1 + \mu_2}{2}\right) - \frac{1}{2}\mu_1 - \frac{1}{2}\mu_2 - \frac{8}{3}\left(\frac{\mu_2 + \mu_3}{2}\right) + \frac{9}{2}\left(\frac{\mu_2 + 2\mu_3}{3}\right) + \frac{1}{6}\mu_2 - \mu_3$$

or, simplifying

$$\varphi[v] = \tfrac{1}{2}\mu_1 + \tfrac{5}{6}\mu_2 + \tfrac{2}{3}\mu_3$$

Alternatively, we can use formula (14.24), which gives us

$$\alpha_1 = \int_0^1 t \; dt$$

$$\alpha_2 = \int_0^1 (t + t^2)dt = \tfrac{5}{6}$$

$$\alpha_3 = \int_0^1 2t^2 dt = \tfrac{2}{3}$$

and so

$$\varphi[v] = \tfrac{1}{2}\mu_1 + \tfrac{5}{6}\mu_2 + \tfrac{2}{3}\mu_3$$

as before.

14.2.14 Example. Let μ_1, μ_2, μ_3 be as before, and let

$$v = \mu_1 + \mu_2(1 - \cos \mu_3)$$

Clearly, $v \in pNA$. We have, by (14.24)

$$\alpha_1 = \int_0^1 dt = 1$$

$$\alpha_2 = \int_0^1 (1 - \cos t)dt = 1 - \sin 1 \cong .16$$

$$\alpha_3 = \int_0^1 t \sin t \; dt = \sin 1 - \cos 1 \cong .30$$

Thus

$$\varphi[v] = \mu_1 + .16\mu_3 + .30\mu_3$$

We might consider one further space, and this is the space bv′NA generated by functions of the form

$$v = f \circ \mu,$$

where, as before, μ is a measure in NA^+ and f is a function of bounded variation, continuous at 0 and 1, and satisfying $f(0) = 0$. We specifically forbid discontinuities at $\mu(S) = 0$ and $\mu(S) = 1$ to avoid the problem of the "μ-near-unanimity" game discussed above.

The strict mathematical development of a value φ on the space bv′NA is quite complex; readers are invited to read [14.2] for details. Essentially, it is based on the fact that any function $v \in$ bv′NA can be expressed as the sum of a "differentiable" part v_p, and a "singular" part v_s. The differentiable part, v_p, can be shown to belong to the space pNA, which we discussed above, while the singular part, v_s, belongs to the space s′NA, which can best be defined by saying that it is the closed subspace of BV generated by functions of the form

$$v = f \circ \mu$$

where μ is a *signed measure* in NA and f is a step-function on the unit interval, continuous at 0 and 1, and satisfying $f(0) = 0$. We have, then, the following theorem:

14.2.15 Theorem. There exists a unique value on the space s′NA satisfying A1–A4.

Proof. Essentially, the proof of uniqueness is obtained by noting that, if $v = f \circ \mu$ as above, then the symmetry axiom will require that

$$\varphi[v] = f(1)\mu \tag{14.26}$$

Thus, we have

$$\varphi\left[\sum f_i \circ \mu_i\right] = \sum f_i(1)\mu_i \tag{14.27}$$

by the linearity axiom. For other members of s′NA, uniqueness is obtained by noticing that the space s′NA is internal, and thus the monotonicity axioms imply continuity. Thus there can exist at most one value on this space. To prove existence, it is necessary to show that the value given by (14.27) does in fact satisfy the axioms.

It remains, now, to obtain a value on the space bv′NA. To do this, we notice that, just as any function of bounded variation (of a real variable) can be expressed as the sum of an absolutely continuous function and a singular function in a unique manner, so also any set function in bv′NA can be expressed as the sum of a function in s′NA and one in pNA in a unique manner. Thus

$$\text{bv′NA} = \text{pNA} \oplus \text{s′NA} \tag{14.28}$$

where the symbol $\oplus$ denotes that this is a direct sum of two vector spaces.

Now, a unique value (satisfying the axioms) exists on each of the two spaces pNA and s'NA. Using φ_1 and φ_2 to denote these, we can then write, for any $v \in bv'NA$

$$\varphi[v] = \varphi_1[v_p] + \varphi_2[v_s]$$

where v_p and v_s are the differentiable and singular parts of v. Since these are uniquely defined by v, it will follow that φ given by (14.27), is well defined for all $v \in bv'NA$. Moreover, φ will satisfy the Axioms A1–A4. Thus,

14.2.16 Theorem. There exists a unique value on the space $bv'NA$ satisfying the Axioms A1–A4.

14.3.   Internal Telephone Billing Rates — An Example

As a possible application of nonatomic game theory, let us consider the problem of allocating the costs of telephone calls made from a certain large institution. For cost accounting purposes, it is desirable to charge the calls to the several projects or contracts that the institution is undertaking; this may be especially important in case of a research institute or university, where several projects may be going on independently of each other, and "fairness" in billing is desired.

Given the present policy of the telephone companies, long-distance telephone calls can be purchased in any of several ways. The "normal" method is direct distance dialing (DDD), according to which each call is billed individually. Individual home owners and small companies usually obtain their long-distance calls in this manner as it is relatively inexpensive for small users. As the number of calls increases, however, the institution will normally find it more economical to obtain either a direct line to a distant city (e.g., Washington, where it wishes contact with funding agencies) or Wide Area Telephone Service (WATS), which, for a sizable initial payment, allows a certain number of calls within a given zone; calls above this basic number are billed individually.

In reality, there are two problems here. First, the institution must determine the system configuration, which will minimize total cost; second, it must allocate these costs internally to the several projects using the telephone system. The first is not, strictly speaking, a game-theoretic problem; its solution is, however, necessary if we are to solve the game-theoretic problem of cost allocation.

Essentially, it may be found that, for a given choice of configuration (e.g., two WATS lines plus one direct line to another city), total costs will depend not only on the total number (and length) of calls made to each destination, but also on the times at which these calls are made. This is due to the fact that, if too many calls are made simultaneously, the WATS lines will be unable to accommodate them all; hence, some will be routed as DDD calls, with a consequently higher cost. Typically, we can divide the day into 24 time periods and let μ_{ij} denote the total minutes of calls in time period j to destination i. Then the cost to the institution, given the kth configuration, is given by

$$C = f_k(\mu_{11}, \ldots, \mu_{mn})$$

where the function f_k is absolutely continuous and piecewise, at least, continuously differentiable.

For optimal choice of the system configuration, costs will be

$$g(\mu_{11}, \ldots, \mu_{mn}) = \min_k f_k(\mu_{11}, \ldots, \mu_{mn})$$

and once again g will be absolutely continuous in the mn variables μ_{ij}, with — at least — one-sided derivatives everywhere.

We are now in a position to allocate the costs. Given the nature of the institution, the following seem reasonable criteria for allocation:

1. Total billings to the several accounts should be just enough to cover the total costs of the service.
2. All calls to a given destination in a given time period should be billed at the same rate, regardless of the person making them.
3. All calls should be charged at least a non-negative amount.

It may, of course, be seen that these correspond to three of the four axioms, which characterize the value for nonatomic games. We may therefore represent this as a nonatomic game in which each moment of telephone call is to be treated as a player. The value of this nonatomic game can then be obtained by the use of formula (14.25).

Thus, if we set

$$g_{ij} = \frac{\partial g}{\partial \mu_{ij}}$$

then we will have

$$\alpha_{ij} = \int_0^1 g_{ij}(t\mu_{11}, \ldots, t\mu_{mn}) dt$$

as the per-minute rate of calls to the ith destination in time period j. Thus, account S would be charged a total amount

$$\varphi[v](S) = \sum \alpha_{ij} s_{ij}$$

where s_{ij} is the total length (in minutes) of calls to the ith destination in the jth time period, chargeable to account S.

Table 14.1 shows examples of these rates, to two different destinations, over the day. As can be seen, these are relatively low during the night hours, then increase during the day to reach a peak when the frequency of calls is greatest.

Table 14.1: Examples of rates charges per minute of call (from Ithaca, New York) to Two Different Destinations, by Hour of Day[a].

Hour	Washington, D.C.	Syracuse, NY
12–8 a.m.	8.4	8.7
8–9	20.3	19.5
9–10	23.0	22.3
10–11	24.1	23.1
11–12 noon	25.2	21.4
12–1 p.m.	20.8	19.1
1–2	23.3	21.0
2–3	26.8	22.6
3–4	26.1	21.9
4–5	23.5	19.7
5–11	13.4	14.2
11–12 midnight	8.4	8.7

[a]Data taken from Billera, L. J., Heath, D. C., & Raanan, J. C. (1978). *Operations Research, 27*, 963.

Problems

1. Let λ and μ be two nonatomic measures such that $\lambda(I) = \mu(I) = 1$, and let

   $$v = 3\mu + \sin^2(\lambda + 2\mu)$$

   Find $\varphi[v]$.

2. Let g be a function of bounded variation on the interval $[0,1]$, continuous at 0 and 1, and satisfying $g(0) = 0$. Let $\|g\|$ be the total variation of the function g over $[0,1]$, Prove that, if μ is a measure with $\mu(I) = 1$

   $$\|g \circ \mu\| = \|g\|$$

3. (Infinite market games.) Consider a situation in which each "player" s, $0 \le s \le 1$, in a game has a certain quantity $a(s)$ of a commodity as initial endowment and has a utility $u(x,s)$ for an amount x of this commodity. Assuming side payments are possible, we then obtain the characteristic function

   $$v(S) = \max\left\{ \int_I u(x(s), s)d\lambda(s) \right\}$$

   subject to

   $$\int_I x(s)d\lambda(s) = \int_I a(s)d\lambda(s)x(s) \ge 0$$

   where λ is Lebesgue measure. Consider the case where

$$a(s) = 8s$$

$$u(x, s) = \sqrt{x+1} - 1 \quad \text{for all} \quad s$$

(a) Show that v can be expressed in closed form as a function of λ and μ, defined by

$$\mu(S) = \int_S 2t \; dt$$

Use this closed form, together with the (14.24) and (14.25), to compute the value $\varphi[v]$.

(b) Show that this value lies in the core of the game v.

4. (The asymptotic approach.) The asymptotic approach, due to Kannai, consists in partitioning the space I of players into a finite collection

$$\Pi = \{S_1, S_2, \ldots, S_m\}$$

of disjoint measurable sets. A "quotient" game V_Π can then be defined by

$$V_\Pi(T) = V\left(\bigcup_{j \in T} S_j\right)$$

for every $T \subset M = \{1, 2, \ldots, m\}$.

A sequence $\Pi_1, \Pi_2, \ldots$ of partitions is said to separate points if (a) each Π_{k+1} is a refinement of Π_k; (b) given $x \neq y \in I$, there is some k: such that x and y are in different elements of Π_k.

For $A \in \mathscr{B}$, let $\Pi_1 = \{A, I - A\}$, and let $= \{\Pi_1, \Pi_2, \ldots\}$ be a sequence that separates points. Define

$$\varphi[V](A) = \lim_k \left\{\sum_i \varphi_i[V_{\Pi_k}]\right\}$$

where the sum is taken over all i such that the sets $S_{ik} \in \Pi_k$ are subsets of A. If this limit exists and is independent of the particular sequence chosen, then this is the *asymptotic value of v on the set A, $\hat{\varphi}[V](A)$.*

Prove that, if $v = f \circ \mu$, where μ is a nonatomic measure, and f is continuously differentiable, with $f(0) = 0$, and $f(1) = 1$, then

$$\hat{\varphi}[V](A) = \mu(A) \quad \text{for all} \quad A \in \mathscr{B}.$$

[*Hint*: Show that the partitions will, for sufficiently large k, satisfy

$$\mu(S) < \varepsilon \quad \text{for all} \quad S \in \Pi_k$$

Show, then, that, if $w = v'v''$

$$g(q) = h'(q)h''(q) + O(\varepsilon)$$

where g, h', h'' are the multilinear extensions (MLEs) of $w_{\Pi_k}, v'_{\Pi_k}, v''_{\Pi_k}$. Use the fact that continuous functions can be uniformly approximated by polynomials.]

5. (Extensions of games.) Define a *fuzzy subset* of I as a function from $I = [0, 1]$ into itself. A *measurable fuzzy set* is a Borel-measurable function from I to itself. Let $\mathscr{F}$ be the collection of all measurable fuzzy sets on I. A *fuzzy set function* is a function v^* from $\mathscr{F}$ to the reals.

Monotonicity of fuzzy set functions is defined in the obvious manner, i.e., v^* is monotone if $v^*(f) \leq v^*(g)$ whenever $f \leq g$.

Prove that there exists a unique mapping, which assigns, to each $v \in \mathrm{pNA}$, a fuzzy set function v^* satisfying

(1) $(\alpha v + \beta w)^* = \alpha v^* + \beta w^*$
(2) $(uv)^* = u^* v^*$
(3) $u^*(f) = \int_I f d\mu$
(4) v monotonic $\Rightarrow v^*$ monotonic

whenever v, $w \in \mathrm{pNA}$, α, $\beta \in R$, $\mu \in \mathrm{NA}$, and $f \in \mathscr{F}$.

[*Hint*: (a) Show that if $v = P(\mu_1, \ldots, \mu_m)$, where the μ_i, are measures, then by (1)–(3), $v^*(f) = P(\beta_1, \ldots, \beta_n)$ where

$$\beta_i = \int_I f d\mu_i$$

Then show that (4) guarantees continuity of the mapping, and use the fact that the polynomials of measures are dense in pNA, to prove uniqueness.

(b) To show existence, let $v_m (m = 1, 2, \ldots)$ be a sequence of finite approximations to the game v, as described in Problem 4, and let v_m^* be the MLE of v_m. Show that, for $v \in \mathrm{pNA}'$, the v_m^* converge to a function v^*, which has the desired properties.

(c) Show, finally, that if χ_S is the characteristic function of the set S (i.e., $\chi_S(s) = 1$ if $s \in S$, and 0 otherwise) then

$$v^*(\chi_S) = v(S)$$

Thus v^* is truly an extension of v.]

Chapter 15

Games without Side Payments

15.1. Introduction

Throughout the last five chapters, we have made the implicit assumption that utility is freely transferable among members of a coalition. This is, in particular, possible in the presence of an "ideal money"—i.e., a commodity whose utility is directly proportional to quantity, and independent of any other assets, or liabilities, which a player may have. Whatever the realism of such a situation, its principal effect is to simplify enormously the description of the game (in characteristic function form). In fact, a single number $v(S)$ is fully adequate to describe the coalition's possibilities: it can obtain this total amount in some manner, and then make side payments of money to satisfy any complaints that its members might make.

In general, unfortunately, the situation is not this simple. Even if money is available in large amounts (which is not always the case), we find that the several players' utility for money is not linear (e.g., the utility of the 1st dollar is much greater than that of the 1001st), and not always independent of their other assets. In some cases, side payments may even be forbidden (e.g., by antitrust or antikickback laws). This being so, we must represent each coalition's possibilities, not by a single number, but rather by a set: the set of all points that the coalition can obtain for its members (cf. our treatment of the two-person bargaining game in Chapter 9).

An important concept here will be Pareto-optimality. This was discussed, in the two-person context, in Chapter 9. More generally, we will say that the point $x = (x_1, \ldots, x_n)$ is Pareto-optimal over the set C if there is no $y \in C$ such that $y_i \geq x_i$, for all i, with inequality holding for at least one value of i. We will also say that x is Pareto-optimal for the coalition S over the set C, if there is no $y \in C$ such that $y_i \leq x_i$ for all $i \in S$, with inequality holding for at least one $i \in S$.

For any $S \subset N$, we shall let $\sum_S$ denote the set of all jointly correlated mixed strategies for the coalition S. Then, an (expected) outcome π will be determined by a choice of strategies $\sigma^{S_1} \in \sum_{S_1}, \sigma^{S_2} \in \sum_{S_2}, \ldots, \sigma^{S_m} \in \sum_{S_m}$, for any partition $\{S_1, S_2, \ldots, S_m\}$, of the player set N. There are, now, two possible meanings to the statement that "the coalition can obtain [some point] for its members." These correspond to the notions of α-effectiveness and β-effectiveness.

15.1.1 Definition. The coalition S is α-effective for the vector x if there is a strategy $\sigma^S \in \sum_S$ such that, for any $\sigma^{N-S} \in \sum_{N-S}$

$$\pi_i(\sigma^S, \sigma^{N-S}) \geq x_i \quad \text{for all} \quad i \in S$$

15.1.2 Definition. The coalition S is β-effective for x if, for any strategy $\sigma^{N-S} \in \sum_{N-S}$, there is some $\sigma^S \in \sum_S$ such that

$$\pi_i(\sigma^S, \sigma^{N-S}) \geq x_i \quad \text{for all} \quad i \in S$$

It is easy to see that α-effectiveness implies β-effectiveness. The converse, however, is not generally true. It certainly holds for finite two-person games (this is the minimax theorem), but even for infinite two-person games, it fails to hold. Assuming finiteness, it may be seen that the two types of effectiveness are equivalent if utility can be transferred linearly among members of S.

Whichever definition of effectiveness is used, the set $v(s)$ can then be described as the set of all payoff vectors for which S is effective. We have then

15.1.3 Definition. An n-person game without side payments is a pair (v, H) where H is a compact subset of R^N and v is a function that assigns to each $S \subset N$ a closed convex subset of R^N satisfying

  (i) $v(S)$ is non-empty.
 (ii) If $x \in v(S)$ and $y_i \leq x_i$ for all $i \in S$, then $y \in v(S)$.
(iii) If $S \cap T = \emptyset$, then $v(S) \cap v(T) \subset v(S \cup T)$.
 (iv) $x \in v(N)$ if and only if $x \leq y$ for some $y \in H$.

In this definition, (ii) is a "comprehensiveness" condition: if S can obtain x, then it can *a fortiori* obtain the smaller amounts y. Condition (iii) is superadditivity in a new setting. Condition (iv) guarantees that $v(N)$ is not too large; this, together with (iii), letting $T = N - S$, guarantees that $v(S)$ is not too large.

With this in mind, we can now define domination $x \vdash y$, through the coalition S, if and only if $x_i > y_i$ for all $i \in S$ and $x \in v(S)$. It is then possible to define von Neumann–Morgenstern solutions (stable sets) and the core of a game.

Some work has been done on stable sets: Stearns [15.15] has proved that all three-person games have stable sets, and has moreover catalogued all such sets. On the other hand, he has also given a seven-person game with no stable sets (cf. the case of games with side payments, where a 10-person game without stable sets has been constructed).

15.1.4 Example. Three players are to decide, by majority vote, the location of a certain facility. Each of the three wishes the facility to be located as close as possible (in a physical sense) to his own "ideal" location. Utility is, essentially, equal to the negative of the distance, i.e., player i's utility is

$$u_i(p, q) = -\sqrt{(p - \alpha_i)^2 + (q - \beta_i)^2} \tag{15.1}$$

where (p, q) is the location chosen for the facility and (α_i, β_i) is player i's ideal location. No side payments are permitted. It is required that (p, q) lies inside a compact convex region Ω; it is assumed that $(\alpha_i, \beta_i) \in \Omega$ for $i = 1, 2, 3$.

Figure 15.1 shows the situation: the points P_1, P_2, P_3, represent the three players' ideal positions, which in the general case will form a triangle. Let us set

$$d_{ij} = \sqrt{(\alpha_i - \alpha_j)^2 + (\beta_i - \beta_j)^2} \qquad (15.2)$$

as the "distance" between players i and j.

We find now that the coalition N can of course obtain any point in Ω. It is clear, however, that only points in the triangle need be considered. Consider a point such as a, in Figure 15.1, outside the triangle. If b is the perpendicular projection of a onto the nearest side (in this case $P_1 P_2$), it is clear that $u^i(b) > u^i(a)$ for $i = 1, 2, 3$. In other words, all three players prefer b to a. Similarly, given any point outside the triangle, there will be, in the triangle, some point that all players prefer. On the other hand, given a point such as c in the triangle, it may be seen that things are quite different: we cannot move to another point without decreasing at least one player's

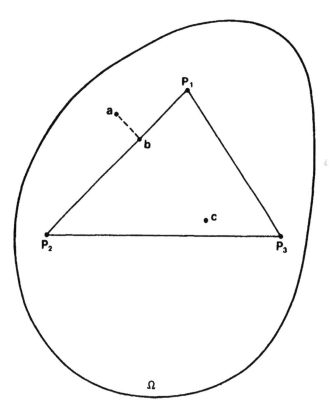

Figure 15.1.

utility. Thus the points in the triangle are said to be Pareto-optimal: it is impossible to increase one player's utility without decreasing that of another.

We are now in a position to describe the sets $v(s)$. For N, we find that a point (x_1, x_2, x_3) can be attained if, for each i, $x_i = u_i(p, q)$, where (p, q) lies in the triangle $P_1 P_2 P_3$. Moreover, by (ii) in Definition 15.1.3, any point (y_1, y_2, y_3) belongs to $v(N)$ if $y_i \leq u_i(p, q)$ for some such (p, q). Thus $v(N)$ will consist of:

(a) the set H, which is the image, under transformation (15.1), of triangle $P_1 P_2 P_3$ and
(b) all x such that $x \leq y$ for some $y \in H$.

For the two-person coalitions, the situation is somewhat simpler. Coalition $\{i, j\}$ $(i \neq j)$ can of course obtain any point $(p, q) \in \Omega$. From their point of view, however, the points on the line segment $P_i P_j$ are Pareto-optimal. Now, if (p, q) lies on the segment $P_i P_j$, we clearly have

$$u^i(p, q) + u^j(p, q) = -d_{ij}, \quad u^i, u^j \leq 0$$

Thus the set $v(\{i, j\})$ consists of all points satisfying

$$x_i + x_j \leq -d_{ij}, \quad x_i, x_j \leq 0$$

Finally, a single player can enforce nothing, he can be sure only that he will receive at least $-m_i$, where

$$m^i = \max \left\{ \sqrt{(p - \alpha_i)^2 + (q - \beta_i)^2} \right\}$$

subject to $(p, q) \in \Omega$. Thus, $v(\{i\})$ consists of all points satisfying

$$x_i \leq -m_i$$

We might look, now, for the von Neumann–Morgenstern solutions of this game. Letting

$$s = \frac{d_{12} + d_{13} + d_{23}}{2}$$

be the semiperimeter of the triangle, we note that the three circles centered at P_1, P_2, and P_3 with radii $s - d_{23}$, $s - d_{13}$, $s - d_{12}$, respectively, are mutually tangent, meeting at the points X, Y, Z, respectively, on the sides of the triangle (see Figure 15.2). These three points give the payoffs (in utility)

$$(d_{23} - s, d_{13} - s, w_3), \quad (d_{23} - s, w_2, d_{12} - s),$$
$$(w_1, d_{13} - s, d_{12} - s)$$

Figure 15.2.

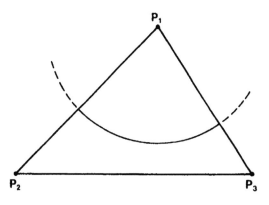

Figure 15.3.

where the w_i can be evaluated (using trigonometry) but certainly satisfy

$$w_i \leq d_{jk} - s$$

with equality holding only in the special case that P_i lies on the line segment $P_j P_k$. It may be verified that these three points form a three-point solution to the game. (In the special case that P_i lies on the segment $P_j P_k$, then $X = Y = Z = P_i$, and the three-point solution degenerates to a single point, which is still a solution — and, moreover, the unique point in the core.)

Thus, our three-person game has a three-point solution, similar to the symmetric solution for the three-person game of Example 11.1.2. It will also have the analogue of the discriminatory solutions studied there (see Figure 15.3). For, let K be any circle, centered at P_1 such that (a) its radius is greater than $s - d_{23}$, and (b) it does not intersect (though it may touch) the line segment P_2P_3. Then the arc of K, which lies inside the triangle, is also a von Neumann–Morgenstern solution.

As can be seen, the behavior of this game is very similar to that of the game with side payments given in Example 11.1.2. The reader should beware: such behavior cannot generally be expected of games without side payments.

15.2.   The Core

We now study the core of games without side payments. Just as for games with side payments, we define the core to be the set of all undominated imputations, i.e., $x \in C(v)$ if and only if:

(a) $x \in v(N)$,
(b) there is no $S \subset N$, $S \neq \varnothing$, and $y \in v(S)$, such that $y_i > x_i$, for all $i \in S$.

Clearly, the existence (or, rather the nonemptiness) of the core depends somehow on the relation among the several sets $v(S)$, and $v(N)$ in particular. For games with side payments, it was seen that this relation can best be expressed in terms of balanced collections. For the present games, we find that some similar results hold. We state these in the form of a theorem:

15.2.1 Theorem. A sufficient condition for the game v to have a nonempty core is that the inclusion

$$\bigcap_{j=1}^{m} v(S_j) \subset v(N) \tag{15.3}$$

holds for any balanced collection $\{S_1, \ldots, S_m\}$.

Before we attempt to prove Theorem 15.2.1, we might wish to compare it with Theorem 10.5.10, dealing with side payments games. The restriction to minimal balanced collections in 10.5.10 is not important: we saw that the condition of 10.5.10 will hold for all balanced collections if it holds for the minimal ones; the same statement is clearly true for the condition of Theorem 15.2.1, since increasing a collection $\{S_1, \ldots, S_m\}$ will merely decrease the set $\cap v(S)$.

More to the point is the fact that the conditions in the two theorems are, in fact, equivalent. In fact, let f be the characteristic function of a game with side payments, then

$$v(S) = \left\{ x \mid \sum_{i \in S} x_i \leq f(S) \right\} \tag{15.4}$$

expresses this as a game without side payments.

Suppose, now, that f satisfies the hypothesis of 10.5.10. Let $\mathscr{B} = \{S_1, \ldots, S_m\}$ be balanced, and suppose

$$x \in \bigcap_{j=1}^{m} v(S_j)$$

Then

$$\sum_{i \in S_j} x_i \le v(S_j), \qquad j = 1, \ldots, m$$

and, if $\mathscr{B}$ has the balancing vector $(\lambda_1, \ldots, \lambda_m)$, we have

$$\sum_{j=1}^{m} \lambda_j \sum_{i \in S_j} x_i \le \sum_{j=1}^{m} \lambda_j f(S_j),$$

or

$$\sum_{i \in N} \left(\sum_{\substack{j \\ i \in S_j}} \lambda_j \right) x_i \le \sum \lambda_j f(S_j)$$

and so

$$\sum_{i \in N} x_i \le \sum \lambda_j f(S_j)$$

This gives us, then

$$\sum_{i \in N} x_i \le f(N)$$

and so $x \in v(N)$. Thus $\cap v(S_j) \subset v(N)$.

Conversely, let $\mathscr{B}$ be a minimal balanced collection with balancing vector λ such that

$$\sum_{j=1}^{m} \lambda_j f(S_j) > f(N)$$

We can then obtain a vector x satisfying

$$\sum_{i \in S_j} x_i = f(S_j), \quad j = 1, \ldots, m \tag{15.5}$$

for each $S_j \in \mathscr{B}$. This can be done because, as we saw in Theorem 10.5.8, the minimal balancing vectors are extreme points of the linear program (10.5.3)–(10.5.5). This will mean that the coefficient matrix of the system (15.5) has linearly independent rows, and so the system always has a solution.

Letting x satisfy (15.5), we see that $x \in v(S_j)$ for each $S_j \in \mathcal{B}$. However,

$$\sum_{i \in N} x_i = \sum_{i \in N}\left(x_i \sum_{\substack{j \\ i \in S_j}} \lambda_j\right) = \sum_j \lambda_j \sum_{i \in S_j} x_i = \sum_j \lambda_j f(S_j) > f(N)$$

and so $x \notin v(N)$.

We see, thus, that the condition of Theorem 15.2.1 is a direct generalization of that in Theorem 10.5.10: in case v is a game with side payments, the two coincide. It is of interest, however, to note that Theorem 15.2.1 is not a full generalization of 10.5.10: in fact, for games with side payments, the condition is both necessary and sufficient; for games without side payments, the condition is merely sufficient.

Finally, we note that, if the balanced collection $\mathcal{B}$ is actually a partition, then the superadditivity condition (ii) of Definition 15.1.3 will actually imply (15.3). Thus (15.3) need only be verified for minimal balanced collections which are not partitions.

15.2.2 Example. We saw above that, for three-person games, the only minimal balanced collection, other than partitions, is $\{\{1, 2\}, \{1, 3\}, \{2, 3\}\}$. Thus, a super additive three-person game will have a nonempty core if

$$v(\{1, 2\}) \cap v(\{1, 3\}) \cap v(\{2, 3\}) \subset v(N)$$

Consider, now, the game given by

$$v(\{1, 2\}) = \{x | x_1 \le 4, x_2 \le 3\}$$
$$v(\{1, 3\}) = \{x | x_1 \le 4, x_3 \le 3\}$$
$$v(\{2, 3\}) = \{x | x_2 \le 3, x_3 \le 3\}$$
$$v(\{i\}) = \{x | x_i \le 0\}, \quad i = 1, 2, 3$$
$$v(N) = \{x | x_1 + x_2 + 2x_3 \le 10\}$$

It is easy to check that v is superadditive. Next, note that

$$x = (4, 3, 3)$$

belongs to each set $v(\{i, j\})$. But $x \notin v(N)$, and so the condition of Equation 15.3 is not satisfied. Nevertheless, v has a nonempty core: we note that any of the points (6, 4, 0), (4, 0, 3), or (0, 4, 3) belongs to the core. Thus the condition is not necessary.

We proceed now with the proof of Theorem 15.2.1. The proof essentially is that given by Scarf [15.11], and is based on the notion of a *primitive set*.

15.2.3 Definition. Let $A = (a_{ij})$ be an $m \times n$ matrix, $m \le n$. We say that the $j_1, j_2, \ldots, j_m$ columns of A form a *primitive set* if:

(a) in the submatrix $\tilde{A}$, consisting of these columns, the minimal entry in each row appears in a different column, and
(b) there is no column j in the matrix A such that, for each i

$$a_{ij} > \min_{1 \le k \le m} a_{i, j_k}$$

15.2.4 Example. Consider the 3×7 matrix

$$A = \begin{pmatrix} 5 & 4 & 2 & 4 & 5 & 3 & 1 \\ 1 & 0 & 4 & 3 & 2 & 5 & 2 \\ 8 & 2 & 5 & 1 & 0 & 2 & 8 \end{pmatrix}$$

In this matrix, columns 1, 4, and 6 form a primitive set. Indeed, the submatrix

$$\tilde{A} = \begin{pmatrix} 5 & 4 & \underline{3} \\ \underline{1} & 3 & 5 \\ 8 & \underline{1} & 2 \end{pmatrix}$$

has the minimal entry of each row (underlined) in a different column. Moreover, there is no column of A that is, componentwise, larger than the column

$$\begin{pmatrix} 3 \\ 1 \\ 1 \end{pmatrix}$$

of row minima in matrix $\tilde{A}$.

It should be noted, moreover, that 1, 4, 6 is not the only primitive set of columns; columns 1, 4, and 5 also form a primitive set, as do also columns 1, 3, and 6.

15.2.5 Definition. For a given matrix A, two primitive sets of columns are *adjacent* if they have $m - 1$ columns in common.

Returning to Example 15.2.4, we notice that the set $\{1, 3, 6\}$ is adjacent to $\{1, 4, 6\}$. In turn, $\{1, 4, 6\}$ is adjacent to $\{1, 4, 5\}$. On the other hand, $\{1, 3, 6\}$ and $\{1, 4, 5\}$ are not adjacent.

It is, of course, clear that a given matrix A might have no primitive sets of columns. Even if a matrix has a primitive set, there may be no primitive set adjacent to this. Nevertheless, adjacent primitive sets frequently exist. Assuming that a primitive set exists, we can remove one of the columns, and see whether any of the remaining columns in the matrix can be introduced so as to form the adjacent set. The situation is of course very similar to that found in linear programming, where, at each pivot-step, one of the basic variables is removed from the basis (i.e., the set of basic variables) and replaced by one of the non-basic variables, to form an adjacent basis, i.e., give an adjacent basic feasible point of the program. We shall, below, see the importance of this similarity.

Once again returning to Example 15.2.4, let us consider the primitive set {1, 4, 6}, giving rise to the matrix

$$\tilde{A} = \begin{pmatrix} 5 & 4 & 3 \\ 1 & 3 & 5 \\ 8 & 1 & 2 \end{pmatrix}$$

Let us suppose we wish to remove column 6 (the third column of $\tilde{A}$) from this set. Which column should be introduced to replace it? Or, to put it another way, we wish to form an adjacent primitive set, {1, 4, j}, where $j \neq 6$. What choice we have for j? The reader may verify directly that the only possible replacement is $j = 5$, giving the new primitive set {1, 4, 5}. Similarly, if we chose to remove column 4, the only possible replacement would be column 3, giving the primitive set {1, 3, 6}. On the other hand, if, from {1, 4, 5}, we remove column 4, we find there is no possible replacement: no adjacent primitive set is to be found in the "direction."

Of course, there are matrices for which more than one replacement may exist. In the "very degenerate" case where all columns are equal, we find that any set of m columns is primitive, so there is a wide choice of replacement. This, however, happens only because of equalities among the entries, as the following lemma shows:

15.2.6 Lemma. Let the matrix $A = (a_{ij})$ be nondegenerate in the sense that no two entries in any row are equal, and suppose $J = \{j_1, j_2, \ldots, j_m\}$ is a primitive set. Remove the column j_k from J. Then there is *at most* one column $j \notin J$, such that $J - \{j_k\} \cup \{j\}$ is a primitive set, i.e., j_k has at most one replacement.

Proof. Without loss of generality, we may assume $J = (1, 2, \ldots, m)$, and for each $i = 1, \ldots, m$

$$\min_{1 \leq j \leq m} a_{ij} = a_{ii}$$

In other words, the minimal entries in the rows of $\tilde{A}$ are on the main diagonal. Suppose we wish to remove column 1 from J. Suppose that j, $m + 1 \leq j \leq n$, is a permissible replacement for 1, giving rise to the new primitive set

$$J' = \{2, \ldots, m, j\}$$

Now we cannot have $a_{1j} < a_{11}$, as this would mean $a_{i1} > \min_{l \in J'} a_{il}$ for all i, contradicting the primitiveness of J'. Thus, we must have $a_{1j} > a_{11}$, and this, together with the primitiveness of J, means that $a_{ij} < a_{ii}$ for some $i \neq 1$. There can be only one such i (as otherwise column j would have two of the row minima in set J'). Call this k:

$$a_{kj} = \min_{l \in J'} a_{kl}$$

For every i other than 1 or k, we will still have

$$a_{ii} = \min_{l \in J'} a_{il}$$

and, since column j has the smallest entry in row k, column k must have the smallest in row 1

$$a_{1k} = \min_{l \in J'} a_{1l}$$

We see, thus, that column k must have the second-smallest entry in row 1

$$a_{1k} = \min_{2 \le l \le m} a_{1l}$$

and is uniquely determined. Then column j must satisfy

$$a_{1j} > a_{1k} \tag{15.6}$$

$$a_{kj} < a_{kk} \tag{15.7}$$

$$a_{ij} > a_{ii}, \quad i \ne 1, k \tag{15.8}$$

There may, of course, be several columns satisfying these constraints. Of all such, we must choose the one that maximizes a_{kj}. For, suppose some other column q also satisfies these constraints, with $a_{kq} > a_{kj}$. In that case, we would have

$$a_{iq} > \min_{l \in J'} a_{il}$$

for all i — contradicting the primitiveness of J'. Thus, the column j is unique, as it must be the one that maximizes a_{kj}, subject to (15.6)–(15.8).

In general, then, we have the following rules:

15.2.7 Rules for Column Replacement. Let J be a primitive set of columns for A, and let $I \in J$ be the column to be removed. For each $j \in J$, let $i(j)$ be the row whose minimal element is in column j.

Let $k \in J$ have the second smallest element (over J) in row $i(l)$.
Consider all $j \notin J$ satisfying

$$a_{i(l), j} > a_{i(l), k} \tag{15.9}$$

$$a_{i(k), j} < a_{i(k), k} \tag{15.10}$$

$$a_{i(q), j} > a_{i(q), q} \quad q \ne k, l \tag{15.11}$$

Among all such, choose that j which maximizes $a_{i(k), j}$. This is the replacement column.

It may be, of course, that there is no column j satisfying (15.9)–(15.11). In that case, no replacement is possible for column l. Suppose, however, that j has been chosen according to rules (15.9). Then

$$J' = J - \{l\} \cup \{j\}$$

is certainly a primitive set. For, condition (a) is satisfied by construction. To prove (b), suppose there was some column q satisfying

$$a_{iq} > \min_{p \in J'} a_{ip}$$

for all i. In that case, either $a_{i(k), q} \geq a_{i(k), k}$, which contradicts the primitiveness of J, or $a_{i(k),j} < a_{i(k),q} < a_{i(k),k}$, which contradicts the maximality of $a_{i(k),j}$.

At this point, the reader should notice the similarity between the present process of choosing adjacent primitive sets and the pivot-step of linear programming. In the present case, one of the columns in the set J is dropped; this will give rise to a unique replacement (if any). In pivot steps, a nonbasic variable is chosen. This will determine (see rules 3.5.2) a basic variable to be removed from the basis. In the nondegenerate case, where no ties are permitted, this basic variable is uniquely determined.

In both processes, of course, it may be that the step is impossible. This happens, in the present process, in case no column satisfies (15.9)–(15.11); in linear programming, it will happen if there are no positive entries in the chosen column — which means that the constraint set is unbounded.

If we know that the constraint set is bounded, then a pivot-step is always possible. Similarly, for the present process, there is one case in which we can guarantee that a replacement column always exists. A matrix $A = (a_{ij})$ is in *normal form* if, for each $i = 1, \ldots, m$

$$a_{ii} = \min_{1 \leq j \leq n} a_{ij} \tag{15.12}$$

and, for each $j = 1, \ldots, m$, and each $k = m + 1, \ldots, n$

$$a_{ik} < a_{ij} \quad \text{if} \quad i \neq j \tag{15.13}$$

Thus, the entries in A are bounded *below* by the main-diagonal entries, and *above* by the off-diagonal entries of the submatrix $\tilde{A}$, which consists of the first m columns. We have then

15.2.8 Lemma. Let A be a matrix in normal form, let J be a primitive set of columns, and let $l \in J$ be such that the $m - 1$ columns in $J - \{l\}$ are not all among the first m columns of A. Then there exists a replacement column for l.

Proof. In the set J, let k be the column with the second smallest entry in row $i(l)$. Clearly, $k \neq i(l)$, as then it would have the smallest entry in row $i(l)$. Therefore, k

cannot be from among the first m columns of the matrix — as then $a_{i(l),k}$ would be larger than $a_{i(l)q}$ for all $a \in J$, $q \neq l$, $q \geq m+1$. Thus, $k \geq m+1$.

Now, k contains the minimal entry (in J) in some row q. This means column $q \notin J$, for certainly $a_{qq} < a_{qk}$. But, then, we find that

$$a_{qq} < a_{qk},$$
$$a_{iq} > \min_{j \in J} a_{ij}, \quad i \neq q$$

and this guarantees that there will be a replacement column.

Let us now consider the linear system

$$By^T = c \tag{15.14}$$

$$y \geq 0 \tag{15.15}$$

where B is an $m \times n$ matrix, $m \leq n$, and y a vector $(y_1, \ldots, y_n)$. There is, generally, no guarantee that such a system is soluble. Suppose, however, that B has the form

$$B = \begin{pmatrix} I & \vdots & D \end{pmatrix}$$

so that its first m columns form an identity matrix. Suppose, moreover, that $c \geq 0$. Then the system will have at least the solution $y_i = c_i$, for $i = 1, \ldots, m$, $y_j = 0$ for $j = m+1, \ldots, n$. Suppose, further, that all entries $b_{ij} \geq 0$, and there is, for each j, at least one i such that $b_{ij} > 0$. Then the set satisfying (15.14) and (15.15) is bounded. If all these conditions hold, we shall say that *the system is in l-standard form*. We have then, the following:

15.2.9 Theorem. Let A and B be $m \times n$ matrices, $m < n$. Let A be in normal form, let the system (15.14)–(15.15) be in l-standard form, and assume nondegeneracy in both cases. Then there exists a primitive set J for the matrix A, and a b.f.p. y for the system (15.14) and (15.15), such that y_j is basic if and only if $j \in J$.

Proof. We note first of all that any b.f.p. of (15.14) and (15.15) has exactly m basic variables, and so the two sets have at least the same size. For a given b.f.p. y, we shall let K be the set of all indices, k, such that y_k is nonbasic. Since there is a total of n variables in the system and m are basic, it follows that K has $n - m$ indices.

We consider, now, objects of the form $(y; J)$, where y is a b.f.p. and J is a primitive set. We shall say such an object has the label $K \cup J$. It is clear that we are looking for a pair $(y; J)$ whose label is $N = \{1, 2, \ldots, n\}$. We shall say two pairs $(y; J)$ and $(y'; J')$ are adjacent if

$$either\ y = y'; \quad J \text{ is adjacent to } J';$$
$$or\ y \text{ is adjacent to } y'; \quad J = J'$$

As we saw before, the point $y_i^0 = c_i$ for $i = 1, \ldots, m$; $y_j^0 = 0$ for $j \geq m + 1$ is a b.f.p. for the system. Its basic variables are $y_1, \ldots, y_m$; hence

$$K^0 = \{m + 1, m + 2, \ldots, n\}$$

Let j be the column (in A), which maximizes a_{ij} subject to $m + 1 \leq j \leq n$. It is easy to see that

$$J^0 = \{2, 3, \ldots, m, \, j\}$$

is a primitive set. We have, then, an initial pair (y^0, J^0), whose label is

$$K^0 \cup J^0 = N - \{1\}$$

Thus, the initial pair $(y^0; J^0)$ is "almost totally labeled," i.e., it is missing only the index 1. On the other hand, the label j appears twice, once in J^0 and once in K^0. Thus we can remove j from one of these two sets. It is not possible to remove it from J^0 (this being the one case in which Lemma 15.2.8 does not apply), but it can certainly be removed from K^0: we take the variable y_j, which is nonbasic at the b.f.p. y^0, and bring it into the basis i.e., obtain an adjacent b.f.p. y^1 for which y_j is basic. This can be done because the constraint set is bounded. Thus we have a new pair $(y^1; J^0)$, adjacent to $(y^0; J^0)$. We have, now

$$K^1 = K - \{j\} \cup \{k\}$$

where y_k was the variable, basic in y^0, which is nonbasic in y^1. The label for $(y^1; J^0)$ is, then

$$J^0 \cup K^1 = J \cup K \cup \{k\} = \{N\} - \{1\} \cup \{k\}$$

There are now two possibilities. One is that $k = 1$. If so, then $N - \{1\} \cup \{k\} = N$, and our process terminates: $(y^1; J^0)$ is as desired. If, however, $k \neq 1$, then $N - \{1\} \cup \{k\} = N - \{1\}$, and we still have an "almost totally labeled" pair. In this case, it is index k, which appears twice.

Since $k \in J^0$, but $k \neq j$ we can remove it from J^0. The remaining elements of J^0 are not all among $\{1, 2, \ldots, m\}$, and so there exists a replacement column for k — call this column l. Then $J^1 = J^0 - \{k\} \cup \{l\}$ is a primitive set, and $(y^1; J^1)$ is a new pair, adjacent to $(y^1; J^0)$. Its label is

$$K^1 \cup J^1 = N - \{1\} \cup \{l\}$$

Again, there are two possibilities. If $l = 1$, then $K^1 \cup J^1 = N$, and the pair $(y^1; J^1)$ is as desired. If not, then $K^1 \cup J^1 = N - \{1\}$, and index l appears in both sets. We then remove it from K^1, by finding a b.f.p. y^2, adjacent to y^1, for which y_l is basic.

Continuing in this manner, we find that we obtain a sequence of adjacent pairs

$$(y^0; J^0)$$
$$(y^1; J^0)$$
$$(y^1; J^1)$$
$$(y^2; J^1)$$
$$(y^2; J^2)$$
$$\vdots$$
$$(y^t; J^t)$$
$$(y^{t+1}; J^t)$$
$$(y^{t+1}; J^{t+1})$$

At alternating steps, one or the other of y and J is changed while the other is kept fixed. Since there are only a finite number of possible pairs, one of three things will eventually happen:

(α) some pair $(y; J)$ is fully labeled;
(β) it is impossible to take a step;
(γ) some pair will be repeated.

We will show that only (α) is possible.

To see that (γ) is not possible, we note that, at each step, the resulting pair $(y; J)$ will be labeled at least by the set $\{2, 3, \ldots, n\} = N - \{1\}$. This is because we always perform the step so as to remove whichever index appears twice, and so long as 1 does not appear in $J \cup K$ then some j appears in both J and K. Thus, until termination, each pair $(y; J)$ obtained has $N - \{1\}$ as its label.

Now, if we restrict ourselves to the set of all pairs with label $N - \{1\}$, we see that each pair has at most two adjacent pairs. Indeed, there is a unique $j \in K \cap J$, and this can be removed either from K (by changing y through a pivot step) or from J (by the column replacement method). In either case, the replacement, if it exists, is unique due to nondegeneracy. We can think of this, then, as a sequence of "rooms," each of which has at most two "doors." The first room of the sequence, however, the pair $(y^0; J^0)$, has only one door. It is clear, then, that no repetition is possible. For, the first room to be repeated would be entered through a third door — unless it be (Y^0, J^0), which would then be entered through a second door. This is impossible, and so there can be no repetition.

An alternative way of explaining the above is to say that a sequence of rooms, each of which has at most two doors, while the first has only one door, must eventually lead to some other room with only one door — or to the outside of the house.

We show, next, that (β) can never happen — a replacement is always possible. In fact, at each step, either y or J is to be changed. If it is y that must be changed, then a step is always possible, given that the constraint set for the system (15.14) and (15.15) is bounded. Suppose, then, that it is J that must be changed.

We know that $1 \notin J$. By Lemma 15.2.6, then, the replacement can be carried out unless $J = \{2, 3, \ldots, m, l\}$, and l is the column to be replaced. But there is only one primitive set J containing $\{2, 3, \ldots, m\}$, and that is J^0. Hence

$$J = J^0 = \{2, 3, \ldots, m, \ j\}$$

Since it is column j that must be replaced, then $J \cup K = N - \{1\}$, with $j \in K$, and since K has $m - n$ elements

$$K = \{m + 1, \ m + 2, \ldots, n\}$$

But this is K^0—associated with the original b.f.p. y^0. Thus the given pair is $(y^0; J^0)$. But this is impossible as it would mean a repetition.

We see, then, that neither (β) nor (γ) is possible. The process must, therefore, terminate in case (α): a totally labeled pair $(y; J)$ will be obtained. This is the desired pair.

We are now in a position to prove Theorem 15.2.1. We shall do this, first of all, by assuming that each of the sets $v(S)$ has only a finite number of Pareto-optimal points. In other words, there is, for each S, $\varnothing \neq S \neq N$, a finite set $H(S)$ such that $x \in v(S)$ if and only if there is $y \in H(S)$ with

$$y_i \geq x_i \quad \text{for all} \quad i \in S$$

All told, now, there will be a very large but finite number, h, of points in the several sets $H(S)$. Let us then form an $n \times h$ matrix A. Each column of A consists of the coordinates of one of these h points, with the following conditions.

(1) For $i = 1, \ldots, n$, the ith column is a point $y \in H(\{i\})$.
(2) For points $y \in H(S)$, the coordinates y_i, $i \notin S$, are arbitrary. These will simply be made very large, the only restrictions being that all these large numbers be distinct, and that the $n(n-1)$ arbitrary numbers in the first n columns be larger than those in the last $h - n$ columns.
(3) There is no need to include any point y with any coordinate y_i smaller than a_{ii}. All such may be discarded.

When this has been done, it is clear that the resulting matrix A is in normal form. Let us, next, form a second $n \times h$ matrix B whose entries are

$$b_{ij} = \begin{cases} 0 & \text{if} \quad i \notin S_j \\ 1 & \text{if} \quad i \in S_j \end{cases}$$

where S_j is the coalition associated with the jth column of A. (In other words, the jth column of A is some vector y in some set $H(S)$; S_j is precisely this S.) Then the first n columns of B form an identity matrix, and there is at least one positive entry in each column. Thus the system

$$Bz^T = (1, 1, \ldots, 1)^T, \quad z \geq 0$$

is in l-standard form. By Equation 15.11, there exists a pair $(z; J)$ such that z is a b.f.p., J is a primitive set, and z_j is basic for all $j \in J$. But in that case, the sets $S_j, j \in J$, form a minimal balanced collection for N, with balancing vector z.

Set now

$$x_i = \min_{j \in J} a_{ij}$$

We will have, of course, $x_i \leq a_{ij}$ for each i and each $j \in J$, and we conclude that $x \in v(S_j)$ for each j. Moreover, $x_i \geq a_{ii}$ for each i.

Under the hypothesis of Equation 15.3, we have

$$\bigcap_{j \in J} v(S_j) \subset v(N)$$

and so $x \in v(N)$. Let $\tilde{x}$ be a Pareto-optimal point in $v(N)$ such that $\tilde{x} \geq x$. We claim that $\tilde{x} \in C(v)$.

Suppose $\tilde{x} \notin C(v)$. This would mean that there was some $S \subset N$, $S \neq \varnothing$, $S \neq N$, and some $y \in v(S)$ such that $y_i > \tilde{x}_i \geq x_i$ for all $i \in S$. There is clearly no loss of generality in assuming that $y \in H(S)$. We note that we must have $y_i > x_i \geq a_{ii}$ for each $i \in S$, and thus y is not one of those points discarded in forming the matrix A. Thus y must be a column of A. But then we would have $y_i > x_i$ for all $i \in S$, and also $y_i > x_i$, for $i \notin S$, because such components y_i are "very large," and x_i cannot be very large as $x \in v(T)$ for some T with $i \in T$ (otherwise the collection would not be balanced). This would contradict the primitiveness of the set J.

This proves the nonemptiness of the core under the finiteness assumption mentioned above, namely, that each $v(S)$ be determined by a finite set $H(S)$. In general, of course, this is not the case; in fact, it may be seen that this will usually give rise to nonconvex sets $v(S)$. However, we can always approximate $v(S)$ in this manner.

More exactly, let us set, for $S \subset N$

$$R^S = \{x | x_j = 0 \quad \text{if} \quad j \notin S\}$$

as the S-coordinate subspace of R^N, and define

$$\tilde{v}(S) = v(S) \cap R^S$$

By condition (ii) of Definition 15.1.3, we can see that

$$v(S) = \tilde{v}(S) + R^{N-S}$$

where the sum of two sets means the set of all possible sums of elements of these sets. Thus $v(S)$ and $\tilde{v}(S)$ determine each other uniquely.

The set $\tilde{v}(S)$, now, will be closed and bounded above, and will have the property that, if $y \in \tilde{v}(S)$ and $x \leq y$, with $x \in R^S$, then also $x \in \tilde{v}(S)$. Thus $\tilde{v}(S)$ is determined by its Pareto-optimal points. By taking a large number of these points, and using them as $H(S)$, an approximation to $v(S)$ will be obtained (see Figure 15.4). This approximation can be made as good as desired by taking sufficiently many of these points. Moreover, the approximation will be from below: the approximating set $v'(S)$ will be a subset of the true $v(S)$.

Assume, then, that the hypothesis of Equation 15.3 holds. Then we approximate the game v in this fashion, getting a sequence of better and better approximations to the sets $v(S)$, $S \neq N$. For each such approximation v', we have

$$\bigcap_{j=1}^{m} v'(S_j) \subset \bigcap_{j=1}^{m} v(S_j) \subset v'(N)$$

since $v'(S) \subset v(S)$, but $v'(N) = v(N)$. Thus a sequence of points in the cores of the approximations v' will be found. Since the Pareto-optimal subset of $v(N)$ is assumed compact, this sequence will have an accumulation point x^*, which lies in the core of v.

15.2.10 Example. Consider the following three-person game: any single player can obtain nothing at all, i.e.

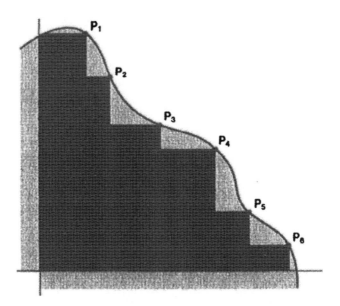

Figure 15.4.

$$v(\{i\}) = \{x \mid x_i \le 0\}, \quad i = 1, 2, 3$$

For the two-player sets, $\{1, 2\}$ can obtain the points $(7, 5, x_3)$ and $(3, 9, x_3)$. Coalition $\{1, 3\}$ can obtain $(4, x_2, 1)$ and $(2, x_2, 8)$. Finally, $\{2, 3\}$ can obtain $(x_1, 6, 2)$ and $(x_1, 1, 7)$. We make no assumption as to $v(N)$, except that it must satisfy the usual superadditivity conditions, and also

$$v(\{1, 2\}) \cap v(\{1, 3\}) \cap v(\{2, 3\}) \subset v(N)$$

To find a point in the core, we form the two 3×9 matrices

$$A = \begin{pmatrix} 0 & 98 & 96 & 7 & 3 & 4 & 2 & 90 & 89 \\ 100 & 0 & 95 & 5 & 9 & 92 & 91 & 6 & 1 \\ 99 & 97 & 0 & 94 & 93 & 1 & 8 & 2 & 7 \end{pmatrix}$$

and

$$B = \begin{pmatrix} 1 & 0 & 0 & 1 & 1 & 1 & 1 & 0 & 0 \\ 0 & 1 & 0 & 1 & 1 & 0 & 0 & 1 & 1 \\ 0 & 0 & 1 & 0 & 0 & 1 & 1 & 1 & 1 \end{pmatrix}$$

The reader should note how A was formed: very large numbers (ranging between 89 and 100) were assigned to players outside the several coalitions. B is simply the incidence matrix of the players in the several coalitions. The linear system is given by

$$z_1 + z_4 + z_5 + z_6 + z_7 = 1$$
$$z_2 + z_4 + z_5 + z_8 + z_9 = 1 + \varepsilon$$
$$z_3 + z_6 + z_7 + z_8 + z_9 = 1 + \varepsilon^2$$

where ε is small. (This avoids degeneracy.)

We start with the initial b.f.p.

$$z^0 = (1, 1 + \varepsilon, 1 + \varepsilon^2, 0, 0, 0, 0, 0, 0)$$

with basic variables z_1, z_2, z_3. As our initial primitive set, we choose columns 2, 3, and 8 of matrix A. Note that $J = \{2, 3, 8\}$ is the only possible primitive set, which includes columns 2 and 3, as column 8 maximizes a_{ij}. We thus have

$$K^0 = \{4, 5, 6, 7, 8, 9\}, \quad J^0 = \{2, 3, 8\}$$

The index 8 appears twice: we can, therefore, make the variable z_8 basic to obtain

$$z' = (1, \varepsilon - \varepsilon^2, 0, 0, 0, 0, 0, 1 + \varepsilon^2, 0)$$

We have, therefore

$$K' = \{3,4,5,6,7,9\}, \quad J^0 = \{2,3,8\}$$

Now, 3 appears twice. We, therefore, remove column 3 from J^0, replacing it with column 9. Then

$$K' = \{3,4,5,6,7,9\}, \quad J' = \{2,8,9\}$$

We bring z_9 into the basis, now, to obtain the new b.f.p.

$$z^2 = (1, \varepsilon - \varepsilon^2, 0, 0, 0, 0, 0, 0, 1 + \varepsilon^2)$$

Now

$$K^2 = \{3,4,5,6,7,8\}, \quad J = \{2,8,9\}$$

Continuing in this fashion, we obtain

$$K^2 = \{3,4,5,6,7,8\}, \quad J^2 = \{2,4,9\}$$
$$K^3 = \{2,3,5,6,7,8\}, \quad J^2 = \{2,4,9\}$$
$$K^3 = \{2,3,5,6,7,8\}, \quad J^3 = \{4,8,9\}$$
$$K^4 = \{2,3,5,6,7,9\}, \quad J^3 = \{4,8,9\}$$
$$K^4 = \{2,3,5,6,7,9\}, \quad J^4 = \{4,5,8\}$$
$$K^5 = \{2,3,4,6,7,9\}, \quad J^4 = \{4,5,8\}$$
$$K^5 = \{2,3,4,6,7,9\}, \quad J^5 = \{5,6,8\}$$
$$K^6 = \{1,2,3,4,7,9\}, \quad J^5 = \{5,6,8\}$$

At this point, we can stop: the variable z_1 has finally become nonbasic, and the pair $(z^6; J^5)$ is totally labeled. The balanced collection, corresponding to columns 5, 6, and 8, is, of course, $\{1, 2\}$, $\{1, 3\}$, $\{2, 3\}$. These three columns form the matrix

$$\tilde{A} = \begin{pmatrix} \underline{3} & 4 & 90 \\ 9 & 92 & \underline{6} \\ 93 & \underline{1} & 2 \end{pmatrix}$$

and the three row minima give us the point

$$x = (3, 6, 1)$$

which, as may be checked, belongs to all these $v(S)$, and hence also to $v(N)$. If, now, x is Pareto-optimal in $v(N)$, it belongs to the core. If not, there is some $x' \in V(N)$, $x' \geq x$, Pareto-optimal. This will certainly be in the core.

15.3. Market Games

An important class of *n*-person games is that dealing with oligopolistic market situations. In the simplest case, we assume that the *n* players are traders, able to deal in commodities so as to increase their utilities. These commodities have, in themselves, no value, other than the utility that they represent.

More precisely, we assume that there are here *m* commodities, $C_1, C_2, \ldots, C_m$. Each of the *n* players (traders) is given an *initial endowment*, i.e., player *i* has a commodity bundle

$$w^i = (w^i_1, w^i_2, \ldots, w^i_m) \tag{15.16}$$

at the beginning of the game (trading period). Each of the players has a personal utility function

$$u^i : R^m \to R$$

so that $u^i(x_1, \ldots, x_m)$ is the utility—to player *i*—of the bundle $(x_1, \ldots, x_m)$.

Let *S* be a coalition. If this coalition forms, we assume that they can redistribute their endowments in any way desired, i.e., they may obtain any *S*-tuple of bundles

$$X(S) = (x^i)_{i \in S} = (x^i_1, x^i_2, \ldots, x^i_m)_{i \in S}$$

satisfying

$$\sum_{i \in S} x^i_j = \sum_{i \in S} w^i_j \qquad \forall j = 1, \ldots, m \tag{15.17}$$

$$x^i_j \geq 0 \qquad \forall i \in S, \ j = 1, \ldots, m \tag{15.18}$$

Any such *S*-tuple of bundles will be called a *feasible allocation* for *S*.

We can now represent this as a game without side payments: for each *S*, *v(S)* is the set of all $(y_1, \ldots, y_n)$ for which there exists a feasible allocation $\overline{X}(S)$ satisfying

$$y_i \leq u^i(x^i_1, \ldots, x^i_m) \tag{15.19}$$

for each $i \in S$.

Generally speaking, there is no guarantee that the sets *v(S)* obtained in this manner will be convex. If they are not convex, there are two alternatives: we can dispense with the need for convexity (which had originally been included in the definition of a game) or we can redefine *v(S)* to equal, not the set of all *y* satisfying (15.19), but rather the convex hull of that set. Either of these alternatives leads to unfortunate complications. There is one important case, however, when convexity of *v(S)* is guaranteed:

15.3.1 Theorem. If the functions u^i are concave, the sets *v(S)* defined by (13.3.4) are convex.

Proof. Let y, $\tilde{y} \in v(S)$ and $0 \leq \lambda \leq 1$. We will have, for $i \in S$

$$y_i \leq u^i(x^i), \quad \tilde{y}_i \leq u^i(\tilde{x}^i)$$

where $X = (x^i)$ and $\tilde{X} = (\tilde{x}^i)$ are feasible allocations for S. But then

$$\hat{X} = \lambda X + (1 - \lambda)\tilde{X}$$

is also a feasible allocation for S, and, by concavity

$$u^i(\hat{x}^i) \geq \lambda u^i(x^i) + (1 - \lambda)u^i(\tilde{x}^i)$$

and so, for each i,

$$\lambda y_i + (1 - \lambda)\tilde{y}_i \leq u^i(\hat{x}^i)$$

Thus $\lambda y + (1 - \lambda)\tilde{y} \in v(S)$, and so $v(S)$ is convex.

In general, the assumption of concavity will simplify our analysis enormously. The following theorem shows a further consequence.

15.3.2 Theorem. If the functions u^i are concave, the game v has a nonempty core.

Proof. Let $\{S_1, S_2, \ldots, S_k\}$ be a balanced collection with balancing vector $(\lambda_1, \ldots, \lambda_k)$, and assume

$$y \in \bigcap_{j=1}^{k} v(S_k)$$

For each S_j, there is a feasible allocation

$$\overline{X}^j(S_j) = (^j x^i) i \in s_j$$

such that

$$y_i \leq u^i(^j x^i) \quad \forall i \in S_j$$

Set, now, for each $i \in N$,

$$\hat{x}^i = \sum_j \lambda_j (^j x^i)$$
$$i \in S_j$$

It is not difficult to verify that

$$\hat{X} = (\hat{x}^i)_{i \in N}$$

is a feasible allocation for N. Now by concavity,

$$u^i(\hat{x}^i) \geq \sum_j \lambda_j u^i(^j x^i)$$
$$i \in S_j$$

and, since

$$\sum_j \lambda_j = 1$$

$$i \in S_j$$

this gives us

$$y_i \leq u^i(\hat{x}^i)$$

and so $y \in v(N)$. Thus, by Theorem 15.2.1, the game has a nonempty core.

The core is, of course, an equilibrium concept: points in the core are in equilibrium, in the sense that no coalition has both the reason and the capacity to disrupt them. Economists are interested, however, in a somewhat stronger notion of equilibrium — one created by a price system.

Let us suppose that a system of prices is given for our commodities. This is not generally meant to imply the existence of money; rather, the ratios of these prices represent rates at which the commodities can be interchanged. Let

$$p = (p_1, p_2, \ldots, p_m)$$

be the price vector. In this case, each player will want to exchange his initial bundle w^i for a new bundle x^i with greater utility. More precisely, player i will look for a bundle x^i such as to

$$\text{Maximize} \quad u^i(x_1^i, x_2^i, \ldots, x_m^i) \tag{15.20}$$

subject to

$$\sum_{j=1}^m p_j x_j^i \leq \sum_{j=1}^m p_j w_j^i \tag{15.21}$$

$$x_j^i \geq 0, \quad j = 1, \ldots, m \tag{15.22}$$

There is, of course, a question as to the existence and uniqueness of this maximum. Assuming that all prices p_j are positive, the constraint set (15.21) and (15.22) is compact, and thus continuity of u^i will guarantee the existence of the maximum. Uniqueness is a much more complicated question, but not really that important. What is important is that, in general, the allocations $\overline{X} = (x^i)_{i \in N}$ obtained in this manner will not be feasible. In some cases, however, they may be; this is the interesting case.

15.3.3 Definition. A *competitive* equilibrium is a pair $\langle p; \overline{X} \rangle$, where $p = (p_1, \ldots, p_m)$ is a price vector, and $\overline{X} = (x^i)_{i \in N}$ is an allocation satisfying

$$\sum_{i \in N} x_j^i = \sum_{i \in N} x_j^i, \quad j = 1, \dots, m$$

$$x_j^i \geq 0, \qquad\qquad i = 1, \dots, n, \; j = 1, \dots, m$$

and such that, for each i

$$u^i(x^i) = \max u^i(\xi^i)$$

subject to

$$\sum_{j=1}^{m} p_j \xi_j^i \leq \sum_{j=1}^{m} p_j w_j^i$$

$$\xi_j^i \geq 0, \quad j = 1, \dots, m$$

We can see the importance of an equilibrium: a set of prices is given. Each player then trades so as to maximize his utility, subject to the obvious budget constraint that the total market value of his final bundle be no greater than that of his initial endowment. If the prices have been "properly" chosen, i.e., if they are equilibrium prices, this process will just clear the market — there will be just enough of each commodity to satisfy every trader's demand, and no more. If, on the other hand, the "wrong" set of prices is chosen, then there will be some imbalance in the market: there will be an excess demand for some commodities (which had been priced too low) and an excess supply of some other commodities (which are priced too high). The practical importance of this, for an economy based on a free market, is easy to understand; the obvious question is whether such an equilibrium exists.

15.3.4 Theorem. Assume that the functions u^i are continuous, concave, and monotone nondecreasing in each variable. Then a competitive equilibrium exists.

Proof. Let us assume first that the u^i are strictly concave. For a given $p = (p_1, \dots, p_m) \geq 0$ such that

$$\sum_{j=1}^{m} p_j = 1$$

let $x^i(p) = (x_1^i, \dots, x_m^i)$ be such as to

$$\text{Maximize} \quad u^i(\xi_1, \dots, \xi_m)$$

subject to

$$\sum_{j=1}^{m} p_j \xi_j \leq \sum p_j w_j^i$$

$$\xi_j \geq 0, \quad j = 1, \ldots, m$$
$$\xi_j \leq M$$

where M is a larger number

$$M > \sum_{i \in N} w_j^i \quad \text{for} \quad j = 1, \ldots, m$$

By compactness, such a maximizing x^i will exist, and by strict concavity, it will be unique. Moreover, due to the monotonicity, it is easy to see that we must have

$$\sum_{j=1}^m p_j x_j^i = \sum_{j=1}^m p_j w_j^i$$

and, moreover

$$x_j^i = M \quad \text{if} \quad p_j = 0$$

Finally, as in the proof of Theorem 4.4.2, $x^i(p)$ depends continuously on p. Let, now

$$f_j(p) = \sum_{i \in N} x_j^i(p) - \sum_{i \in N} w_j^i$$

be the access *demand* for C_j. We are looking for a p such that all the excess demands vanish. Define

$$g_j = \max\{0, f_j\} \quad \text{and} \quad p_j' = \frac{p_j + g_j}{1 + \sum_{k=1}^m g_k}$$

It is easy to see that f_j, g_j, and hence p_j' depend continuously on p. Moreover, we will have

$$\sum_{j=1}^m p_j' = \frac{\sum p_j + \sum g_j}{1 + \sum g_k} = 1$$

since $\sum p_j = 1$, and also

$$p_j' \geq 0$$

since $p_j \geq 0$ and $g_j \geq 0$. Thus the transformation

$$T(p) = p'$$

is a continuous mapping of the unit simplex into itself, and must therefore have a fixed point p^*

$$T(p^*) = p^*$$

We claim that $p*$ is an equilibrium price vector.

To see this, we note first of all that all $p_j^* > 0$. For, suppose some $p_j = 0$. Then, for each i, $x_j^i = M$, and so

$$f_j(p) = nM - \sum_{i \in N} w_j^i > 0$$

so that

$$g_j > 0 \quad \text{and} \quad p_j' \frac{p_j + g_j}{1 + \sum g_k} > 0$$

We note next that it is not possible for all f_j to be positive, nor for all to be negative. In fact, we have

$$\sum_j p_j f_j = \sum_j p_j \left(\sum_i x_j^i - \sum_i w_j^i \right) = \sum_i \left(\sum_j p_j x_j^i - \sum_j p_j w_j^i \right)$$

and this vanishes since each of the parentheses in this last expression vanishes. Inasmuch as all $p_j^* > 0$, we conclude that either all $f_j = 0$ or some $f_j > 0$ while some other $f_l < 0$, and hence $g_l = 0$.

Assume, then, that $f_j > 0$ while $f_l < 0$. Then

$$p_l' = \frac{p_l + g_l}{1 + \sum g_k} = \frac{p_l}{1 + \sum g_k} < p_l$$

since $p_l > 0$, but the denominator is greater than 1. We conclude that we must have

$$f_j(p^*) = 0 \quad \text{for all} \quad j$$

We must next show that our constraints $\xi_j \leq M$, which we introduced above to guarantee existence of a maximum, are not a problem. In fact, for any j, we know that

$$\sum_{i \in N} x_j^i(p^*) = \sum_{i \in N} w_j^i$$

and, as all $x_j^i \geq 0$, we will have, for each i,

$$x_j^i \leq \sum_{i \in N} w_j^i < M$$

Suppose that the true maximum — i.e., the maximum without the restriction $\xi_j \leq M$, were at some point with $z_j^i > M$. Then we would find, by concavity, that, for all t, $0 < t < 1$,

$$u^i((1-t)x^i + tz^i) > u^i(x^i)$$

but, for small values of t,

$$(1-t)x^i + tz^i \leq M$$

and this would contradict the maximality of $u^i(x^i(p))$. Thus, we conclude that the pair $(p^*; X)$ where

$$X = (x^i(p^*))_{i \in N}$$

is a competitive equilibrium.

This completes the proof in case the u^i are strictly concave. If they are merely concave, we may approximate them by

$$\tilde{u}^i(x) = u^i(x) - \varepsilon \sum_{j=1}^{m} (M - x_j)^2$$

where $\varepsilon > 0$ is small. It is easy to see that $\tilde{u}^i$ is strictly concave, and, moreover, it is monotone over the sets $0 \leq x_j \leq M$. For each value of ε, this game will have a competitive equilibrium; as $\varepsilon \to 0$, these equilibria will have at least one accumulation point, which is a competitive equilibrium for the original game.

It should be noticed that, in this case, it is possible for some $p_j^* = 0$. This generally represents a saturation phenomenon: beyond a certain point, commodity C_j gives no further utility to the players. Thus it becomes, in effect, a surplus commodity, and its price vanishes.

We prove, finally, an important relation between the two equilibrium concepts, which we have discussed: the competitive equilibrium always belongs to the core. The exact statement of this is as follows:

15.3.5 Theorem. If (p^*, X) is a competitive equilibrium for a market game, then the payoff $y = (y_1, \ldots, y_n)$

$$y_i = u^i(x_1^i, \ldots, x_m^i)$$

belongs to the core.

Proof. Suppose $y \notin C(v)$. Then there is some $S \subset N$, $S \neq 0$, and some allocation $Z(S)$, feasible for S, such that

$$u^i(z_1^i, \ldots, z_m^i) > y_i$$

for all $i \in S$.

Now, x^i maximizes u^i subject to the constraints (15.21) and (15.22), and so z^i cannot satisfy them. By feasibility, it must satisfy $z_j^i > 0$, and so we must have

$$\sum_{j=1}^{m} p_j^* z_j^i > \sum_{j=1}^{m} p_j^* w_j^i$$

Summing this over all $i \in S$, we have

$$\sum_{i \in S} \sum_{j=1}^{m} p_j^* z_j^i > \sum_{i \in S} \sum_{j=1}^{m} p_j^* w_j^i$$

or

$$\sum_{j=1}^{m} p_j^* \left(\sum_{i \in S} z_j^i - \sum_{i \in S} w_j^i \right) > 0$$

but this is impossible as all $p_j^* \geq 0$, and the parentheses cannot be positive. This contradiction proves that y is undominated, and therefore belongs to the core.

15.3.6 Example. Consider a three-person game with two commodities; initial endowments are

$$w^1 = (29, 0), \quad w^2 = (16, 0), \quad w^3 = (0, 29)$$

Let all three players have the same utility function

$$u(x_1, x_2) = x_1^{.8} + x_2^{.9}$$

It can be verified that the equilibrium price vector is

$$p^* = (.392, .608)$$

with corresponding allocation

$$x^1 = (15, 9), \quad x^2 = (10, 4), \quad x^3 = (20, 16)$$

and utility payoffs

$$y^* = (15.95, 9.79, 23.11)$$

This point certainly lies in the core of the game.

It should be noted, now, that y^* is not the only point in the core of the game. Players 1 and 2 can, respectively, obtain the amounts $29^{.8} = 14.79$ and $16^{.8} = 9.19$, respectively. They can, however, gain nothing by trading. Thus any Pareto-optimal payoff that gives them exactly these utilities, or slightly more, will be undominated. Thus the equilibrium payoff is not, in general, the only point in the core; the core is usually considerably larger.

15.4. Approaches to the Value

As seen in the previous sections, certain satisfactory results have been obtained in generalizing the notions of stable sets and core to games without side payments. As regards the value, results are not nearly so pleasant.

Generally speaking, the situation is as follows. For the side-payments case, the Shapley value is generally a well-established concept. For two-person games, the Nash value is also well established. As it happens, these two coincide for two-person games with side payments (at least in the fixed threat case). What we would like is something that generalizes both the Nash and Shapley values for n-person games without side payments. There seem to be several ways of doing this; none, unfortunately, seems wholly satisfactory.

We saw in Chapter 9 that, if the utility transformation (9.3) is made, the Nash solution to the resulting game is obtained simply by an equal division of the possible increments in utility, and this would be the case even if side payments were allowed. One possible approach, in the n-person case, is to look for some linear transformation of utilities, which will somehow allow us to solve the game as though it were a game with side payments.

More precisely, let v be a game without side payments. We can transform this into a game with side payments w by setting

$$w(S) = \max \sum_{i \in S} x_i \tag{15.23}$$

subject to

$$x \in v(S) \tag{15.24}$$

And $\varphi[w]$, the Shapley value of this game, can then be obtained. Now, if this $\varphi[w]$ is feasible, i.e., if it belongs to $v(N)$, it is easy to see that it must be Pareto-optimal and, moreover, individually rational, i.e.

$$\varphi_i[w] \geq w(\{i\}) = \max\{x_i | x \in V(\{i\})\}$$

Unfortunately, $\varphi[w]$ is not usually feasible; it does not lie in $v(N)$.

Let us now choose a "utility scale" vector

$$\lambda = (\lambda_1, \ldots, \lambda_n) \geq 0$$

and, in effect, "rescale" each player's utility by a factor λ_i. We would then obtain a new game w_λ, given by

$$w_\lambda(S) = \max\left\{\sum_{i \in S} \lambda_i x_i | x \in v(S)\right\} \tag{15.25}$$

and obtain the value $\varphi[w_\lambda]$. It may the happen that the vector $\hat{\varphi}[v; \lambda]$

$$\hat{\varphi}_i = \frac{\varphi_i[w_\lambda]}{\lambda_i}, \quad i = 1, \ldots, n \tag{15.26}$$

is feasible; this clearly will depend on the particular vector λ chosen. If, indeed, $\hat{\varphi}[v; \lambda] \in v(N)$, then we say this it is a value for the game v.

The following theorem, given without proof, is due to Shapley:

15.4.1 Theorem. For a game v satisfying Definition 15.1.3, there exists at least one vector $\lambda = (\lambda_1, \ldots, \lambda_n) > 0$ such that

$$\hat{\varphi}(v; \lambda) \in v(N)$$

Shapley's proof of this theorem uses a fixed point argument. There is then, in general, no uniqueness: a game has at least one value, but may have several such.

An alternative approach, due to Harsanyi, is based on the idea that, for a game with side payments, the coalition S has the amount

$$C_S = \sum_{T \subset S} (-1)^{s-t} v(T) \tag{15.27}$$

available as a "dividend" for its members. In fact, Lemma 12.1.7 showed that

$$v(S) = \sum_{T \subset S} C_T \tag{15.28}$$

and the value is then given by

$$\phi_i[v] = \sum_{\substack{S \subset N \\ i \in S}} C_S \frac{1}{s} \tag{15.29}$$

Thus, in effect, the Shapley value divides this dividend equally among members of S; the total value to a player is the sum of the dividends, which he receives from the several coalitions which include him.

For a game without side payments, let us, first of all, choose a vector $(\mu_1, \ldots, \mu_n) > 0$. Then, for each $i \in S$, let

$$\Delta_i(\{i\}) = \max\{x_i | x \in v(\{i\})\} \tag{15.30}$$

We then proceed to assign dividends $\Delta_i(S)$ inductively to each player from each coalition, as follows: if $i \notin S$, then $\Delta_i(S) = 0$. Assuming that $\Delta_i(T)$ has been defined for all $T \subset S$, $T \neq S$, let us set, for $i \in S$

$$x_i(S) = \sum_{\substack{T \subset S \\ T \neq S}} \Delta_i(T) \tag{15.31}$$

The point $x(S)$ is, more or less, what the members of coalition S have already received from its proper subcoalitions. Then S may have some surplus — or some deficit — to make up.

Let r be given by

$$r(S) = \max\{t \,|\, x(S) + t\mu \in v(S)\} \tag{15.32}$$

This r may be positive or negative; it will, however, always exist given Definition 13.1.3. Then we set

$$\Delta_i(S) = \mu_i r(S) \tag{15.33}$$

It is easy to see, now, that y, given by

$$y = x(N) + \Delta(N) \tag{15.34}$$

belongs to $v(N)$ and is, moreover, Pareto-optimal, except in the degenerate special case that the set $v(N)$ has "vertical walls," i.e., boundary points at which the surface of $v(N)$ contains some line segment parallel to one of the coordinate axes.

It may be seen, of course, that y will depend on the particular choice of μ. To correspond to the Nash value, in the two-person case, it is necessary that the hyperplane

$$\pi = \left\{ x \,\left|\, \sum_{i=1}^{m} \frac{x_i - y_i}{\mu_i} = 0 \right. \right\}$$

be a support to the set $v(N)$ at the point y. In fact, it can be shown that there exists at least one $\mu = (\mu_1, \ldots, \mu_n)$ such that this holds.

The reader should note that the μ_i obtained here correspond, more or less, to the reciprocals of the λ_i in Shapley's process described above. (Of course, the two procedures are different and therefore it is not generally true that $\mu_i = \lambda_i^{-1}$.)

A third possible approach consists in generalizing the notion of multi-linear extension. As in Section 15.2, let us set

$$\tilde{v}(S) = v(S) \cap R^S \tag{15.35}$$

For a vector $(q_1, \ldots, q_n)$, $0 \le q_i \le 1$, we can then set

$$F(q_1, \ldots, q_n) = \sum_{S \subset N} P_S(q) \tilde{v}(S) \tag{15.36}$$

where

$$P_S(q) = \prod_{i \in S} q_i \prod_{i \notin S} (1 - q_i) \tag{15.37}$$

and the sum on the right side of (15.36) represents the set of all possible sums of terms $P_S(q)u(S)$ with $u \in \tilde{v}(S)$; alternatively, we have

$$F(q_1, \ldots, q_n) = \left\{ \sum_{S \subset N} P_S(q)u(S) | u(S) \in \tilde{v}(S) \right\} \tag{15.38}$$

Then F is a point-to-set mapping and can be thought of as the MLE of v, for the no-side-payments case.

Let us suppose, now, that the Pareto-optimal surface of $F(q)$ can be characterized by some equation, i.e., suppose it is the set of all x satisfying

$$G(q_1, \ldots, q_n; x_1, \ldots, x_n) = 0 \tag{15.39}$$

and suppose G is differentiable. In that case, the system of differential equations

$$\frac{dx_i}{dt} = -\frac{\partial G}{\partial q_i} \bigg/ \frac{\partial G}{\partial x_i} \tag{15.40}$$

$$q_i(t) = t \quad \text{for all} \quad i, \ 0 \le t \le 1 \tag{15.41}$$

$$x_i(0) = 0 \quad \text{for all} \quad i \tag{15.42}$$

is meaningful. Assuming it has a solution $x(t)$, for $0 \le t \le 1$, it may be seen that, for each t, $x(t)$ lies on the Pareto-optimal subset of $F(t, t, \ldots, t)$. Thus we see that $x(1)$ is Pareto-optimal in $F(1, \ldots, 1)$, which coincides with $v(N)$, and is also a possible value for the game v. Unfortunately, questions relating to the existence and uniqueness of this value remain open.

15.4.2 Example. Player 1 has a pound of coffee, player 2 has a coffee urn, and player 3 has a pound of sugar. We assume that player 1 likes his coffee without sugar, player 2 prefers it with sugar — but is willing to drink it without — and that there is no use for sugar then as a coffee sweetener.

Under these assumptions, a single player can obtain nothing at all; neither can coalitions {1, 3} and {2, 3}. On the other hand, {1,2} can produce a certain quantity (say, 3 gallons) of coffee brew, which has utility 100 to player 1, or 25 to player 2. (Of course, they can divide the coffee any way they wish.) The three players together can obtain the same three gallons of brew, with the additional advantage that sugar is now available for those who want it. Thus this will have utility of 100 units in total, which the three can divide as they wish. The characteristic function is then

$$v(\{i\}) = \{x | x_i \le 0\}, \quad \text{for} \quad i = 1, 2, 3$$
$$v(\{i, 3\}) = \{x | x_i \le 0, x_3 \le 0\}, \quad \text{for} \quad i = 1, 2$$
$$v(\{1, 2\}) = \{x | x_1 + 4x_2 \le 100, x_1 \le 100, x_2 \le 25\}$$
$$v(N) = \left\{ \begin{array}{l} x_1 + x_2 + x_3 \le 100, \\ x_i + x_j \le 100, \quad \text{for} \quad i \ne j \\ x_i \le 100, \quad \text{for} \quad i = 1, 2, 3 \end{array} \right\}$$

If we apply the Shapley technique, we find that, with $\lambda = (1, 1, 1)$, we have

$$w(\{i\}) = 0$$
$$w(\{1, 3\}) = w(\{2, 3\}) = 0$$
$$w(\{1, 2\}) = w(N) = 100$$

which has the value (50, 50, 0). Thus

$$\hat{\varphi}[v; \lambda] = (50, 50, 0)$$

which is feasible, and thus the desired value.

Using, instead, the Harsanyi technique, we have also $\mu(1, 1, 1)$. Then, of course, $\Delta_i(S) = 0$ for all one-player sets, and for $S = \{1, 3\}$ or $\{2, 3\}$. Then $x(\{1, 2\}) = (0, 0, 0)$, and so $r(\{1, 2\}) = 20$. Thus

$$\Delta_1(\{1, 2\}) = \Delta_2(\{1, 2\}) = 20$$

Then $x(N) = (20, 20, 0)$ and we will have $r(N) = 20$. Thus

$$\Delta_i(N) = 20, \quad i = 1, 2, 3$$

and

$$y = (40, 40, 20)$$

is the corresponding value. It is easily checked that the plane

$$(x_1 - 40) + (x_2 - 40) + (x_3 - 20) = 0$$

or $x_1 + x_2 + x_3 = 100$, is indeed a support to $v(N)$.

Finally, we may apply the MLE approach. This gives rise to a complicated system of differential equations, but leads to the solution

$$x(1) = (51.7, 47.8, 0.5)$$

Thus, the three approaches give three different "values" to the same game. All three seem, somehow, unsatisfactory.

15.5.   The Bargaining Sets

We consider, finally, the problem of generalizing the several concepts introduced in Chapter 13: bargaining sets, kernel, nucleolus. It turns out that both the kernel and the nucleolus are defined through an explicit interpersonal comparison of utilities, which while not entirely intuitive, was at least permissible so long as there was a common unit of utility. For games without side payments, it is difficult to see how these ideas can be generalized.

Given a coalition structure $\mathcal{T} = \{T_1, \ldots, T_m\}$, an i.r.p.c. is a pair $\langle x; \mathcal{T} \rangle$ where x is Pareto-optimal in the set

$$\bigcap_{j=1}^{m} v(T_j)$$

and x does not lie in the interior of any $v(\{i\})$. Similarly, a c.r.p.c. has the additional property that x does not lie in the interior of any $v(S)$, $S \subset T_j \in \mathcal{T}$. Objections and counterobjections can be defined just as in Section 13.1, and the several sets $\mathcal{M}, \mathcal{M}_1, \mathcal{M}^{(i)}, \mathcal{M}_1^{(i)}$, etc., are generalized without difficulty. Unfortunately the proof of Theorem 13.1.9 does not generalize, as this seems to depend directly on a free transferability of utility.

15.5.1 Example. Consider the three-person game of Example 15.I.4. For two-person coalitions, we find that the following are the only elements of $\mathcal{M}_1^{(i)}$:

$$\mathcal{T} = \{\{1, 2\}, \{3\}\}, \quad x = (d_{23} - s, d_{13} - s, w_3)$$
$$\mathcal{T} = \{\{1, 3\}, \{2\}\}, \quad x = (d_{23} - s, w_2, d_{12} - s)$$
$$\mathcal{T} = \{\{1\}, \{2, 3\}\}, \quad x = (w_1, d_{13} - s, d_{12} - s)$$

corresponding to the three points X, Y, Z, which, as we saw above, are the points of tangency of three mutually tangent circles centered at the vertices of the triangle (see Figure 15.5).

If the three-person coalition forms, we have

$$\mathcal{T} = \{N\}, \quad x = (d_{23} - t, d_{13} - t, d_{12} - t)$$

corresponding to the point T, which is the center of a small circle, tangent to all three of the large circles mentioned above.

15.5.2 Example. Consider the three-person game with characteristic function

$$v(\{i\}) = \{x \mid x_i \le 0\}, \quad i = 1, 2, 3$$
$$v(\{1, 2\}) = \{x \mid x_1 + 2x_2 \le 32\}$$
$$v(\{1, 3\}) = \{x \mid 2x_1 + x_3 \le 32\}$$
$$v(\{2, 3\}) = \{x \mid x_2 + 2x_3 \le 32\}$$
$$v(\{1, 2, 3\}) = \{x \mid x_1 + x_2 + x_3 \le 30\}$$

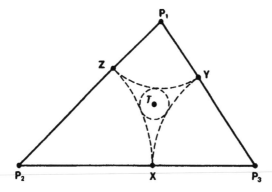

Figure 15.5.

It can be seen that, for $\mathcal{T} = \{\{1,2\},\{3\}\}$, the only x with $\langle x; \mathcal{T} \rangle \in \mathcal{M}_1^{(i)}$ is

$$x = (32/3, 32/3, 0)$$

Similar results hold for the other coalition structures which include a two-player coalition.

Suppose, next, that $\mathcal{T} = \{N\}$. At least one of the three inequalities

$$4x_1 - x_2 < 32, \qquad 4x_2 - x_3 < 32, \qquad 4x_3 - x_1 < 32$$

must hold, as otherwise we would have $x_1 + x_2 + x_3 \geq 32$, and we would have $x \notin v(N)$.

Suppose that

$$4x_1 - x_2 < 32$$

In this case, 1 would object against 2 with $\langle y; \mathcal{U} \rangle$, where $\mathcal{U} = \{\{1,3\},\{2\}\}$ and

$$y = (x_1 + \varepsilon, 0, 32 - 2x_1 - 2\varepsilon)$$

If 2 wishes to counter-object, he must give player 3 the amount y_3. But then he can receive at most z_2, where

$$z_2 = 32 - 2y_3 = 4x_1 - 32 - 4\varepsilon$$

But $x_2 > 32 - 4x_1$, and so $z_2 < x_2$. We conclude 2 has no counterobjection. The same considerations hold if $4x_2 - x_3 < 32$ or if $4x_3 - x_1 < 32$. Thus, there is no $\langle x; \mathcal{T} \rangle \in \mathcal{M}_1^{(i)}$ if $\mathcal{T} = \{N\}$.

15.5.3 Example. Let v be a three-person game, with

$$v(\{i\}) = \{x|x_i \leq 0\}, \quad i = 1, 2, 3$$
$$v(\{1, 2\}) = \{x|x_1 \leq 1, x_2 \leq 2\}$$
$$v(\{1, 3\}) = \{x|x_1 \leq 2, x_3 \leq 1\}$$
$$v(\{2, 3\}) = \{x|x_2 \leq 1, x_3 \leq 2\}$$
$$v(\{1, 2, 3\}) = \{x|x_1 + x_2 + x_3 \leq 1\}$$

In this case, we see that, for $\mathscr{T} = (\{1, 2\}, \{3\})$, there are no $\langle x; \mathscr{T} \rangle \in \mathscr{M}_1^{(i)}$. In fact, there is only one Pareto-optimal point here, and that is

$$x = (1, 2, 0)$$

Here, it is clear that 1 has an objection $\langle (2, 0, 1); \{\{1, 2\}, \{3\}\} \rangle$ against 2, and 2 has no counterobjection.

Similar considerations hold for $\mathscr{T} = \{\{1, 3\}, \{2\}\}$ or $\mathscr{T} = \{\{1\}, \{2, 3\}\}$.

If $T = \{N\}$, there are four stable values of x: these are $(2, 1, 0)$, $(0, 2, 1)$, $(1, 0, 2)$, and $(1,1,1)$. For all these, and no others, $\langle x; \{N\} \rangle \in \mathscr{M}_1^{(i)}$. Note that these four points are the core of v.

As can be seen from Examples 15.5.2 and 15.5.3, there is generally speaking no guarantee that, for a given $\mathscr{T}$, there will be some $\langle x; \mathscr{T} \rangle \in \mathscr{M}_1^{(i)}$. This is unfortunate as the existence theorem — Theorem 13.1.9 — is one of the most important in the theory of games with side payments. The obvious question is whether the definitions can be modified so as to obtain some existence properties.

It we study Examples 15.5.2 and 15.5.3, we see that the difficulties arising in the two examples are really of two different types. The problem in Example 15.5.3 is quite simply that there are not enough Pareto-optimal points. If, e.g., {1, 2} forms, this coalition really has no choice; it can only obtain the payoff (1, 2, 0). But this means that player 2 is receiving too much; he can never hope to protect his share. The situation is quite different in Example 15.5.2: there, the coalition N can divide 30 utiles in any way it wishes, and yet cannot avoid objections from some of its members.

It is possible to avoid difficulties of the type encountered in Example 15.5.3 by a simple expedient. Instead of insisting that, for an i.r.p.c. $\langle x; \mathscr{T} \rangle$, the payoff x be Pareto-optimal in the set $\cap v(T_j)$, we merely ask that it lie on the boundary of each of the sets $v(T_j)$. If we do this, we find that, for this example, the sets $\mathscr{M}_1^{(i)}$ are no longer empty, i.e., we will find that, for $\mathscr{T} = \{\{1, 2\}, \{3\}\}$, the point $\langle x; \mathscr{T} \rangle$ with $x = (1, 1, 0)$ belongs to $\mathscr{M}_1^{(i)}$. Of course, an obvious question arises as to how the point (1, 1, 0) can be stable when (1, 2, 0) is available; the answer, perhaps, is that this type of situation is too degenerate to give wholly satisfactory results.

Even with this modification, of course, the game of Example 15.5.2 will still have no stable i.r.p.c.s $\langle x; \mathscr{T} \rangle$ with $\mathscr{T} = \{N\}$. A modification of the bargaining set, due to Asscher, corrects this situation. As is Section 10.1, we shall say $i \gg j$, for an i.r.p.c. $\langle x; \mathscr{T} \rangle$, if i and j are partners in $\mathscr{T}$, i has an objection against j, and j has no counterobjection.

15.5.4 Definition. The i.r.p.c. $\langle x; \mathscr{T} \rangle$ belongs to the bargaining set $\mathscr{M}_1^{(i)}$ if, for every pair i, j of partners in $\mathscr{T}$, either

(a) $i \not\gg j$ or
(b) $i \gg j$, and there exists a chain of players $k_0 = j, k_1, \ldots, k_q = i$ such that

$$k_0 \gg k_1 \gg \cdots \gg k_{q-1} \gg k_q$$

It may be seen that, for games with side payments, the set $\tilde{\mathscr{M}}_1^{(i)}$ reduces to $\mathscr{M}_1^{(i)}$. In fact, Lemma 13.1.3 tells us that case (b) above cannot happen. Thus only case (a) is left: no $i \gg j$.

15.5.5 Theorem. Let v be an n-person game and let $\mathscr{T}$ be a coalition structure. Then there is at least one x such that $\langle x; \mathscr{T} \rangle \in \tilde{\mathscr{M}}_1^{(i)}$.

Proof. Let us define a new relationship, $\rightarrow$, by saying that $i \rightarrow j$ if there is some chain $k_0 = i, k_1, \ldots, k_q = j$ such that

$$k_0 \gg k_1 \gg \cdots \gg k_q$$

but no chain $l_0 = j, l_1, \ldots, l_p = i$ such that

$$l_0 \gg l_1 \gg \cdots \gg l_p.$$

Then it is not difficult to see that the relation $\rightarrow$ is acyclic. It will follow, just as in the proof of 11.1.9, that at any $\langle x; \mathscr{T} \rangle$, and for any $T_i \in \mathscr{T}$, there is some $i \in \mathscr{T}_k$ such that no $j \rightarrow i$.

Continuing as in the proof of 13.1.9, we will assume that each $v(\{i\})$ has the form $\{x | x_i \leq 0\}$, and let $X(\mathscr{T})$ be the set of all vectors, which are on the (upper) boundary of $v(S_j)$ for each $S_j \in \mathscr{T}$. The proof then generalizes directly to show that there is at least one x such that $\langle x; \mathscr{T} \rangle \in \tilde{\mathscr{M}}_1^{(i)}$.

15.5.6 Example. Let us consider once again the game of Example 15.5.2, with $\mathscr{T} = \{N\}$, see Figure 15.6.

As we saw above, there is no x with $\langle x; \{N\} \rangle \in \mathscr{M}_1^{(i)}$. In other words, at each x, there is some pair i, j with $i \gg j$. It is not possible (in this game) to have $i \gg j$ and $j \gg i$; thus, the desired points x must have a cycle $i \gg j \gg k \gg i$. This will hold if the three inequalities

$$4x_1 - x_2 < 32, \quad 4x_2 - x_3 < 32, \quad 4x_3 - x_1 < 32$$

hold simultaneously. This will be true in the interior of a small triangle, with vertices (10.9, 11.4,7.7), (7.7,10.9,11.4), and (11.4,7.7,10.9). This triangle, then is the desired set, $\tilde{\mathscr{M}}_1^{(i)}$.

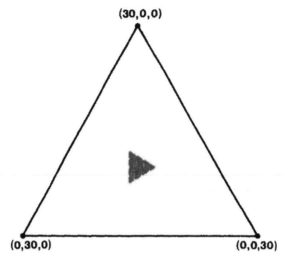

(30,0,0)

(0,30,0)   (0,0,30)

Figure 15.6.

Problems

1. Consider a three-person economy with three commodities, C_1, C_2, and C_3. Each of the three players has the utility function

$$u(x, y, z) = 1 - e^{-x-2y-z}$$

where x, y, and z are the amounts of the commodities C_1, C_2, and C_3, respectively. Assuming that the three players have the initial endowments

$$\omega^1 = (3, 2, 0), \quad \omega^2 = (2, 0, 3), \quad \omega^3 = (0, 1, 4)$$

respectively, characterize the sets $v(S)$ for each of the possible coalitions $S \subset N$.

2. For the economy of Problem 1, find a competitive equilibrium (prices and quantities). Verify that the corresponding imputation is in the core of the game.

3. For the economy (market game) of Problem 1, find the λ form of the Shapley value.

4. Let Ω be a compact set (of alternatives). Assume each $i \in N$ has a utility function u^i, defined and continuous over Ω, and that these utility functions are such that, for any three points x, y, and $z \in \Omega$, one of the following conditions holds:
   (1) $u^i(x) \geq \min\{u^i(y), u^i(z)\}$ for all $i \in N$
   (2) $u^i(y) \geq \min\{u^i(x), u^i(z)\}$ for all $i \in N$
   (3) $u^i(z) \geq \min\{u^i(x), u^i(y)\}$ for all $i \in N$

Define a simple game with player set N over the alternative set Ω by giving the winning coalitions, and specifying that each winning coalition can enforce any point of Ω (whereas losing coalitions can guarantee only that the outcome will be a point in Ω.)

Prove that, if the game is proper (i.e., the complement of any winning coalition loses, and any subset of a losing coalition loses), then the core $C(v)$ is nonempty.

[*Hint*: (a) Show that, for such a game, domination is an acyclic relation.

(b) Show that, if Ω is finite, there must be, by (a), an undominated imputation.
(c) If Ω is infinite, show that, for any finite $\Lambda \subset \Omega$, the set

$$\Omega - \text{dom } \Lambda = \{x | x \in \Omega; \text{ no } y \in \Lambda \text{ dominates } x\}$$

is nonempty. Use the compactness of Ω to show that $C(v)$ is nonempty.]

5. An n-person game without side payments need not have a stable set. Consider the simple seven-player game v, where H is the convex hull of the five points

$$c = (2, 0, 2, 0, 2, 0, 1)$$
$$p_1 = (1, 1, 2, 0, 0, 0, 0)$$
$$p_2 = (0, 0, 1, 1, 2, 0, 0)$$
$$p_3 = (2, 0, 0, 0, 1, 1, 0)$$
$$0 = (0, 0, 0, 0, 0, 0, 0)$$

and the minimal winning coalitions are $\{1, 3, 5\}$, $\{1, 2, 7\}$, $\{3, 4, 7\}$, $\{5, 6, 7\}$. Any winning coalition can enforce any imputation, whereas the other coalitions are effective only for the imputations in which their members receive 0. Then v has no stable set solutions.

(a) The core of v is simply $\{c\}$.
(b) If L_i is the segment $[p_i, c]$, then the set of imputations undominated by c is $L_1 \cup L_2 \cup L_3$. If V is stable it must contain at least one point, $q_i \neq c$, from each L_i.
(c) The points q_i must satisfy $q_1^7 = q_2^7 = q_3^7$. Hence V contains only one point other than c from each L_i. Hence V is not stable; v has no stable set solutions.

6. Let Γ be a game in extensive or normal form, and let v be the β-form of the characteristic function. Let $C(v)$ be the core of v (also known as the β-core of Γ). Define a supergame of Γ as an infinite sequence of repetitions of Γ, and define a *cooperative supergame c-strategy* as a joint strategy f for the n players in the supergame (i.e., a function defined over all information sets, for each repetition of the game). (Note that choices on each play of the supergame can be made to depend on the result of previous plays.)

For a supergame c-strategy f, let

$$p_r^i(f), \quad r = 1, 2, \ldots$$

be the expected payoff to player i in the rth play of the game, and set

$$P_r^i(f) = \frac{1}{r} \sum_{k=1}^{r} p_k^i(f).$$

The c-strategy f is summable if, for each i, the limit

$$Q^i(f) = \lim_{r \to \infty} P_r^i(f)$$

exists and is finite.

A summable supergame c-strategy f is a *strong equilibrium c-point* if there is no coalition S for which there exists a supergame c-strategy g satisfying

(a) $g^{N-S} = f^{N-S}$ (where f^T represents the restriction of the function f to points at which members of T play), and

(b) there exist numbers $\varepsilon_i > 0$ for all $i \in S$ such that

$$\lim_{k \to \infty} \text{Prob}\{\text{for some } r \geq k, \ P_r^i(g) \geq Q^i(f) + \varepsilon_i \text{ for all } i \in S\} > 0.$$

Prove Aumann's theorem: the imputation $x \in C(v)$ if and only if $x = (Q(f))$ for some strong equilibrium point f.

Chapter 16

Spatial Games

16.1. The Core and Near-Core Solution Concepts

In their original development of n-person games, von Neumann and Morgenstern represented all outcomes as payoff vectors, i.e., points in n-dimensional *utility space*. This representation in utility space continued even after new models for n-person games, e.g., games without side payments, were developed. This is the representation which we have used, up to this point, in this book.

An alternative representation of outcomes in terms of *policy space* is possible, however. In this model, outcomes of the games are policies, which are assumed to lie in some low-dimensional Euclidean space; utilities to the players are usually defined in terms of distance from their most preferred, or ideal, points. Such a representation is most useful when the outcomes correspond to public goods (so that several individuals might obtain similar benefits from the same outcome) and so have been used mainly in connection with applications to political science. We refer the reader to Black [16.1], Plott [16.14], Owen [16.12], Ordeshook [16.11], and Riker and Ordeshook [16.15], for some of the earlier developments.

16.1.1 Definition. An *n-person voting game* is a set $N = \{1, 2, \ldots, n\}$ of players, called *voters,* together with a collection W of subsets of N, the *winning coalitions.* All other subsets of N are called *losing coalitions.* It is assumed that

  (i) $N \in W$
 (ii) $\varnothing \notin W$
(iii) If $S \in W$, and $S \subset T$, then $T \in W$

The game will be called *proper* if, for any S and $T \in W$, $S \cap T \neq \varnothing$. It will be called *decisive* if, for any S, either S or $N - S \in W$.

We shall, in this section, deal exclusively with proper, decisive voting games. Some of the results given here are valid for other voting games, but it should not be assumed that this is so in general.

16.1.2 Definition. An *m-dimensional spatial game* is a voting game, together with a collection of n points P^i, $i = 1, \ldots, n$, in m-dimensional Euclidean space R^m. Point P^i is voter i's *ideal point.* The convex hull of the points P^i is the *Pareto-optimal set.*

Heuristically, the situation is simple enough. The voters are to choose among alternative policies (or candidates) by voting; a winning coalition can enforce any point on which its members agree. It is assumed that player i would be happiest with a policy occupying point P^i; failing that, he would like a position as close as possible to P^i, i.e., we assume that i's utility is a decreasing function of the distance between the achieved outcome and P^i.

The points of space R^m are called *positions* or *policies*. Given two policies, x and y, we define a partition of N as follows

$$Q(x, y) = \{i | d(x, P^i) < d(y, P^i)\}$$

$$Q(y, x) = \{i | d(x, P^i) > d(y, P^i)\}$$

$$I(x, y) = \{i | d(x, P^i) = d(y, P^i)\}$$

where d represents the usual (Euclidean distance between two points. The interpretation is that each voter prefers points which are closer to his ideal point; hence, $Q(x, y)$ is the set of voters who prefer x to y; $I(x, y)$ is the set of voters indifferent between x and y. Thus, x *dominates* y if $Q(x, y) \in W$.

The *core* is (as always) the set of undominated policies.

The following will be an important concept:

16.1.3 Definition. Let $(N, W, \{P^i\})$ be an m-dimensional spatial game. The hyperplane

$$\sum_k a_k x_k = b \qquad (16.1)$$

is a *median hyperplane* for this game if the two sets

$$\left\{ i | \sum_k a_k P^i_k > b \right\}$$

and

$$\left\{ i | \sum_k a_k P^i_k < b \right\}$$

are both losing coalitions.

Essentially, a median hyperplane is a hyperplane such that the ideal points to either side of it (i.e., in the open half-spaces) form a losing coalition. For a decisive game, this can only happen if at least one of the ideal points lies on the hyperplane. No two median hyperplanes can be parallel, but there is one median hyperplane in each direction in space; i.e., given any hyperplane in space, there will be exactly one median hyperplane parallel to this. Note also that, for a decisive game, if each *open* half-space contains only a losing coalition of ideal points, then each *closed* half-space contains a winning coalition.

It is not difficult to see that, when $m = 1$, so that the voters can, in fact, be ordered along a line, there will be a point in the core, obtained as follows.

If the linear ordering is strict (no two ideal points coincide) there will be a voter such that the set of all voters to either side of him forms a losing coalition. For decisive proper games, he will be unique — the median or swing voter. If the linear ordering is not strict, so that several voters' ideal points coincide, the swing voter need not be unique, but there will be a unique swing position. This position can easily be seen to lie in the core. (Note that, since the space here has dimension 1, the median "hyperplane" is nothing other than the median point.)

For $m = 2$, and *a fortiori* for greater m, the core is usually empty. We see this in Figure 16.1, where points A, B, and C represent the ideal points for voters 1, 2, and 3 respectively. (In this three-voter game, we assume each coalition of two or more voters wins.)

To see that the core is empty, we consider point x which does not lie on line AB. It is easy to see that y, the foot of the perpendicular to AB through x, is closer to both A and B than is x. Thus $Q(y, x) = \{1, 2\}$, and we conclude that y dominates x. But y does not lie on AC, and so y is dominated by z — the foot of the perpendicular to AC through y. In turn, z is dominated by w, etc. Since no point can lie on all three lines AB, AC, and BC simultaneously (unless AB, AC, and BC are concurrent) we conclude that the core is empty.

In fact, it is possible for a two-dimensional array of voters to have a non-empty core. Figure 16.2 illustrates this, for a five-voter game (in which all coalitions of three or more voters win). The point here is that one of the ideal points — P^5 in this case — is very centrally located, at the intersection of the two line segments $P^1 P^3$ and $P^2 P^4$. The effect of this is that every line that passes through P^5 is a median line: there are at most two ideal points on each side of such a line. Conversely, all median lines pass through P^5.

(The above analysis suggests that core points exist only in degenerate cases, i.e., two or more coincident points, three or more collinear points, etc. In fact that is the case for symmetric games, where each voter has only one vote. In general weighted voting games, however, if one voter has many votes, a core point may exist even under nondegeneracy. The obvious example occurs when one voter (a dictator) has more than half the votes. In such case the dictator's position is clearly a core point.)

Figure 16.1.

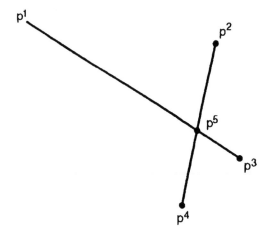

Figure 16.2.

16.1.4 Theorem. A necessary and sufficient condition for a spatial game to have a nonempty core is that all median hyperplanes pass through one same point. That point (the intersection of all median hyperplanes) is necessarily the ideal point of at least one voter. Moreover, this point is not only undominated; it actually dominates all other points.

Proof. Suppose that point x does not lie on a median hyperplane, H — as is the case with x and the hyperplane (line) AB in Figure 16.1. In this case, the point y, at the foot of the normal line (perpendicular) to H through x, must dominate x, for $Q(y, x)$ contains at least all j whose ideal points lie in the closed half-space consisting of H and the opposite side of H from x. Since H is a median hyperplane, that set is a winning coalition. Thus x is not in the core.

Conversely, suppose x lies on all median hyperplanes. Then x will dominate all other points. For, if y is any other point, the hyperplane through x normal to the line xy is a median hyperplane, and so x dominates y.

We do not prove one last statement in the theorem, namely that if x lies on all median hyperplanes, it is necessarily the ideal point of at least one player. This is left as an exercise for the reader.

We see, then, that the core is frequently empty for higher-dimensional games. This deprives us of one of the most appealing solution concepts. We look, then, for points which come "as close as possible" to being in the core. Three approaches seem reasonable.

(a) *The Copeland Winner*: The core consists of points which are undominated by any other. Absent such a point, we might look for a point X^* which is dominated by "as few" points as possible. This idea was originally advanced by Copeland [16.2]. In the context of Euclidean space, we take this "as few" to mean a set with minimal m-dimensional Lebesgue measure (i.e., area or volume).

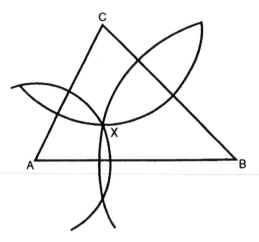

Figure 16.3.

Figure 16.3 shows the same configuration as 16.1. Three circles are drawn, passing through x, and centered at the ideal points A, B, and C, respectively. Now, x is dominated by all points in the interior of the three cigar-shaped areas which are bounded by these circles. It dominates all other points, except for a set of area 0, namely, points on the three circular arcs. It can be shown (see Problem 6) that the total shaded area is then

$$H(x) = \angle CAB \ d(A, x) + \angle ABC \ d(B, x) + \angle BCA \ d(C, x) - 2T$$

where the angles are measured in radians, and T is the area of the triangle. This being so, it will follow without much difficulty that $H(x)$ is minimized by choosing x to be the weighted arithmetic mean, with weights equal to the three angles, of points A, B, and C, i.e.

$$X^* = \frac{\alpha A + \beta B + \gamma C}{\pi} \qquad (16.2)$$

where α, β, γ are the measures (in radians) of the three angles of the triangle ABC; if in degrees, the denominator should be changed to 180°. The X^* obtained in this manner is the Copeland winner.

For more than three voters, calculation of the Copeland winner can be quite complicated, especially if the dimension m is large. The reader is referred to Grofman et al. [16.7] and to Owen and Shapley [16.13] for a discussion of this problem.

(b) *The Yolk*: As discussed above, the core point, if it exists, must lie at the intersection of all the median hyperplanes. When there is no such point, we might look for the point which is closest to all such planes, in the sense of minimizing its distance from them, i.e., a point x which gives

$$r = \min_{x} \max_{K} \ d(x, K) \qquad (16.3)$$

where $d(x, K)$ is the (perpendicular) distance from x to hyperplane K, and the maximum is taken over all median hyperplanes. Such an x is the *yolk center*, and the ball of radius r centered at x (the smallest ball which touches all median hyperplanes) is known as the *yolk*. This concept was first introduced by McKelvey [16.8]; its principal properties have been developed in Feld *et al.* [16.5].

In Figure 16.4, the same configuration of 16.1 is used once again. The yolk here is the circle inscribed in triangle ABC. Its center — the incenter of the triangle — can be found, as is well known from elementary Euclidean geometry, at the intersection of the three angle bisectors.

For more complex configurations, locating the yolk is more complicated. We note that there are an infinite number of median hyperplanes, one in each direction in space. However, only a finite number of hyperplanes need be considered in searching for x. Since distance from a hyperplane is a linear function of position, it follows that the yolk can be found by linear programming. (The problem is that it is by no means clear which hyper-planes need to be considered.) Readers are referred to McKelvey [16.8], Ferejohn *et al.* [16.6] for details.

(c) *The Minimal Response Point*: We mentioned above that the core point (if it exists) will dominate all other points. In its absence, we might look for the point which "comes closest" to dominating all others.

Let x be a policy, and suppose y (an alternative policy) dominates x (or at least, suppose x does not dominate y). If the core is empty, y can certainly be dominated, by a third alternative, say, called z. If there is such a z close to x, then we can say that x "comes close" to dominating y. If there is no such z near x, then x is far from dominating y. We look, then, for the z closest to x (in the sense of Euclidean distance)

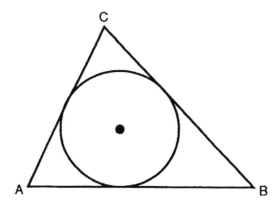

Figure 16.4.

which dominates y. In fact, there will usually be no closest z, since domination is (topologically) an open relationship. Nevertheless, an infimal distance will exist. Let, then

$$s(x, y) = \inf\{d(x, z)|z \text{ dominates } y\} \qquad (16.4)$$

We may think of x as the position chosen by an incumbent politician, whereas y is that chosen by a challenger, seeking to unseat him (or her). Then $s(x, y)$ is the "change of position" necessary for the incumbent to repel this challenger. Of course, the challenger is also able to choose (alternatively, the incumbent's opposition is free to choose a challenger), and so we might expect that, given x, y will be chosen so as to maximize this $s(x, y)$. Thus, let

$$\sigma(x) = \max_{y} s(x, y) \qquad (16.5)$$

be the *finagle radius* of position x. The *minimal response point* is that x which minimizes the finagle radius $\sigma(x)$, while the minimal value, δ, of $\sigma(x)$ is the *minimal response distance*. This point and distance, under the names *finagle point* and *finagle radius*, respectively, were introduced in Wuffle *et al.* [16.18]. Their properties have not yet been studied in detail.

In Figure 16.5, three mutually tangent circles have been drawn, centered at points A, B, and C respectively. (The three points of tangency are, coincidentally, the three points of tangency of the inscribed circle (yolk) described above.) A small, fourth circle, J, has been drawn, tangent to these three circles. The center, x, and radius, δ, of J are the minimal response point and the minimal response distance, respectively.

For other configurations, calculation of the minimal response point is considerably more complicated: it reduces to a nonlinear programming problem.

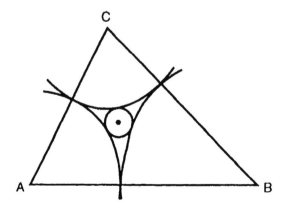

Figure 16.5.

16.1.5 Example. To see how these points compare, consider the 3-person voting game in which every coalition with 2 or more voters is a winning coalition. Let the three voters' ideal points be at $A = (0, 0)$, $B = (0, 1)$, and $C = (\sqrt{3}, 0)$, respectively. It may be seen that this triangle has angles of $\alpha = 90°$, $\beta = 60°$, and $\gamma = 30°$; the sides are $a = 2$, $b = \sqrt{3}$ and $c = 1$ units long.

(a) Calculation of the Copeland winner is easy enough; we need merely apply formula (16.2), which yields

$$X^* = \frac{90A + 60B + 30C}{180} = (0.289, 0.333)$$

(b) To calculate the yolk, we note that the semiperimeter of the triangle is $(3 + \sqrt{3})/2 = 2.366$ units long. From elementary geometry we know that the radius of the inscribed triangle is given by $r^2 = (s-a)(s-b)(s-c)/s$, so in this case $r = 0.366$. Thus the yolk is a circle of 0.366 units radius, tangent to the three sides of the triangle. In particular, it must be tangent to both the coordinate axes, and it follows that its center will be the point (0.366, 0.366).

(c) Calculating the minimal response point is slightly more complicated. The three mutually tangent circles (as in Figure 16.5) have radii $s-a = 0.366$, $s-b = 0.634$ and $s-c = 1.366$, respectively. Now, the condition for four circles to be mutually tangent is that the square of the sum of the four curvatures (i.e., the reciprocals of the radii) be equal to twice the sum of their squares. Hence, the small circle which is tangent to all three of these will have a radius equal to $\delta = 0.095$. This is the minimal response distance. The minimal response point, the center of this circle, can be obtained by noticing that it must be at distance $s-a+\delta$, $s-b+\delta$, and $s-c+\delta$ from the three vertices; it is in fact the point (0.311, 0.341).

16.2.   The Modified Power Index

Alternatively, we might try to define a power index for spatial games. With no spatial considerations, of course, there are well-known power indices (the Shapley value or Banzhaf-Coleman index) which can be applied directly to the set W of winning coalitions for a given game. (We merely define $v(S) = 1$ if $S \in W$, and $v(S) = 0$ otherwise.) The idea, however, is to modify such an index so as to take account of the geometry of the points P^i.

As discussed in Section 12.1, above, the Shapley value can be obtained by taking an arbitrary permutation (ordering) of the n players, and assigning to player i the marginal contribution $v(S \cup \{i\}) - v(S)$ which i makes to the coalition S of all players who precede him or her in the ordering. The value is then the arithmetic mean of these $n!$ marginal contributions.

In case of a voting game, this can be simplified by introducing the idea of a *pivot*:

16.2.1 Definition. Let (N, W) be a voting game, and let $\ll$ be a strict linear ordering of N. For $i \in N$, let $S\,(i; \ll)$ be the set of all j such that $j \ll i$. Then, i is a pivot for this ordering if and only if $S\,(i; \ll)$ is a losing coalition, while $S\,(i; \ll) \cup \{i\}$ is a winning coalition.

As a very simple example, if W consists of all coalitions with k or more elements, then the pivot is always the kth element in the ordering. More generally, the pivot is the first element (in the given ordering) to complete a winning coalition. It is not difficult to see that, by conditions (i–iii) of Definition 16.1.1 above, every ordering (permutation) of the set N has a unique pivot. If the game is decisive, then the pivot for ordering $\ll$ will also be the pivot for the reverse ordering $\gg$ (where $a \gg b$ if and only if $b \ll a$).

If, now, we define $v(S) = 1$ for every $S \in W$, and $v(S) = 0$ otherwise, we see that every permutation will have a unique player i (the pivot) for whom the marginal contribution $v(S \cup \{i\}) - v(S)$ is 1. For all others, this contribution is 0. Thus for a voting game the value Φ_i is simply the number of permutations for which i is a pivot, divided by $n!$. This can be interpreted as saying that Φ_i is the probability that i will be the pivot, assuming a uniform probability distribution over the set of all permutations of N.

Suppose, however, that the points P^i are introduced. It seems reasonable that the geometry of these points should affect the probabilities assigned to the various permutations. As an example, consider a 3-person game (where W consists of all coalitions of two or more voters), where the ideal points are as in Figure 16.6.

In such a situation, since players 1 and 2 are very close together, with 3 much further away, the two orderings 1–3–2 and 2–3–1 (corresponding to situations where 3 would agree with one of 1 and 2 against the other) would seem to be much more difficult to obtain, and so we would be justified in assigning lower probabilities to these two orderings than to the other four. But with the given voting rule (any two or more players form a winning coalition), the pivot is always the second player in any order. Thus Φ_3, the probability that 3 will be the pivot, would be considerably less than $1/3$. Heuristically, this is not unreasonable: since voters 1 and 2 have their ideal points close together, they will tend to agree on most issues, and 3's opinion will usually have little effect on the outcome.

Now, how can we best develop this idea? A priori, there are many possibilities. To choose among these, we might consider several *desiderata:*

Figure 16.6.

D1. If P^i and P^j are close together, then *ceteris paribus,* permutations in which i and j are close together should have higher probability than those in which these two players are far apart.

D2. The probabilities should be invariant under rigid motions of the configuration of ideal points.

D3. The probabilities should be continuous functions of the ideal points, except possibly for discontinuities when two or more of the ideal points coincide, when three or more lie on the same line, etc.

D4. If two ideal points P^i and P^j coincide, the effect should be as if the two players, i and j, were consolidated into a single player.

Now, there may be many schemes which satisfy D1–D4, above. The following is due to Shapley [1975]. First, Shapley defines *issues space* as the dual, R^{m*}, of the policy space (i.e., the space of linear functional over policy space R^m).

Let $(N, W, \{P^i\})$ be a spatial game, and let f be an issue. Then f defines an ordering of the voters, in which i precedes j if and only if $f(P^i) \leq f(P^j)$. This ordering need not be a strict order (it is quite possible that $f(P^i) = f(P^j)$ for two distinct voters i and j), but, in the great majority of cases, it will be strict. (The condition for such "ties" is that the gradient of the functional f be orthogonal to the line segment $P^i P^j$, and, since there are only finitely many voters, such ties will happen only in degenerate cases.)

16.2.2 Example. Consider the configuration of Figure 16.7, and let f be the linear functional given by $f(P) = v \cdot P$, where v is shown by the arrow, and $\cdot$ is the usual inner product (dot product). We can project points P^i $(i = 1, 2, 3)$ by a perpendicular projection, onto the (directed) line l, which is parallel to vector v. Their images are then Q^i. Note that, on the line, these points are in the order Q^2, Q^1, Q^3. Thus, f induces the order 2, 1, 3 on the voters.

It is now easy to see that, if f and g are issues, with $g = rf$, $r > 0$, then f and g induce the same order on a given set of points $\{P^i\}$. It follows that, to obtain all the orders induced by issues, it suffices to consider one issue in each direction of the space R^{m*}.

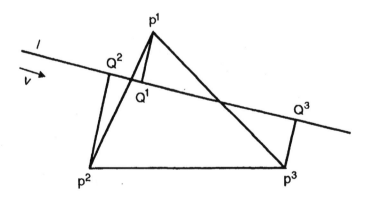

Figure 16.7.

Let K, then, be the unit sphere in R^{m*}, i.e., the set of all linear functional f with norm $|f| = 1$. Let μ be Lebesgue $(m-1)$-dimensional measure on K. For $i \in N$, define

$$A_i = \{f \in K \mid i \text{ is pivot in the order induced by } f\} \qquad (16.6)$$

Then the spatial index of power Ω is defined by

$$\Omega_i[W, \{P^j\}] = \mu(A_i)/\mu(K) \qquad (16.7)$$

Equivalently, we might define Ω_i as the probability that i is pivot, when permutations are chosen according to a uniform distribution over the sphere K of linear functional with unit norm.

The following is important:

16.2.3 Theorem. Assume that all ideal points P^i are distinct. Then the measure μ of that subset of K consisting of linear functional which do not induce an ordering with a unique pivot is 0.

Proof. We know that if an ordering is strict, then it must have a unique pivot. Hence, if f does not induce an ordering with a unique pivot, the ordering is not strict, i.e., there are distinct voters i and j such that $f(P^i) = f(P^j)$. But this can only happen if the gradient of f is orthogonal to the line $P^i P^j$.

Since there are only n voters, the gradient of f must be orthogonal to one of $n(n-1)/2$ lines, i.e., it must lie in one of $n(n-1)/2$ linear subspaces of R^m, each of dimension $m-1$. Thus f must lie in the union of $n(n-1)/2$ linear subspaces of R^{m*}, each of dimension $m-1$. The intersection of this union with K will have dimension $m-2$, and therefore its $(m-1)$-dimensional measure is 0.

16.2.4 Corollary. For any game $(N, W, \{P^i\})$

$$\sum_{i \in N} \Omega_i = 1$$

Proof. This follows directly from the fact that almost every f (i.e., except for a set of measure 0) gives rise to a unique pivot. Thus the sum of the measures of the sets A_i must be equal to the measure of set K. (It should perhaps be proved that each H_i is measurable; we will dispense with this proof.)

16.2.5 Example. Consider the three-voter game represented by Figure 16.7, where W consists of all coalitions with two or more voters.

In this case, we have $m = 2$. Then K is the unit "sphere" in R^{2*}, i.e., the circumference of a circle, with measure $\mu = 2\pi$. Each f can be represented by an angle, namely, the direction of its gradient.

Let the issue f have the gradient v as shown in the figure. We saw above that this issue induces the order 2, 1, 3. Now, it is not too difficult to see that the (necessary and sufficient) condition for f to induce this order is that v go *from left to right,* and that the perpendicular to v, through P^1, lie *inside* the angle $P^2 P^1 P^3$. This angle has measure α radians, and thus the measure of the set which induces this order is α. On the other hand, the issue $-f$, whose gradient is the opposite vector $-v$, will induce the opposite order, 3, 1, 2. Thus this order is also induced by a set of measure α. Now, these are the only two orders in which voter 1 is the pivot. We conclude that

$$\mu(A_1) = 2\alpha$$

In a similar way, we can see that $\mu(A_2) = 2\beta$, and $\mu(A_3) = 2\gamma$, where β and Y are the measures of the other two angles in the triangle. Since $\mu(K) = 2\pi$, we conclude that

$$\Omega_1 = \alpha/\pi, \quad \Omega_2 = \beta/\pi, \quad \Omega_3 = \gamma/\pi$$

Since α, β, and γ are the three angles of a triangle, their sum is π, and we see that

$$\Omega_1 + \Omega_2 + \Omega_3 = 1$$

as Corollary 16.2.5 said.

In general, it is interesting to see how this index measures voting power. In a two-out-of-three majority voting game, the power assigned to the three voters is proportional to the angles in the triangle formed by their ideal points. Thus, in a situation such as that of Figure 16.6, voter 3's ideal point is very far from those of the other two voters; the corresponding angle is consequently very small. It follows that voter 3 has very little power there. On the other hand, in a situation such as that of Figure 16.8, voter 2 is a "central" voter: his position is almost directly between that of the other two voters. The corresponding angle is large (obtuse) and so voter 2's power is quite large. Heuristically, 2 seems to be in position to play off 1 and 3 against each other in such a case, forming coalitions with one or the other with little danger that they will act together against him.

In Example 16.2.6, we note that, given this particular voting game, the power index depends only on the three angles of the triangle; i.e., it will be the same for any similar triangle. This suggests that, in general, this might be so for all voting games. In fact, the following theorem tells us that this is indeed the case.

16.2.6 Theorem. For a given voting game (N, W), the power index Ω is invariant under similarity transformations of the spatial configuration $\{P^i\}$.

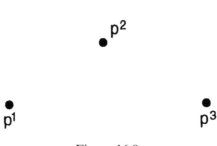

Figure 16.8.

Proof. Any similarity transformation of Euclidean space can be obtained as a composition of four basic types of transformations: *translations, rotations* centered at the origin, *reflections* in a mirror passing through the origin, and *dilatations* centered at the origin. (A dilatation is a transformation which transforms each point X into point rX, where r is a fixed non-zero scalar.)

Under a translation, each point P^i is transformed into point $P^{i\prime} = P^i + C$, where C is a fixed vector in R^m. Let f be an issue. Since f is a linear functional, $f(P^{i\prime}) = f(P^i) + f(C)$, and so f induces the same order on the points $P^{i\prime}$ as on the points P^i. Thus, Ω will be invariant under translations.

Under a rotation or reflection, each P^i is transformed into point $P^{i\prime} = MP^i$, where M is an orthogonal matrix in space R^m. Let, then, M^* be the adjoint transformation in the dual space, i.e., M^*f is the issue defined by $M^*f(P) = f(MP)$ for all $P \in R^m$. Clearly, M^*f induces the same order on points P^i as f does on the points $P^{i\prime}$. Now, in the dual space, orthogonal transformations map K into itself, and preserve Lebesgue measure. Thus, Ω will be invariant under rotations centered at the origin or reflections in a mirror through the origin.

Under a dilatation, each P^i is transformed into $P^{i\prime} = rP^i$ where r is a nonzero scalar. If f is an issue, it is a linear functional, and so $f(P^{i\prime}) = rf(P^i)$. Thus, if $r > 0$ (a *direct* dilatation), f induces the same order on the points $P^{i\prime}$ as on points P^i, and hence Ω is invariant under this transformation. If, on the other hand, $r < 0$ (an *inverse* dilatation), we see that f induces the same order on the points rP^i, as $-f$ does on points P^i. But this transformation of R^{m*} (f going into its negative $-f$) maps K into itself and preserves Lebesgue measure. Thus Ω will be invariant under any dilatation (direct or inverse) centered at the origin.

We see that Ω is, then, invariant under these four basic types of similarity transformations of the points $\{P^i\}$. Thus it must be invariant under any composition of such transformations. But every similarity transformation is of this type. Hence, Ω is invariant under all similarity transformations.

The following theorem is also important:

16.2.7 Theorem. Let $(N, W, \{P^i\})$ be an m-dimensional spatial game, and suppose that the n points P^i are all contained in some k-dimensional linear subspace, G, of R^m ($k < m$). Then the power index Ω is the same for this game, whether we think of the P^i as points of R^m or as points of G.

Proof. Let G^* be the dual subspace, i.e., the space of all linear functionals g whose gradient lies in G, and let H^* be its orthogonal complement, i.e., the space of all linear functionals h such that $h(x) = 0$ for all $x \in G$. Then R^{m*} is the direct sum of G^* and H^*, i.e., every linear functional f can be expressed in a unique manner in the form $f = g + h$, where $g \in G^*$ and $h \in H^*$. Let L be the perpendicular projection of R^{m*} into G^*, i.e., if $f = g + h$ is as above, then $L(f) = g$.

Now, since all $P^i \in G$

$$f(P^i) = g(P^i) + h(P^i) = g(P^i) = L(f(P^i))$$

Thus, the order induced by f is necessarily the same as that induced by $L(f)$.

Now, if $f = g + h$ as above, let $g' = g/|g|$, the normalized form of g, and let $M(f) = g'$. Then g' induces the same order as g, and so $M(f)$ induces the same order as f. It follows that, if A_i is the set of issues in $K \cap G^*$ which induce orders with voter i as pivot, then $M^{-1}(A_i)$ is the set of issues in K which induce orders with i as pivot.

We note, finally, that L', as a transformation of the $(m-1)$-sphere K into a $(k-1)$-sphere $K \cap G^*$, does not preserve measure. However, if A is a measurable subset of $K \cap G^*$, then the ratio $v(A)/v(K \cap G^*)$, where v is Lebesgue $(k-1)$-measure, is the same as the ratio $\mu(M^{-1}(A))/\mu(K)$. (This can best be seen if we think of a sphere of radius R in 3-space, If a section of such a sphere is bounded (only) by two meridians of longitude, then the surface area of this section is equal to $2R$ times the angle between the two meridians, while the part of the equator which lies inside the section is equal to R times that same angle.) Thus, the power index Ω is the same regardless of whether we consider all issues in K or only issues in $K \cap G^*$, i.e., regardless of whether we treat the P^i as points of R^m or as points of the subspace G.

For games in characteristic function form, the Shapley value can be thought of as a possible outcome of the game. For spatial games, on the other hand, the power index Ω is not a possible outcome — outcomes are policies. To obtain an outcome from Ω, we define the center of power:

16.2.8 Definition. The center of power of the spatial game $(N, W, \{P^i\})$ is the point

$$P^* = \sum_{i \in N} \Omega_i P^i \tag{16.8}$$

where Ω is the spatial power index.

16.2.9 Example. Consider once again the 3-voter configuration of Example 16.2.5. The power index here is given by

$$\Omega = (\alpha/\pi, \beta/\pi, \gamma/\pi)$$

where α, β, γ are the three angles of the triangle. Thus the center of power is

$$P^* = \frac{\alpha P^1 + \beta P^2 + \gamma P^3}{\pi} \tag{16.9}$$

16.2.10 Example. Consider the five-voter configuration of Figure 16.2. Assume, once again, that W consists of all coalitions with 3 or more voters.

For this configuration, we note that P^5 lies on the line segment $P^1 P^3$. Thus, for any issue f, $f(P^5)$ cannot be smaller than both $f(P^1)$ and $f(P^3)$. It follows that no f can induce an ordering in which voter 5 precedes both voters 1 and 3. Nor, for similar reasons, can f induce an ordering in which 5 follows both 1 and 3. In other words, if f induces a strict ordering on the voters, 5 must lie *between* 1 and 3 in that ordering.

Note, however, the P^5 also lies on the segment $P^2 P^4$. Thus, if f induces a strict ordering, 5 must lie between 2 and 4 in that ordering.

Thus, in any strict ordering induced by an issue, we see that voter 5 must be preceded by exactly one of voters 1 and 3, and also by exactly one of voters 2 and 4.

In other words, 5 will necessarily be third in any such ordering. But, for this W, the pivot in any ordering is the third voter. It follows that 5 will be the pivot for the orderings induced by almost all issues. Hence the power index here is

$$\Omega_1 = \Omega_2 = \Omega_3 = \Omega_4 = 0, \quad \Omega_5 = 1$$

In other words, voter 5 here (due to his central position) has all the power. The center of power is then

$$P^* = P^5$$

It is of interest to note that, for Example 16.2.10, the center of power P^* coincides with the Copeland winner X^*, as given by equation (16.2). Note also that in 16.2.11, the center of power is in fact the unique core point. As mentioned above, the core of a decisive spatial game, if nonempty, consists of exactly one point which is necessarily the Copeland winner. Thus for this example also, the center of power and the Copeland winner coincide. The question is whether this is merely a coincidence.

In fact, the following is true:

16.2.11 Theorem. Let $(N, W, \{P^i\})$ be a proper decisive m-dimensional spatial game. Suppose $m = 1$ or 2. Then the Copeland winner X^* is unique, and coincides with the center of power P^*.

For dimension $m = 1$, the theorem is merely a restatement of the fact, mentioned above, that the position of the median (pivot) voter is the unique core point. For $m = 2$, we omit the proof of this theorem, which is quite lengthy. An outline of the proof is given in Problem 7 at the end of this chapter. We note, however, that the Copeland winner is the point minimizing a function which, however, is not itself easy to determine. Thus a direct computation of the Copeland winner can be quite complicated. On the other hand, computation of the power index Ω is relatively straightforward in many cases. If nothing else, a numerical approximation — calculating the pivot in a reasonably large number of directions — can be done quite quickly with computers. Since (at least for dimension 2) the two problems are equivalent, we can simplify the Copeland winner calculation considerably.

16.2.12 Example. Let $N = \{1, 2, 3, 4\}$, and let W consist of all coalitions with three or more voters, and also of coalitions S with two voters, such that $1 \in S$. (We can think of voter 1 as having 2 votes, while each of the other players has 1 vote; 3 votes are needed to win.) Let the voters' ideal points be at $P^1 = (-1, 0)$, $P^2 = (1, 0)$, $P^3 = (0, \sqrt{3})$, and $P^4 = (0, 1/\sqrt{3})$. Thus, the first three voters form an equilateral triangle; voter 4 is at the centroid of the triangle (see Figure 16.9).

We note that, if the gradient of issue f goes from upper left to lower right, and its perpendicular lies in the angle $P^3 P^1 P^4$, then f induces the order 3–1–4–2; if its perpendicular lies in the angle $P^4 P^1 P^2$, f induces 3–4–1–2 with the same perpendiculars, but from lower right to upper left, f induces the reverse orders 2–4–3–1 and 2–1–4–3 respectively. In each of these cases, voter 1 is the pivot. These issues cover an angle equal to twice angle $P^3 P^1 P^2$, or $120°$. Thus, $\Omega_1 = 1/3$.

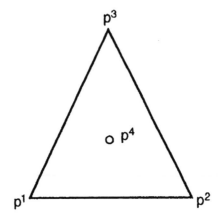

Figure 16.9.

Next, note that, if the gradient of f goes from lower left to upper right, and its perpendicular lies inside the angle $P^1 P^2 P^4$, then f induces the order 1–2–4–3. With the same perpendicular, but from right to lower left, f induces the order 3–4–2–1. In both these cases, 2 is the pivot. Since the angle $P^1 P^2 P^4$ measures $30°$, and 2 is a pivot in a set measuring twice this angle, or $60°$, we find $\Omega_2 = 1/6$. By symmetry, $\Omega_3 = 1/6$ also.

Finally, since this index must assign total power 1 to all the voters, we find $\Omega_4 = 1/3$.

The center of power is, then, given by

$$P^* = (2P^1 + P^2 + P^3 + 2P^4)/6$$

or

$$P^* = (-0.167, 0.481)$$

This center of power is also the Copeland winner, X^*.

16.2.13 Example. Let (N, W), be as in Example 16.2.10, and let the ideal points now be $P^1 = (0, 1/\sqrt{3})$, $P^2 = (0, \sqrt{3})$, $P^3 = (1, 0)$, and $P^4 = (-1, 0)$. Note that the difference is that in this example, the strong voter is at the centroid of the triangle (in the previous example, he or she was at one of the vertices) (see Figure 16.10).

It is not too difficult to see that, since P^1 is in the interior of the convex hull of the other three points, it is not possible for any issue to induce an ordering in which voter 1 is either first or last, i.e., voter 1 will always be in either second or third position in the order. But this means that voter 1 will always be the pivot. Thus, the index of power is

$$\Omega_1 = 1, \quad \Omega_2 = \Omega_3 = \Omega_4 = 0$$

and the center of power is here point P^1. In fact, P^1 is here the unique point in the core.

Figure 16.10.

Problems

1. In two dimensions, in a 3-person majority game (any coalition of 2 or more players wins), let the three players' ideal points be at (0, 0), (1, 3), and (3, 0).
   (a) Find the yolk for this game.
   (b) Find the spatial power index and the center of power.

2. In two dimensions, in a 5-person majority game (any coalition of 3 or more players wins), let the five players' ideal points be at (0, 0), (0, 1), (1, 2), (3, 1), and (3, 0).
   (a) Find the yolk for this game.
   (b) Find the spatial power index and the center of power.

3. In a spatial game, let the yolk have center 0 and radius σ. Show that a sufficient condition for a point Y to dominate point X is that

$$d(0, Y) < d(0, X) + 2\sigma$$

   where d is the usual (Euclidean) distance.

4. For a decisive spatial game, show that, if the core is nonempty, it consists of exactly one point. Moreover, in such case, the point in the core dominates all other points in the space.

5. For a spatial game, the *minimum response point* is the point X which gives the minimum in

$$\sigma = \min_{X} \max_{Y} \inf_{Z} d(X, Z)$$

   where the minimum is taken over all policies X, the maximum is taken over all policies Y, and the infimum is taken over all policies Z such that Z dominates Y. The distance σ is the *minimum response distance* or *finagle radius*.

(a) Show that, for a three-person majority game in two dimensions, the minimal response point and minimal response distance are respectively the center and radius of a circle which is tangent to the three circles K_1, K_2, and K_3 which are centered at the three ideal points P^1, P^2, and P^3, respectively, and tangent to each other.

(b) Find the minimal response point for the game in exercise 1, above.

6. Prove (16.1.2). (Use Figure 16.3, together with the fact that, if two circles centered at A and B, respectively, intersect at the two points P and X, then the angles PAB and XAB are congruent. Then the area of the hexagon $PAQCRB$ must be twice the area of ABC. Use also the fact that the area of the intersection of two sets is equal to the sum of their areas, minus the area of their union.)

7. Prove Theorem 16.2.9
   (a) For any angle Θ, let $U_\Theta = (\cos\Theta, \sin\Theta)$, and let f_Θ be the issue defined by $f_\Theta(X) = U_\Theta \cdot X$. Then, for all but a finite number of Θ, $0 \le \Theta \le 2\pi$, the order induced by f_Θ has a unique pivot. Call this pivot, $P(\Theta)$.
   (b) The orders induced by issues f_Θ and $f_{\Theta+\pi}$ have the same pivot. Moreover

$$H(X) = 2 \int_0^\pi (U_\Theta \cdot (P(\Theta) - X))^2 d\Theta$$

   (c) Define the function

$$G(X) = \int_0^\pi (P(\Theta) - X) \cdot (P(\Theta) - X) \, d\Theta$$

   Then $G(X)$ is minimized at the Copeland winner, X^*.
   (d) The difference $H(X) - G(X)$ is constant. Hence H and G are minimized at the same point.

Chapter 17

Other Applications

17.1. Introduction

In this final chapter we shall consider certain applications of game theory. We will not try to give a very rigorous, complete treatment of these topics, but rather explain how game theory can be applied to them.

17.2. Single-Item Auctions

Consider a situation in which an *auctioneer* wishes to sell an item. There are n individuals called *bidders*. Each bidder i values the item at a price θ_i: this is known as her *type*. These types are independent random variables, with uniform distribution over the interval [0, 1]. Each individual knows her own type, but nothing else. Bidder i sends a *message*: a bid, b_i. The auctioneer receives a vector $\mathbf{b} = (b_1, \ldots, b_n)$ of messages.

17.2.1. First-Price Sealed-Bid Auctions

In a *first-price sealed-bid auction*, each bidder sends her bid b_i in ignorance of the others' bids. Once the bids are known, the high bidder receives the item, and must pay the amount of his bid. The payoff to bidder i is then

$$u_i(\mathbf{b}, (\theta_i)) = \theta_i - b_i \quad \text{if } i \text{ is the high bidder}$$
$$0 \qquad\qquad \text{otherwise} \tag{17.1}$$

Now, how should a given individual, say individual 1, bid? Clearly she should not choose $b_1 > \theta_1$ as she would thus risk a negative payoff, and could never get a positive payoff. Now, if she makes a bid b, the probability that she will be high bidder is equal to the probability that all other $n-1$ bidders make lower bids. Of course 1 does not know how they will bid.

Assume, first, that bidders h, (other than 1) all bid their type: $b_h = \theta_h$. Since the types are uniformly distributed in the unit interval, the probability that a particular $b_h < b$ is simply b. Since the bids are independent, the probability that $b_h < b$ for all h other than 1 will be b^{n-1}. Thus, the probability that 1's bid of b will be high bid is b^{n-1}.

Bidder 1 will then gain $\theta_1 - b$ with probability b^{n-1} and 0 otherwise. Her expected winnings are therefore

$$\pi(b) = (\theta_1 - b)b^{n-1} \tag{17.2}$$

It is straightforward now to maximize this expression: the maximum is found at

$$b_1 = (n-1)\theta_1/n \tag{17.3}$$

There is one difficulty with this analysis, namely, it assumes that all bidders other than 1 will in fact bid their types. In other words, bidder 1 is the only one assumed to act strategically.

Assume, then, that all bidders, other than 1, make the strategic choice $b_h = (n-1)\theta_h / n$. Let 1 make the bid b. Note that, if so, none of them will bid more than $(n-1)/n$, and so it cannot be good for bidder 1 to bid more than this same amount.

Now, the condition $b_h < b$ is equivalent to $\theta_h < bn/(n-1)$. This has probability equal to the smaller of $bn/(n-1)$, and 1. Since this is true for all $h \neq 1$, the probability that individual 1's bid of b be the largest bid is $[bn/(n-1)]^{n-1}$ if $b < (n-1)/n$, and 1 otherwise. Thus the expected payoff to 1, for $b < (n-1)/n$, becomes

$$\pi^*(b) = (\theta_1 - b)[bn/(n-1)]^{n-1}$$
$$= [n/(n-1)]^{n-1}[(\theta_1 - b)b^{n-1}] \tag{17.4}$$

Since this is merely the positive constant $[n/(n-1)]^{n-1}$ times the previous objective function $\pi(b)$, it follows that it must have its maximum at the same point, i.e., bidder 1's optimal reply is still

$$b_1 = (n-1)\theta_1/n$$

This analysis is clearly true for each one of the bidders. We find, then, that the bids

$$b_h = (n-1)\theta_h/n \tag{17.5}$$

constitute a Nash equilibrium for the bidding game.

This result depends on the probability distribution of the vector $\boldsymbol{\theta}$. In a somewhat more general case, we will assume that the several θ_i are identically distributed independent random variables. Let $F(\theta)$ be the cumulative distribution of each one of these. We look for a symmetric equilibrium. This should have the form of a function β: individual h will bid $b_h = \beta(\theta_h)$. It seems reasonable to assume that β should be a monotone, strictly increasing function, so that an inverse function β^{-1} will exist and also be monotone increasing.

Assume all bidders, other than 1, are using the equilibrium strategy β. Let individual 1, with type θ, bid b. This will be the high bid if for each the $n-1$ other bidders, $\beta(\theta_h) < b$, or equivalently, $\theta_h < \beta^{-1}(b)$. Now the probability that these $n-1$ variables all be smaller than a given θ will be the cumulative distribution function

$$G(\theta) = [F(\theta)]^{n-1} \tag{17.6}$$

and so the probability that 1's bid of b be high is given by $G(\beta^{-1}(b)) = [F(\beta^{-1}(b))]^{n-1}$. Thus, 1's expected payoff will be

$$\pi(b) = (\theta - b)\,[G(\beta^{-1}(b))] \tag{17.7}$$

To maximize, we differentiate. Remembering that the derivative of β^{-1} is $1/\beta'$, and letting g be the derivative of G

$$\pi'(b) = -[G(\beta^{-1}(b))] + (\theta - b)\,[g(\beta^{-1}(b))]/[\beta'(\beta^{-1}(b))] \tag{17.8}$$

At the maximizing point, this derivative must be 0, and $b = \beta(\theta)$. Thus

$$0 = -[G(\theta))] + (\theta - \beta(\theta))[g(\theta)]/[\beta'(\theta)]$$

which reduces to the differential equation

$$G(\theta)\beta'(\theta) + g(\theta)\beta(\theta) = \theta g(\theta) \tag{17.9}$$

This has the general solution

$$G(\theta)\beta(\theta) = \int \theta g(\theta)d\theta + c$$

To find constant of integration c we note that for $\theta = 0$, any positive bid will lead to a loss. Thus $\beta(0) = 0$, and so

$$G(\theta)\beta(\theta) = \int_0^\theta x\,g(x)dx$$

$$\beta(\theta) = [G(\theta)]^{-1}\int_0^\theta x\,g(x)dx \tag{17.10}$$

17.2.1 Example. Consider a first-price auction where the common distribution for the several values θ_i is a beta distribution in the interval $[0, 1]$ with parameters $(2, 2)$:

$$F(\theta) = 3\theta^2 - 2\theta^3$$

and let $n = 3$. How much should a bidder offer if he values the item at its maximum, 1? How much if he values it at 0.8?

In this case

$$G(\theta) = F(\theta)^2 = 9\theta^4 - 12\theta^5 + 4\theta^6$$

and

$$g(\theta) = 360\theta^3 - 600\theta^4 + 240\theta^5$$

Application of (17.10) gives us

$$\beta(\theta) = (7.2\theta^5 - 100\theta^6 + 3.429\theta^7)/(90\theta^4 - 120\theta^5 + 40\theta^6)$$

In particular, we find $\beta(1)=0.629$, and $\beta(0.8)=0.576$. Thus the bidders will offer considerably less than their valuation of the item.

17.2.4.   Second-Price Auctions

In a second-price auction, the bidding proceeds as before. Once again, the high bidder wins the item. This time, however, the winner must pay, not her own bid, but rather the second-highest bid.

Once again, we consider bidder 1's strategy. We know her payoff will be

$$\pi(b_1, b_2, \ldots, b_n) = \theta_1 - b_h \quad \text{if } b_1 \text{ is high bid}$$
$$0 \text{ otherwise} \tag{17.11}$$

where b_h is the second-highest bid.

We claim, now, that bidder 1 should in fact bid her value: $b_1 = \theta_1$, as this strategy dominates all others.

In fact, consider a bid $b > \theta_1$. Let b_h be the highest of the remaining bids. Discounting the possibility of ties, there are three possibilities: (1) $b_h > b > \theta_1$; (2) $b > b_h > \theta_1$; (3) $b > \theta_1 > b_h$. We show that, in any one of these cases, the bidder would do at least as well (and in some cases better) by bidding θ_1 instead.

In case (1), bidder 1 is not the high bidder, so she would receive a payoff 0 with either b or θ_1.

In case (2), she would be high bidder with b, and receive the payoff $\theta_1 - b_h < 0$. With bid θ_1 she is not the high bidder and so receives payoff 0.

In case (3), she is the high bidder with either bid, and so receives the payoff $\theta_1 - b_h$ with either b or θ_1.

Thus, the bid θ_1 is never worse than b, and sometimes better.

Suppose, on the other hand, that $b < \theta_1$. Once again discounting the possibility of ties, there are three possibilities: (1) $b_h > \theta_1 > b$; (2) $\theta_1 > b_h > b$; (3) $\theta_1 > b > b_h$.

In case (1), bidder 1 is not the high bidder, so she would receive a payoff 0 with either b or θ_1.

In case (2), she wins with bid θ_1, and receives the payoff $\theta_1 - b_h > 0$. With bid b she is not the high bidder and so receives 0.

In case (3), she is the high bidder with either bid, and so receives the payoff $\theta_1 - b_h$ with either b or θ_1.

We find, then, that with a second-price auction, the strategy of bidding one's value: $b_h = \theta_h$ is a dominant strategy for all players. Note that this does not depend on the probability distribution for the value vector $\boldsymbol{\theta}$: the only thing required for our proof is that ties have probability 0. This is a very strong result, as it means the bidders will not have to carry out complicated calculations: they don't even need to know the distribution.

There remains the question of whether the principal — who wishes to maximize the selling price for the object — should choose a first-price or a second-price auction. We will consider the uniform distribution over [0, 1].

Consider first the first-price auction. For the uniform distribution, we find that the maximum of n variables has cumulative distribution $G(x) = x^n$. The density of this maximum is then $g(x) = nx^{n-1}$ and it has expected value $E[max\ \theta] = n/(n+1)$. Since the bid is $b = (n-1)\theta/n$, the expected selling price is $[n/(n+1)][(n-1)/n] = (n-1)/(n+1)$.

Consider, next, the second-price auction. Let θ_h be the second-highest type. In this case, we note that the event "$\theta_h \leq x$" can be expressed as the union of the disjoint events: (1) "all $\theta_i \leq x$" and (2) "one $\theta_k > x$ and all $n-1$ other $\theta_i \leq x$". The probability of (1) is, as before, x^n. For (2), we note that there are here n possibilities, since the high bidder, k, can be any one of n. Thus, the probability of (2) is $n(1-x)x^{n-1}$. The cumulative distribution of θ_h is therefore

$$G(x) = x^n + n(1-x)x^{n-1} = nx^{n-1} + (1-n)x^n$$

Some small calculations will then give us the expected value $E[\theta_h] = (n-1)/(n+1)$. Since θ_h is the selling price, we find that the two mechanisms give the same expected payoff to the principal.

We conclude that, if the seller's utility for money is in fact equal to the quantity of money (or, more generally, a linear function of money), he should be indifferent between the two types of auction. Of course there is no guarantee that seller's utility is in fact a linear function of money. With more complicated utility functions additional calculations (which we will avoid here) would be necessary.

17.3. Mechanisms

Up to now, we have assumed that players play a game with well-defined rules, and well-defined payoff functions (though in some cases the players might not be certain as to these). In some cases, however, the problem is to design a game so as to induce the players to act in a certain way. We therefore consider a situation with $n+1$ agents: a set $N = \{1, 2, \ldots, n\}$ of players (known as *individuals*), and an additional agent called the *principal*. This principal has an objective function, and will be in a position to modify the individuals' payoff functions so as to maximize (or minimize) this function. (Note that, in the auctions described above, the principal is the seller or auctioneer; in the case of Example 17.3.1 below, the principal is the regulator.) We start with a very simple example, which gives us some idea of how things may be done.

17.3.1 Example. Decoupling: Consider a service company, delivering some type of utility to the public at large. The service can be, e.g., water, gas, electricity. We will specifically consider an electric service company. The players here the company and its customers, though there is an additional player, usually unseen: the *regulator*.

In general these companies are *regulated monopolies*. A regulator determines the rates that they can charge for their services. These rates should be enough to cover both fixed and variable costs, plus a reasonable rate of return to their investors. Now, it is easy enough to calculate the variable costs of the service: purchases of fuel, payments to suppliers and employees, and so forth. What is more problematic is the calculation of payments for fixed costs and investors' return.

Usually, the regulator will calculate a, the variable costs per kilowatt-hour, with little difficulty. He then calculates b, the fixed costs authorized for depreciation on plant elements, interests paid to bondholders, and return to stockholders, as well as c, the expected sales (in kilowatt-hours) during a given time period. Thus he authorizes a rate of

$$x = a + b/c \qquad (17.12)$$

dollars per kilowatt-hour.

Now, if sales in the given time period are indeed equal to c kilowatt-hours, the company will receive

$$xc = ac + b \qquad (17.13)$$

dollars in the time period. Of this, ac will cover the variable costs, while the remaining b will be equal to the income authorized to cover fixed costs and investors' returns.

Unfortunately, things do not always work out as planned. Actual sales will depend not only on the rates but also on the aggregate demand function of the remaining players (the firm's customers), and this is known only probabilistically. Moreover, the firm is generally able (by advertising, say) to induce its customers to increase consumption above the calculated amount c.

Suppose, then, that consumption is equal to d kilowatt-hours, where $d>c$. In that case total receipts will be

$$xd = ad + b + x(d - c) \qquad (17.14)$$

dollars. Thus, the firm receives ad to cover its variable costs, plus b, its authorized returns for fixed costs, plus a last term, $x(d-c)$, which represents an unearned profit to the firm. Clearly it is in the firm's interest to increase demand.

It remains to determine the regulator's objective function. This is not easy to specify, but we should take into account that high consumption will, on the one hand, cause environmental pollution, and, on the other hand, require increased imports of fuel. It seems reasonable, then, that other things being equal, the regulator prefers consumption to be relatively low. Thus the regulator and the electric company have conflicting interests: the regulator prefers low consumption, while the firm will increase its profits with high consumption. Since the firm can communicate with its customers — something that the regulator is not allowed to do — it seems the firm will succeed in increasing consumption.

There is, however, something that the regulator can in fact do: to wit, he can modify the firm's payoff function. This is usually done by a technique known as *decoupling*.

In a decoupling mechanism, the firm's rates are modified with some regularity. Thus, to continue with the above example, suppose that consumption, d, over a given time period was indeed greater than the initial estimate, c. In this case, the firm would have received an unearned profit $k = x(d-c)$. If so, then for the subsequent period the firm would be allowed, not the quantity b, but rather the quantity $b' = b - k$, to cover fixed costs and returns to investors. The rate allowed for the next time period would be

$$x' = a + b/c - k/c \qquad (17.15)$$

dollars per kilowatt hour, rather than $a + b/c$.

(Note that, if $d < c$, so that the firm receives less than b dollars in the given period, then $k < 0$, so that the regulator will allow the firm to increase its rates for the next period.)

Thus any unearned profit in a given time period is taken back from the firm in the subsequent period, and it is no longer in the firm's interest to (induce its customers to) increase consumption. There will no longer be a conflict between the regulator and the firm.

17.3.1. Description of Mechanisms

As mentioned above, we consider a situation with $n+1$ players, the several *individuals* $\{1, 2, \ldots, n\}$, and the *principal*. Each individual i has a type θ_i which determines his utility function, his capabilities, and such. (This θ_i may be a single number, a vector, or something more complicated.) Let Θ_i be the set of possible types for that individual. The individual is aware of his type (he knows what he likes!) but does not know the other individuals' types, knowing only that they are determined according to a known probability distribution. Let Θ be the cartesian product $\Theta_1 \times \Theta_2 \times \ldots \times \Theta_n$. A *type profile* is an n-tuple $\theta = (\theta_1, \ldots, \theta_n) \in \Theta$.

There are k types of assets, which are to be distributed among the individuals. An assignment is a vector $a = (a_1, \ldots, a_k)$. An allocation is an n-tuple $x = (a^1, \ldots, a^n)$ of assignments. There is a compact set A in $n \times k$ space, which consists of the feasible allocations.

As an example, there may exist a vector $C = (c_1, c_2, \ldots, c_k)$ of available assets. In such case, A might consist of all non-negative allocations $(a^1, \ldots, a^n)$ such that $\sum_i a^i = C$. In some cases it may be more complicated than that, possibly consisting only of integer-valued assignments. There may be other types of restrictions.

A *result* is a pair [**x, t**], where **x** is a feasible allocation, and $\mathbf{t} = (t_1, \ldots, t_n)$ is a vector of monetary payments. The principal would like to have the result be a function of the type profile **θ** — usually, a result that satisfies some objective. However, the principal does not know the types; rather, he will have to base his action on some *messages* that the several individuals send.

A *mechanism* is a function **y** which assigns, to each *n*-tuple of messages, a result.

At the beginning of the game, the principal promulgates the mechanism **y**, i.e., the rule which will determine the allocations and payoffs.

Once the mechanism is known, each individual *i* announces a message, m_i, In many cases this is an element of Θ_i, though it may be something different — possibly, something simpler. Frequently, the individual claims that this message is indeed his type, θ_i. There is however nothing to prevent the individual from lying.

Letting $\mathbf{m} = (m_1, \ldots, m_n)$ be the profile (vector) of messages, the mechanism **y** is then used to give a result

$$\mathbf{y(m)} = [\mathbf{x(m)}, \mathbf{t(m)}] \tag{17.16}$$

In other words, individual *i* will receive the assignment $a^i(\mathbf{m})$ and must then pay the monetary amount $t_i(\mathbf{m})$. In many cases, of course, $t_i < 0$, so the individual actually receives some money.

The utility to a given individual is given by

$$U_i(\mathbf{x, t, \theta}) = V_i(a^i, \theta_i) - t_i \tag{17.17}$$

Thus, the utility to an individual of a given allocation depends on his type and assignment, but not on the other players' types or assignments. The value of the monetary payment is in fact equal to the quantity of money (and independent of any assignment received. This type of utility function is said to be *quasi-linear in money* (or simply *quasi-linear*); money is frequently said to be the *numéraire*. The advantage of this assumption lies in the fact that this allows for a linear transfer of utility – monetary side payments are allowed. Thus in general the principal would like to develop a mechanism that maximizes the sum

$$\Sigma_i V_i(a^i, \theta_i)$$

over the set *A*. In other words, if all individuals can explain their utility functions, and are truthful, the mechanism will maximize the total [expected] utility. Any adjustments can then be made through the side payments **t**. We will therefore define a *social welfare function* as a function *f* which assigns, to each type profile $\mathbf{\theta} = (\theta_1, \ldots, \theta_n)$ an allocation $f(\mathbf{\theta}) = (a^1(\mathbf{\theta}), \ldots, a^n(\mathbf{\theta}))$ such that

$$\Sigma_i V_i(f(\mathbf{\theta}), \theta_i) = \max\{\Sigma_i V_i(\mathbf{x}, \theta_i) \mid \mathbf{x} \in A\} \tag{17.18}$$

Generally speaking, the maximization problem implied in the social welfare function may be complicated, but reasonably straightforward. In some cases it could be as easy as a linear program. The problem, however, is that the vector $\boldsymbol{\theta} = (\theta_1, \ldots, \theta_n)$ is unknown: each individual i will deliver a message m_i which he claims is related in some way to his type θ_i but he may find it in his interest to lie. Moreover the individual player may be unable to describe his utility function properly: he knows whether he prefers one allocation to another, but cannot give a total description of his type. The problem faced by the principal consists in obtaining the outcome $f(\boldsymbol{\theta})$ on the basis of the messages $\mathbf{m} = (m_1, \ldots, m_n)$. We give one such instance.

17.4. Public Goods

Consider a situation in which two individuals must pay for a public good. For a level x of this good (i.e., if a sum of money x is invested), the agent $i\,(=1, 2)$ will derive utility

$$V_i(x) = \theta_{i1} x + \theta_{i2} \log x \qquad (17.19)$$

Where θ_{i1} and $\theta_{i2} > 0$, and log denotes the natural logarithm. Assuming payments of t_i from the individuals, we have

$$U_i(x, t_i) = \theta_{i1} x + \theta_{i2} \log x - t_i \qquad (17.20)$$

We assume $\theta_{11} + \theta_{21} < 1$ as otherwise the individuals would be willing to pay for unbounded levels of the good.

The optimal investment level, x^*, will maximize the sum of the U_i and thus be given by $V_1' + V_2' = 1$. This gives us

$$\theta_{11} + \theta_{12}/x^* + \theta_{21} + \theta_{22}/x^* = 1 \qquad (17.21)$$

which reduces to

$$x^* = \frac{\theta_{12} + \theta_{22}}{1 - \theta_{11} - \theta_{21}} \qquad (17.22)$$

This is the *efficient level of investment*. It remains to determine the payments, t_1 and t_2, that the two agents must make. It seems reasonable that they should be functions of the two types. The problem is that each individual knows only her own type, the regulator knows neither type, and, as mentioned above, the individuals may find it profitable to lie about this.

Suppose, however, that the regulator asks each agent to send a message, m_i, a positive number (approximately the level of investment that the agent would be willing to pay for). He agrees that the level of investment will then be

$$f(m_1, m_2) = m_1 + m_2 \qquad (17.23)$$

and that agents 1 and 2 must then pay fees

$$t_1 = m_1 - m_2 \log(m_1 + m_2) \qquad (17.24)$$

$$t_2 = m_2 - m_1 \log(m_1 + m_2) \qquad (17.25)$$

respectively.

This determines a non-cooperative game between the two agents. We therefore look for equilibrium strategies. From (17.20), we have

$$U_1 = \theta_{11}(m_1 + m_2) + \theta_{12} \log(m_1 + m_2) - m_1 + m_2 \log(m_1 + m_2) \qquad (17.26)$$

For m_1 to be an optimal response against m_2, the derivative of U_1 with respect to m_1 must be 0, so

$$\frac{\partial U_1}{\partial m_1} = \theta_{11} + \frac{\theta_{12} + m_2}{m_1 + m_2} - 1 = 0$$

or, equivalently,

$$\frac{\partial U_1}{\partial m_1} = \theta_{11} + \frac{\theta_{12} - m_1}{m_1 + m_2} = 0 \qquad (17.27)$$

For an equilibrium, m_2 must also be an optimal response against m_1, and so

$$\frac{\partial U_2}{\partial m_2} = \theta_{21} + \frac{\theta_{22} - m_2}{m_1 + m_2} = 0 \qquad (17.28)$$

Adding these two equations, we obtain

$$\theta_{11} + \theta_{21} + \frac{\theta_{12} + \theta_{22}}{m_1 + m_2} - 1 = 0$$

and so

$$m_1 + m_2 = \frac{\theta_{12} + \theta_{22}}{1 - \theta_{11} - \theta_{21}} = x^* \qquad (17.29)$$

Thus, at equilibrium, $x = m_1 + m_2$ is an efficient level of investment, and the mechanism described by (17.23)–(17.25) will induce efficiency.

Knowing the equilibrium value of $m_1 + m_2$, it is now straightforward to solve (17.27)–(17.28) for m_1 and m_2, and this then allows us to obtain the values of the t_i.

It should be noticed, of course, that the two agents do not reveal their types: they merely send a message which consists of a single number. This of course presents a problem as, not knowing each other's type, the agents will find it difficult to find the equilibrium strategies (messages). The hope is however that, if an equilibrium is not reached, they might have an incentive to adjust their messages and in this way approach the equilibrium.

It is not always possible to design a mechanism that will (a) induce the individuals involved to make truthful statements, and (b) allow the principal to obtain a socially optimal result. Nevertheless there are cases when this can be done. The following example illustrates this. The reader is encouraged to read the literature for a rigorous treatment and for other examples.

17.4.1 Example. Consider a situation in which the two agents' utilities V_i are given by

$$V_1(x) = 0.4x + 0.24 \, \log \, x$$

$$V_2(x) = 0.35x + 0.18 \, \log \, x$$

where x is money, in thousands of dollars. In this case the efficient level of investment is

$$x^* = (0.24 + 0.18)/(1 - 0.4 - 0.35) = 1.68$$

From (17.27)–(17.28) we now obtain the equilibrium values

$$0.4 + (0.32 - m_1)/1.68 = 0$$

$$m_1 = 0.912$$

$$m_2 = 0.768$$

and then

$$t_1 = 0.912 - 0.768 \, \log(1.68) = 0.594$$
$$t_2 = 0.768 - 0.912 \, \log(1.68) = 0.295$$

Note that the two agents pay only a total of $889, where the total investment is $1680. Presumably the remaining $791 would come from the public coffers.
    Finally, we have

$$U_1(x) = 0.4(1.68) + 0.24 \, \log(1.68) - 0.594 = 0.203$$
$$U_2(x) = 0.35(1.68) + 0.16 \, \log(1.68) - 0.295 = 0.376$$

Thus, agent 1 pays $594 and receives a benefit which she values at $797. Agent 2 pays $295 and receives a benefit which he values at $671.
    Suppose, next, that the two agents send messages of, for example, $m_1 = 1.05$ and $m_2 = 0.75$ respectively. In that case, the total investment will be $1800, and the two agents will have to pay fees

$$t_1 = 1.05 - 0.75 \, \log(1.8) = 0.609$$

$$t_2 = 0.75 - 1.05 \, \log(1.8) = 0.133$$

Their net utilities will be

$$U_1(x) = 0.4(1.8) + 0.24 \ \log(1.8) - 0.609 = 0.252$$

$$U_2(x) = 0.35(1.8) + 0.18 \ \log(1.8) - 0.133 = 0.603$$

Thus, they do substantially better than at equilibrium. This is however a non-cooperative situation. Thus, we must consider *individual incentives*. We assume that the partial derivatives $\partial U_i / \partial m_i$ do, in fact, represent these incentives. We have, then.

$$\frac{\partial U_1}{\partial m_1} = 0.4 + \frac{0.24 - 1.05}{1.8} = -0.05$$

$$\frac{\partial U_2}{\partial m_2} = 0.35 + \frac{0.16 - 0.75}{1.8} = 0.022$$

and we see that agent 1 will find it in his interest to decrease the value of m_1. Agent 2 would like to increase m_2, but not by so much. Thus, assuming each agent, i, modifies his message in proportion to the incentive $\partial U_i / \partial m_i$, total investment will decrease.

In general, we can imagine that this type of situation will happen: the two agents will, originally, send messages which are not efficient. Note, however, that if $m_1 + m_2 > x^*$, then the sum of the two partial derivatives, $\partial U_1/\partial m_1 + \partial U_2/\partial m_2$, will be negative, and so at least one, and possibly both, of the agents will want to decrease her message. Assuming the two agents modify their message in proportion to the respective incentives (derivatives), the sum $m_1 + m_2$ will decrease.

Conversely, if $m_1 + m_2 < x^*$, the tendency will be for the agents to modify their messages so that this sum increases. In either case we can imagine the agents modifying their messages until the efficient level, x^*, is reached.

17.5.   Two-sided Matching

In Example 10.4.5 we discuss the *assignment game*: in this game, there are two disjoint sets of players, M and M'. The only truly effective coalitions are one-player coalitions, and two-player coalitions consisting of one player from each of these two sets. In that example, the two sets consisted of sellers and buyers of houses, respectively; each seller had only one house, and each buyer could buy at most one house. We saw there that for such games the core will always be nonempty, and can in fact be found in an efficient manner by, e.g., the well-known *Hungarian algorithm* (Kuhn [17.10]).

Now, the assignment problem is, in effect, a linear programming problem, and for such games, the solution to the dual problem can be used to obtain at least one point in the core. But the linear programming representation is due to the possibility of side payments: the buyer gives a sum of money to the seller in exchange for the house. But what if side payments were not allowed? Could the core, then, be empty?

It is of course difficult to imagine — at least in our economic system — how a market for houses could function without side payments. There are however other types of two-sided matching systems which might function without side payments. The most obvious is a *marriage market*.

A *marriage market* consists of two sets, $M = \{m_1, \ldots, m_k\}$ (the men) and $W = \{w_1, \ldots, w_l\}$ (the women).

Each man, m_i, has a preference relation, p_i, on the set $W' = W \cup \{0\}$, where "0" represents "remaining unmarried." We write

$$w_j p_i w_h$$

to denote that man m_i would rather marry w_j than w_h.

This p_i is a *strict linear order* relation — it is transitive, asymmetric, and complete:

(1) If $w_j\, p_i\, w_h$ and $w_h\, p_i\, w_r$, then $w_j\, p_i\, w_r$
(2) It is not possible that $w_j\, p_i\, w_h$ and $w_h\, p_i\, w_j$
(3) For any m_i and any two (distinct) w_j and w_h, **either** $w_j\, p_i\, w_h$ **or** $w_h\, p_i\, w_j$

Additionally, for man m_i, we set

$$D_i = \{w_j | 0\, p_i w_j\}$$

These are the women he finds *unacceptable*: m_i would prefer to be single rather than marry any member of D_i.

(Note that, by condition (3), we are explicitly ruling out the possibility of ties. Our treatment could be modified to include these possibilities, but we will not do so here.)

Similarly, each woman w_j has a preference relation, q_j, on $M' = M \cup \{0\}$. This is also transitive, asymmetric, and complete. She also has a set E_j of unacceptable men:

$$E_j = \{m_i | 0\ q_j m_i\}$$

A *marriage assignment*, now, is a **one-to-one** mapping μ from a subset $S \subset M$ **onto** a subset $T \subset W$. Now, $\mu(m_i) = w_j$ means that man m_i marries woman w_j. If $m_i \notin S$, then this man remains single. In a similar way, if woman $w_h \notin T$, then w_h remains single.

Since μ is one-to-one, two different men cannot marry the same woman. Similarly, since μ is a function, the same man cannot marry two different women. On the other hand, several of the men may remain single. Similarly, several of the women may remain single.

Note that, since μ is one-to-one, the inverse mapping μ^{-1} is defined on T. We will extend μ from the set S to the set M by setting $\mu(m_i) = 0$ for $m_i \notin S$. Similarly we will extend μ^{-1} from T to W by setting $\mu^{-1}(w_h) = 0$ for $w_h \notin T$.

We will assume that each individual's preference depends only on his or her spouse. Thus, given two assignments λ and μ, m_i prefers λ to μ if and only if $\lambda(m_i)\, p_i \mu(m_i)$. He is indifferent between λ and μ if and only if $\lambda(m_i) = \mu(m_i)$. (A similar statement holds for women.)

We look now for the **core** of the set of all assignments. It turns out that domination, as in Example (10.4.5) above, can only be through one- or two-person coalitions. An assignment is dominated via a single person if that person is married to someone he or she dislikes. In other words, μ is dominated via the coalition $\{m_i\}$ if $\mu(m_i) \in D_i$, and it is dominated via $\{w_j\}$ if $\mu^{-1}(w_j) \in E_j$.

We will say that an assignment is *individually rational* if it is not dominated via any singleton coalition. We therefore consider only individually rational assignments. For such assignments, domination can only be through two-person coalitions consisting of one man and one woman.

An assignment μ is *unstable* if there exists a two-person coalition $\{m_i, w_j\}$ such that

(1) $w_j p_i \mu(m_i)$
(2) $m_i q_j \mu^{-1}(w_j)$

In other words, both this man and this woman would be happier if married to each other than with their present spouses. The core is the set of stable assignments.

The question then is whether there is at least one stable assignment. In fact, the following algorithm, the *deferred acceptance* method, which is due to Gale and Shapley [17.4], shows that the core is always nonempty. There are two versions of this, one more favorable to the men and the other to the women. We give the one that is more favorable to the men. We will assume that each of the men finds at least one of the women acceptable, and each of the women finds at least one of the men acceptable: none of the D_i is equal to W, and none of the E_j is equal to M. All individuals are unmarried, prior to the start of the algorithm.

The algorithm consists of several stages.

Stage 1(a): Each of the men proposes marriage to his favorite woman (the one that ranks first in his preference order).

Stage 1(b): Any woman who has received at least one proposal from an acceptable man will accept the proposal of the man she most prefers, among those who have proposed to her. She is then *engaged* to that man. She rejects all other proposals.

Any remaining woman is said to be *uncommitted*. Any man whose proposal has been rejected is uncommitted.

Stage 2(a): Any engaged man will remain engaged to that same woman. Any uncommitted man will propose to his first choice among those women who have not previously rejected him, assuming that she is acceptable to him. He may not propose to any woman who has previously rejected or *jilted* him. He will not propose to any woman whom he finds unacceptable.

Stage 2(b): Any engaged woman who receives a proposal from at least one man whom she prefers to her current fiancé will break off her current engagement (*jilt* that man) and accept the proposal of the man she most favors from among those who propose to her. She rejects all other proposals.

Any uncommitted woman who has received at least one proposal from an acceptable man will accept the proposal of the man she most prefers, among those who have proposed to her. She is then *engaged* to that man. She rejects all other proposals.

Any man whose current proposal is rejected remains uncommitted. Jilted men are now uncommitted.

Except as mentioned in 2(b), women will maintain their *status quo*: any engaged woman remains engaged to the same man; any uncommitted woman remains uncommitted.

Stages, 3, 4, etc. will be exactly as Stage 2. The process will continue in this manner until each man *either* (a) is engaged, *or* (b) has been rejected by all the women he finds acceptable.

Since a man is not allowed to propose twice to the same woman, and there are l women, then no man can propose more than l times. Since there are only k men, there will be at most kl proposals. Thus, the process will terminate after not more than kl stages. When the process terminates, all the engaged couples proceed to get married. All uncommitted men and women will remain single.

It remains to prove that the assignment thus obtained is stable.

First of all, the algorithm explicitly states that no man ever proposes to any unacceptable woman, and no woman ever accepts a proposal from an unacceptable man. Thus, the assignment obtained is individually rational.

Suppose, then, that the assignment is unstable. There must then be a pair (m_i, w_j) who prefer each other to their assigned spouses (or unmarried status). This means, of course, that w_j is acceptable to m_i. Since she is preferable to his wife, and men propose to women in order of their preference, m_i must have proposed to w_j at some time. This means that she either rejected his proposal when made, or accepted his proposal but later jilted him. In either of these two cases, it was because she had a proposal from some man, m_h, whom she preferred to m_i. Now she is currently either married to this m_h, or married to some other man after jilting m_h and possibly some other men. However, every time she jilted a man, it was for a preferable man. Since the preference relation is transitive, she prefers her current husband to m_i. Since the relation is asymmetric, she does not prefer m_i to her current husband.

We conclude that the current assignment is stable.

17.5.1 Example. Consider a marriage market with 4 men and 4 women. The men's preferences are as follows:

$$m_1 : \quad w_1, w_2, w_3, w_4$$
$$m_2 : \quad w_2, w_3, w_1, w_4$$
$$m_3 : \quad w_1, w_2, w_3, w_4$$
$$m_4 : \quad w_1, w_3, w_4, w_2$$

and the women's preferences;

$$w_1 : \quad m_2, m_4, m_3, m_1$$
$$w_2 : \quad m_3, m_4, m_1, m_2$$
$$w_3 : \quad m_1, m_2, m_3, m_4$$
$$w_4 : \quad m_1, m_2, m_3, m_4$$

The algorithm will then proceed as follows:

Stage 1(a): m_1 proposes to w_1; m_2 proposes to w_2; m_3 proposes to w_1; m_4 proposes to w_1
Stage 1(b): w_1 accepts m_4; w_2 accepts m_2
Current engagements: $\{m_4, w_1\}$, $\{m_2, w_2\}$

Stage 2(a): m_1 proposes to w_2; m_3 proposes to w_1
Stage 2(b): w_2 accepts m_3 and jilts m_2
Current engagements: $\{m_4, w_1\}$, $\{m_3, w_2\}$

Stage 3(a): m_1 proposes to w_3; m_2 proposes to w_3
Stage 3(b): w_3 accepts m_1
Current engagements: $\{m_4, w_1\}$, $\{m_3, w_2\}$, $\{m_1, w_3\}$

Stage 4(a): m_2 proposes to w_1
Stage 4(b): w_1 accepts m_2 and jilts m_4
Current engagements: $\{m_2, w_1\}$, $\{m_3, w_2\}$, $\{m_1, w_3\}$

Stage 5(a): m_4 proposes to w_3
Stage 5(b): no acceptances
Current engagements: $\{m_2, w_1\}$, $\{m_3, w_2\}$, $\{m_1, w_3\}$

Stage 6(a): m_4 proposes to w_4
Stage 4(b): w_4 accepts m_4
Current engagements: $\{m_2, w_1\}$, $\{m_3, w_2\}$, $\{m_1, w_3\}$, $\{m_4, w_4\}$.

At this point, all men are engaged. The engaged couples are now married. We leave it to the reader to check that this is stable.

There is, of course, an alternative approach, in which, at each stage, the women propose and it is up to the men to accept the proposals they like. In that case, the algorithm would go as follows:

Stage 1(a): w_1 proposes to m_1: w_2 proposes to m_3: w_3 proposes to m_1: w_4 proposes to m_1
Stage 1(b): m_1 accepts w_1; m_3 accepts w_2
Current engagements: $\{m_1, w_1\}$, $\{m_3, w_2\}$

Stage 2(a): w_3 proposes to m_2: w_4 proposes to m_2:
Stage 2(b): m_2 accepts w_3
Current engagements: $\{m_1, w_1\}$, $\{m_3, w_2\}$, $\{m_2, w_3\}$

Stage 3(a): w_4 proposes to m_3
Stage 3(b): m_3 rejects w_4

Stage 4(a): w_4 proposes to m_4
Stage 4(b): m_4 accepts w_4
Current engagements: $\{m_1, w_1\}$, $\{m_3, w_2\}$, $\{m_2, w_3\}$, $\{m_4, w_4\}$.

The algorithm terminates here as all women are now engaged. Once again the reader may check that this assignment is stable.

The marriage problem is one of the simplest types of matching problems. Other types include:

1. Student-college matching

There are applicants, a_i, and colleges c_j. In this case, college c_j has a finite number r_j of openings. Applicants will have preferences over colleges, and colleges, over prospective students. This can be treated much as the marriage problem. The easiest approach is to modify the marriage problem as follows: each of the applicants is treated as man; each of a college's r_j openings is treated as a different woman. The algorithm described above is then carried out as before.

Once again, the algorithm will give a stable matching, with one proviso: the colleges' preferences for an individual do not depend on whether another individual has accepted. This might, however, happen, as colleges will frequently seek diversity in their student population, or might be looking for a football team, which would mean that a particular quarterback is a good choice only if the college has not recruited another quarterback. (Other examples might exist.) In such cases, there is no guarantee of a stable matching.

2. The Roommate Problem

In this problem, there is only one set, M; each member $i \in M$ has preferences over the remaining members of M. The problem is to assign the individuals, two to a room.

In this case, there is no guarantee of a stable matching. In fact, consider a case with 4 individuals. Suppose their preferences are

$$m_1 : \quad m_2, m_3, m_4$$
$$m_2 : \quad m_3, m_1, m_4$$
$$m_3 : \quad m_1, m_2, m_4$$
$$m_4 : \quad m_2, m_3, m_1$$

In this case, there is no stable matching. The reason is that whatever player is matched with the "unlikeable" roommate, m_4, will be able to form a dominating coalition with one of the other two. If for example the matching is $\{ \{m_1, m_2\}, \{m_3, m_4\} \}$, then m_2 and m_3 would both be happier with each other than with their assigned roommates. Thus the assignment is unstable. It is easy to check that the two other possible assignments are also unstable.

Problems

1. In a first-bid auction, assume there are 4 bidders. Let the several bidders' types θ_i be independent random variables with the common (cumulative) distribution function

$$F(\theta) = \theta^2 \quad \text{for} \quad 0 \le \theta \le 1$$

Find the equilibrium bid function $\beta(\theta)$. How much should an individual bid if he values the item at 0.9? If he values it at 0.2?

2. In a first-bid auction, assume there are 5 bidders. Let the types (values) θ have common distribution function $F(\theta) = \sqrt{\theta}$ for θ in the unit interval [0, 1]. Find the equilibrium bid function $\beta(\theta)$.

3. In a first-bid auction, assume there are 3 bidders. Let the types (values) θ have common distribution function $F(\theta) = 2\theta - \theta^2$ for θ in the unit interval [0, 1]. Find the equilibrium bid function $\beta(\theta)$.

4. Consider a *duopoly*, in which two individuals must decide how much to produce of a certain good. The selling price of this good is determined by the demand function

$$p = 10 - m_1 - m_2$$

where m_1 and m_2 are the amounts produced by the two individuals. Each individual has a secret technology which gives rise to the cost function cost function

$$c_i = \theta_{i1} m_i^2 + \theta_{i2} m_i$$

Thus individual 1's utility is given by

$$U_{-1} = (10 - m_1 - m_2)m_1 - \theta_{11} m_1^2 + \theta_{12} m_1$$

with a similar utility function for individual 2.
(a) Find the equilibrium quantities and price, as a function of the θ_{ij}.
(b) Assuming $\theta_{11} = 1$, $\theta_{12} = 2$, $\theta_{21} = 2$, and $\theta_{22} = 2$, find the equilibrium quantities and price. How much profit (utility) would each of the producers receive?
(c) Suppose the two individuals produce $m_1 = 2$ and $m_2 = 1$. What would the incentives be?

5. Modify exercise 4, above, by setting $\theta_{11} = 1.5$, $\theta_{12} = 2$, $\theta_{21} = 2.5$, and $\theta_{22} = 1$. Let the utility functions be the same as above. What would the equilibrium quantities and price be?

6. Modify Example 17.4.1, above, by setting

$$V_1(x) = 0.4x + 0.2 \ \log x$$
$$V_2(x) = 0.4x + 0.15 \ \log x$$

(a) Find the equilibrium messages, m_1 and m_2.
(b) Suppose the messages sent are $m_1 = 0.8$, $m_2 = 0.9$. What would the incentives (partial derivatives) be?

7. Consider a marriage market with five men and six women. The men's preferences are as follows:

$$m_1 : w_1, w_2, w_3, w_4, w_5, w_6$$
$$m_2 : w_3, w_1, w_4, w_6, w_5, w_2$$
$$m_3 : w_3, w_2, w_5, w_1, w_4, w_6$$
$$m_4 : w_1, w_4, w_6, w_2, w_5, w_3$$
$$m_5 : w_1, w_4, w_2, w_3, w_6, w_5$$

and the women's preferences are

$$w_1 : m_3, m_1, m_2, m_5, m_4$$
$$w_2 : m_3, m_4, m_2, m_5, m_1$$
$$w_3 : m_1, m_4, m_2, m_5, m_3$$
$$w_4 : m_2, m_1, m_3, m_4, m_5$$
$$w_5 : m_5, m_1, m_4, m_3, m_2$$
$$w_6 : m_3, m_1, m_4, m_5, m_2$$

Find two stable assignments for this marriage market.

8. Consider a marriage market with five men and five women.

the men's preferences are as follows:

$$m_1 : w_1, w_2, w_3, w_4, w_5$$
$$m_2 : w_3, w_2, w_4, w_5, w_1$$
$$m_3 : w_4, w_1, w_5, w_2, w_3,$$
$$m_4 : w_1, w_4, w_2, w_5, w_3$$
$$m_5 : w_1, w_5, w_3, w_4, w_2$$

and the women's preferences are

$$w_1 : m_3, m_1, m_2, m_5, m_4$$
$$w_2 : m_3, m_4, m_2, m_5, m_1$$
$$w_3 : m_1, m_4, m_2, m_5, m_3$$
$$w_4 : m_2, m_1, m_3, m_4, m_5$$
$$w_5 : m_5, m_1, m_4, m_3, m_2$$

Find two stable assignments for this marriage market.

Appendix

A.1. Convexity

Throughout this book, the notion of convexity has been used extensively. We give below some of the properties of convex functions, though other properties have been given in the text, notably in Chapters 2 and 4.

A.1.1 Definition. A real-valued function $f(x)$, defined over a real linear space, is said to be *convex* iff, for any values x, y of the independent variable, and any such r such that $0 \leq r \leq 1$

$$f(rx + (1-r)y) \leq rf(x) + (1-r)f(y) \tag{A.1}$$

The function f is said to be *strictly convex* if, whenever $x \neq y$ and $0 < r < 1$, strict inequality holds in (A.1)

A.1.2 Definition. A function f is said to be concave (strictly concave) iff $-f$ is convex (strictly convex).

The following theorem gives obvious properties of convex functions. No proof is given.

A.1.3 Theorem. Let f and g be convex functions, and suppose that $c \geq 0$. Then the functions $f + g$, cf, and $\max\{f, g\}$ are all convex.

In a similar way, if f and g are concave, so are $f + g$, cf, and $\min\{f, g\}$.
It is easy to see that a linear function is both concave and convex. Conversely, if a function is both concave and convex, it is linear.

A.1.4 Definition. A set $X \subset V$, where V is a real linear space, is said to be *convex*, iff whenever $x, y \in S$ and $0 \leq r \leq 1$, then

$$rx + (1-r)y \in S \tag{A.2}$$

Thus, a set S is convex if the segment joining two points in S lies entirely in S. This compares with the idea that a function is convex if the chord joining two points on the graph of the function lies entirely above the surface. Thus, the set of points *above* the graph of a convex function is convex (this is often given as the definition of a convex function). This relation between convex sets and convex functions is mirrored in the following theorem, which represents a partial analogue of A.1.3. Its proof is trivial.

A.1.5 Theorem. Let S_α be convex for $\alpha \in A$ (where A is some index set). Then $\cap_{\alpha \in A} S_\alpha$ is also convex.

Very often we are given a nonconvex set K, and we wish to apply some theorem which holds only for convex sets. This makes it necessary for us to associate with K a convex set $H(K)$.

A.1.6 Definition. Let K be any set. Then its *convex hull* $H(K)$ is the intersection of all the convex sets which contain K.

By Theorem A.1.5, the convex hull of any set is convex; it is clear that, if a set is convex, it is its own convex hull. There is, however, a different way of defining the convex hull: i.e., in terms of *convex linear combinations*.

A.1.7 Definition. Let $x^1, x^2, \ldots, x^p$ be p points in some real linear space. The point y will then be said to be a *convex linear combination* of $x^1, x^2, \ldots, x^p$ if there exist real numbers, $c_1, \ldots, c_p$, satisfying

   (i) $c_j \geq 0 \quad j = 1, \ldots p$
   (ii) $\sum c_j = 1$
   (iii) $y = \sum c_j x^j$

The connection between convex linear combinations and the convex hull is shown in the following theorem.

A.1.8 Theorem. For any set K, the convex hull $H(K)$ is precisely the set S of all points y which are convex linear combinations of elements of K.

Proof. It is easy to see that $K \subset S$. Now, S is convex, for suppose

$$y = \sum_1^p c_j x^j, \quad x^1, \ldots, x^p \in K$$

are

$$y' = \sum_{p+1}^{p+q} c_j x^j, \quad x^{p+1}, \ldots, x^{p+q} \in K$$

are both in S, with $(c_1, \ldots, c_p)$ and $(c_{p+1}, \ldots, c_{p+q})$ satisfying (i) and (ii) of A.1.7. Then, if $0 \leq r \leq 1$, we can write

$$y'' = ry + (1-r)y' = \sum_{j=1}^{p}(rc_j)x^j + \sum_{j=p+1}^{p+q}(1-r)c_j x^j$$

and the numbers $(rc_1, \ldots, rc_p, (1-r)c_{p+1}, \ldots, (1-r)c_{p+q})$ satisfy A.1.7(i), (ii). Hence, $y'' \in S$, and so S is convex. It follows that $H(K) \subset S$.

Conversely, let $y = \sum_{j=1}^{p} c_j x^j$, where $x^j \in K$ and the c_j are as above. We prove by induction on p that $y \in H(K)$. In fact, for $p = 1$, $y = x^1$, and so $y \in K \subset H(K)$. Suppose, then, that this is true for $p-1$. We may assume, without loss of generality, that $c_1 > 0$, and so $r = \sum_{j=1}^{p-1} c_j > 0$. We have then

$$y = ry' + (1-r)x^p$$

where

$$y' = \sum_{j=1}^{p-1} \frac{c_j}{r} x^j$$

It is clear that $y' \in S$ and so, by the induction hypothesis, $y' \in H(K)$. But $x^p \in H(K)$. By convexity, it follows that $y \in H(K)$. Hence, $S \subset H(K)$. This proves the theorem.

The idea of a convex hull enables us to assign a convex set to any set. On the other hand, it is sometimes of interest to reduce a convex set to some subset, whose convex hull it is.

A.1.9 Definition. Let S be a convex set, and suppose that $x \in S$. We say that x is an *extreme point* of S if there do *not* exist two points, x', x'' in S, such that $x' \neq x''$ and

$$x = \tfrac{1}{2}(x' + x'')$$

The following theorem is of importance and has been used in the text of this book.

A.1.10 Theorem. A compact convex subset S of n-dimensional euclidean space is the convex hull of its extreme points. Moreover, any $y \in S$ can be expressed as a convex linear combination of at most $n + 1$ extreme points of S.

Proof. It is clear that the second part of the theorem implies the first. We shall, therefore, prove only this second part. We prove it by induction on n. In fact, for $n = 1$, the only compact convex sets are the empty set $\varnothing$, single points, and intervals $[a, b]$. For $\varnothing$ and single points, the theorem is trivial; for sets $[a, b]$, it is clear that a and b are extreme, and every $y \in [a, b]$ is a convex linear combination of a and b.

Assume, then, that the theorem is true for $n-1$. It is clear that it will hold for all $(n-1)$-dimensional sets, even when these are subsets of a higher dimensional space. Let S be as in the hypothesis, and let $y \in S$.

Take any line passing through y. Because of compactness and convexity, the intersection of this line and S is a closed line segment, with end-points y' and y''.

Consider the point y'. It is a boundary point of S, and hence (see Problem 2.1), there is a hyperplane P passing through y' such that the set S lies entirely on or "above" (to one side of) P. Clearly, P is closed and convex, and hence $S \cap P$ is a compact convex set of dimension at most $n-1$. As $y' \in S \cap P$, it can be expressed as a linear combination of at most n extreme points of $S \cap P$.

Suppose that $x \in S \cap P$, and suppose that $x = \frac{1}{2}(x' + x'')$, with x', $x'' \in S$. The set P can be characterized by an equation $L(x) = \alpha$, where L is a linear functional, and we know $L(x) \geq \alpha$ for all $x \in S$. By linearity, we must have $L(x') = L(x'') = \alpha$, and so x', $x'' \in P$. This means that, if x is extreme in $S \cap P$, it is extreme in P.

We have thus shown that y is a convex linear combination of the point y'' and at most n extreme points of S. However, y'' was obtained by taking an arbitrary line through y; this line can always be taken so that y'' is extreme. Thus y will be a linear combination of at most $n + 1$ extreme points of S.

A.2. Fixed Point Theorems

We give below, without proof, two theorems which have been used in the text. The proofs of these theorems are quite lengthy and can be found in the references in the Bibliography.

A.2.1 Theorem. (The Brouwer Fixed Point Theorem) Let S be a compact convex subset of n-dimensional euclidean space, and let f be a continuous function mapping S into itself. Then there exists at least one $x \in S$ such that $f(x) = x$.

This theorem is a topological theorem and thus will apply to any set topologically equivalent to such a set S. (Quite often, it is stated that S is an n-simplex.) The proof in the general case is quite lengthy and requires an extensive background in topology, though for $n = 1$, it is an easy consequence of the intermediate value theorem [it then reduces to showing that the equation $f(x) - x = 0$ has a root in the given interval]. A generalization which is very useful in game theory is the Kakutani fixed point theorem, given below.

A.2.2 Definition. Let f be a function with domain X (a topological space) and such that, for $x \in X$, $f(x)$ is a subset of some topological space Y. Then, f is said to be *upper semicontinuous* (u.s.c.) at a point x_0 if for any sequence $x_1, x_2, \ldots$, converging to x_0, and any sequence of points $y_1, y_2, \ldots$ with $y_i \in F(x_i)$, the limit of the sequence $\{y_n\}$ (if the sequence converges) belongs to $f(x_0)$. The function f is upper semicontinuous if it is u.s.c. at each point of X.

A.2.3 Theorem. (Kakutani) Let S be a compact convex subset of n-dimensional euclidean space, and let f be an upper semicontinuous function which assigns to each $x \in S$ closed convex subset of S. Then there is some $x \in S$ such that $x \in f(x)$.

Bibliography

General The following books contain material of general interest, as opposed to books and papers which are more properly related to one or another chapter of this work and which are given in chapter bibliographies.

1. Dresher, M., L. S. Shapley, and A. W. Tucker, eds., *Advances in Game Theory*, Annals of Mathematics Studies No. 52, Princeton, New Jersey (Princeton University Press), 1964.
2. Dresher, M., A. W. Tucker, and P. Wolfe, eds., *Contributions to the Theory of Games, III*, Annals of Mathematics Studies No. 39, Princeton, New Jersey (Princeton University Press), 1957.
3. Karlin, S., *Mathematical Methods and Theory in Games, Programming and Economics*, Reading, Massachusetts (Addison-Wesley), 1959.
4. Kuhn, H. W., and A. W. Tucker, eds., *Contributions to the Theory of Games, I-II*, Annals of Mathematics Studies Nos. 24, 28, Princeton, New Jersey (Princeton University Press), 1950, 1953.
5. Kuhn, H. W., and A. W. Tucker, eds., *Linear Inequalities and Related Systems*, Annals of Mathematics Studies No. 38, Princeton, New Jersey (Princeton University Press), 1956.
6. Luce, R. D., and H. Raiffa, *Games and Decisions*, New York (Wiley), 1957.
7. Tucker, A. W., and R. D. Luce, eds., *Contributions to the Theory of Games, IV*, Annals of Mathematics Studies No. 40, Princeton, New Jersey (Princeton University Press), 1959.
8. von Neumann, J., and O. Morgenstern, *Theory of Games and Economic Behavior*, Princeton, New Jersey (Princeton University Press), 1944, 1947.

References 1, 2, 4, 5, and 7 above consist of many separate papers and will be referred to, in the chapter bibliographies, as *Annals 24, Annals 28*, etc.

Chapter 1

1. Berge, C., "Topological Games with Perfect Information," *Annals 39*.
2. Gale, D., and F. M. Stewart, "Infinite Games with Perfect Information," *Annals 28*.
3. Gill, R., "The Monty Hall Problem is not a Probability Puzzle (it's a Challenge in Mathematical Modelling)," *Statistica Neerlandica* **65** (February 2011), 58–71.

4. Kuhn, H. W., "A Simplified Two-Person Poker," *Annals 24.*
5. Kuhn, H. W., "Extensive Games and the Problem of Information," *Annals 28.*
6. Morgan, J. P., Chaganty, N. R., Dahiya, R. C., and Doviak, M. J. "Let's Make a Deal: The Player's Dilemma," *Am. Stat.* **45** (1991), 284–287.
7. Nash, J., and L. S. Shapley, "A Simple Three-Person Poker Game," *Annals 24.*
8. vos Savant, M. "Game Show Problem," *Parade* (1990–1991), retrieved from http://marilynvossavant.com/game-show-problem/, December 16, 2012.

Chapter 2

1. Brown, G. W., and J. von Neumann, "Solutions of Games by Differential Equations," *Annals 24.*
2. Dresher, M., and S. Karlin, "Solutions of Convex Games as Fixed Points," *Annals 28.*
3. Farkas, J., "Theorie der Einfachen Ungleichungen," *J. Reine Angew. Math.* **124** (1902), 1–27.
4. Gale, D., H. W. Kuhn, and A. W. Tucker, "On Symmetric Games," *Annals 24.*
5. Motzkin, T. S., H. Raiffa, G. L. Thompson, and R. M. Thrall, "The Double Description Method," *Annals 28.*
6. Robinson, J., "An Iterative Method of Solving a Game," *Ann. of Math.* **54** (1951), 296–301.
7. Shapley, L. S., and R. N. Snow, "Basic Solutions of Discrete Games," *Annals 24.*
8. Weyl, H., "Elementary Proof of a Minimax Theorem due to von Neumann," *Annals 24.*

Chapter 3

1. Dantzig, G., L. R. Ford, and D. R. Fulkerson, "A Primal-Dual Algorithm for Linear Programs," *Annals 38.*
2. Dantzig, G., and D. R. Fulkerson, "On the Max-Flow Min-Cut Theorem of Networks," *Annals 38.*
3. Gass, S. I., *Linear Programming*, New York (McGraw-Hill), 1958.
4. Goldman, A., and A. W. Tucker, "Theory of Linear Programming," *Annals 38.*
5. Tucker, A. W., "Dual Systems of Homogeneous Linear Relations," *Annals 38.*
6. Vajda, S., *The Theory of Games and Linear Programming*, London (Methuen), New York (Wiley), 1956.
7. Wolfe, P., "Determinateness of Polyhedral Games," *Annals 38.*

Chapter 4

1. Bohnenblust, H. F., S. Karlin, and L. S. Shapley, "Games with Continuous Convex Payoff," *Annals 24.*
2. Bohnenblust, H. F., S. Karlin, and L. S. Shapley, "Solutions of Discrete Two-Person Games," *Annals 24.*

3. Borel, E., "Sur les jeux où interviennent le hasard et l'habileté des joueurs," *Eléments de la Théorie des Probabilités*, 3ème edition, Paris (Librairie Scientifique), 1924.
4. Dresher, M., S. Karlin, and L. S. Shapley, "Polynomial Games," *Annals 24*.
5. Duffin, R. J., "Infinite Programs," *Annals 38*.
6. Gross, O., "A Rational Game on the Square," *Annals 39*.
7. Karlin, S., "Operator Treatment of the Minmax Principle," *Annals 24*.
8. Karlin, S., "On a Class of Games," *Annals 28*.
9. Karlin, S., "Reduction of Certain Classes of Games to Integral Equations," *Annals 28*.
10. Restrepo, R., "Tactical Problems Involving Several Actions," *Annals 39*.
11. Shiffman, M., "Games of Timing," *Annals 28*.
12. Sion, M., and P. Wolfe, "On a Game Without a Value," *Annals 39*.

Chapter 5

1. Berkovitz, L. D., and W. H. Fleming, "On Differential Games with Integral Payoff," *Annals 39*.
2. Berkovitz, L. D., "A Variational Approach to Differential Games," *Annals 52*.
3. Berkovitz, L. D., "A Differential Game without Pure Strategy Solutions on an Open Set," *Annals 52*.
4. Blackwell, D., "Multi-Component Attrition Games," *Nav. Res. Legist. Quart.* **1** (1954), 327–332.
5. Dubins, L. E., "A Discrete Evasion Game," *Annals 39*.
6. Everett, H., "Recursive Games," *Annals 39*.
7. Fleming, W. H., "The Convergence Problem for Differential Games," *Annals 52*.
8. Isaacs, R., *Differential Games*, New York (Wiley), 1965.
9. Isbell, J., "Finitary Games," *Annals 39*.
10. Kuhn, H. W., "Extensive Games and the Problem of Information," *Annals 28*.
11. Milnor, J., and L. S. Shapley, "On Games of Survival," *Annals 39*.
12. Mycielski, J., "Continuous Games of Perfect Information," *Annals 39*.
13. Ryll-Nardzewski, C., "A Theory of Pursuit and Evasion," *Annals 52*.
14. Scarf, H. E., "On Differential Games with Survival Payoff," *Annals 39*.
15. Shapley, L. S., "Stochastic Games," *Proc. Nat. Acad. Sci. U.S.A.* **39** (1953), 327–332.

Chapter 6

1. Aumann, R. J., and M. Maschler, *Repeated Games with Incomplete Information: A Survey of Recent Results*, Princeton, New Jersey (Mathematica), 1967.
2. Aumann, R. J., and M. Maschler, *Repeated Games with Incomplete Information: The Zero-Sum Extensive Case*, Princeton, New Jersey (Mathematica), 1968.
3. Blackwell, D., "An Analog of the Minimax Theorem for Vector Payoffs," *Pacific J. Math* **6** (1956), 1–8.

4. Harsanyi, J. C., "Games with Incomplete Information Played by Bayesian Players, I, II, III," *Management Sci.* **14** (1968), 159–182, 320–332, 468–502.

5. Kohlberg, E., "Optimal Strategies in Repeated Games with Incomplete Information," *Int. J. Game Theory* **4** (1975), 7–24.

6. Mertens, J. F., and S. Zamir, "The Value of Two-Person Zero-Sum Games with Lack of Information on Both Sides," *Int. J. Game Theory* **1** (1972), 39–64.

7. Mertens, J. F., and S. Zamir, "Minmax and Maxmin of Repeated Games with Incomplete Information," *Int. J. Game Theory* **9** (1980), 201–215.

8. Ponssard, J. P. "A Note on the L-P Formulation of Zero-Sum Sequential Games with Incomplete Information," *Int. J. Game Theory* **4** (1975), 1–5.

9. Ponssard, J. P., and S. Sorin, "The L-P Formulation of Finite Zero-Sum Games with Incomplete Information," *Int. J. Game Theory* **9** (1980), 99–105.

Chapter 7

1. Arrow, K. J., *Social Choice and Individual Values*, Cowlen Commission Monograph 12, New York (Wiley), 1951.

2. Davidson, D., S. Siegel, and P. Suppes, "Some Experiments and Related Theory on the Measurement of Utility and Subjective Probability," Applied Mathematics and Statistics Laboratory, *Technical Report 1*, Stanford University, 1955.

3. Hausner, M., "Multi-Dimensional Utilities," *Decision Processes* (R. M. Thrall, C. H. Coombs, and R. L. Davis, eds.), New York (Wiley), 1954.

4. Hernstein, J. N., and J. Milnor, "An Axiomatic Approach to Measurable Utility," *Econometrica* **21** (1953), 291–297.

5. Isbell, J., "Absolute Games," *Annals 40*.

6. Luce, R. D., "A Probabilistic Theory of Utility," *Technical Report 14*, Behavioral Models Project, Columbia University, 1956.

7. Suppes, P., and M Winet, "An Axiomatization of Utility Based on the Notions of Utility Differences," *Management Sci.* **1** (1955), 259–270.

Chapter 8

1. Aumann, R. J., "Subjectivity and Correlation in Randomized Strategies," *J. Math. Econ* **1** (1974), 67–96.

2. Harsanyi, J. C, "Approaches to the Bargaining Problem before and after the Theory of Games: A Critical Discussion of Zeuthen's, Hick's and Nash's Theories," *Econometrica* **24** (1956), 144–157.

3. Harsanyi, J. C., "Games with Randomly Disturbed Payoffs: A New Rationale for Mixed Strategy Equilibrium Points," *Int. J. Game Theory* **2** (1973), 1–23.

4. Hart, S., D. Schmeidler, "Existence of Correlated Equilibria," *Math. Oper. Res.* **14** (1989), 18–25.

5. Lemke, C. E., "Bimatrix Equilibrium Points and Mathematical Programming," *Management Sci.* **11** (May 1965).

6. Maynard Smith, J., *Evolution and the Theory of Games*, Cambridge, UK (Cambridge University Press), 1982.

7. Nash, J., "Equilibrium Points in *n*-Person Games," *Proc. Nat. Acad. Sci. U.S.A.* **36** (1950), 48–49.
8. Nash, J., "Non-Cooperative Games," *Ann. of Math.* **54** (1951), 286–295.
9. Selten, R., "Reexamination of the Perfectness Concept for Equilibrium Ponts in Extensive Games," *Int. J. Game Theory* **4** (1975), 25–55.
10. Shapley, L. S., "Some Topics in Two-Person Games," *Annals 52*.
11. van Damme, E., *Stability and Perfection of Nash Equilibria*, Berlin (Springer Verlag), 1987.

Chapter 9

1. Braithwaite, R. B., *Theory of Games as a Tool for the Moral Philosopher*, Cambridge (Cambridge University Press), 1955.
2. Harsanyi, J. C., "Approaches to the Bargaining Problem before and after the Theory of Games: A Critical Discussion of Zeuthen's, Hick's and Nash's Theories," *Econometrica* **24** (1956), 144–157.
3. Nash, J., "The Bargaining Problem," *Econometrica* **18** (1950), 155–162.
4. Raiffa, H., "Arbitration Schemes for Generalized Two-Person Games," *Annals 28*.
5. Rubinstein, A., "Perfect Equilibrium in a Bargaining Model," *Econometrica* **50** (1982), 97–109.
6. Zeuthen, F., *Problems of Monopoly and Economic Warfare*, London (G. Routledge and Sons), 1930.

Chapter 10

1. Bondareva, O. N., "Certain Applications of the Methods of Linear Programming to the Theory of Cooperative Games," *Problemy Kibernet.* **10** (1963), 119–139.
2. McKinsey, J. C. C., "Isomorphism of Games and Strategic Equivalence," *Annals 24*.
3. Owen, G., "The Core of Linear Production Games," *Math. Programming* **9** (1975), 358–370.
4. Shapley, L. S., "On Balanced Sets and Cores," *Naval Res. Logist. Quart.* **14** (1967), 453–460.
5. Shapley, L. S., and M. Shubik, "The Assignment Game I: The Core," *Internal. J. Game Thy.* **1** (1972), 111–130.
6. Tijs, S. H., "On *S*-Equivalence and Isomorphism of Games in Characteristic Function Form," *Internal. J. Game Thy.* **4** (1976), 209–210.

Chapter 11

1. Bott, R., "Symmetric Solutions to Majority Games," *Annals 28*.
2. Gillies, D. B., "Solutions to General Non-Zero-Sum Games," *Annals 40*.
3. Gurk, H., and J. Isbell, "Simple Solutions," *Annals 40*.
4. Kalisch, G. K., and E. D. Nering, "Countably Infinitely Many Person Games," *Annals 40*.

5. Lucas, W. R, "The Proof That a Game May Not Have a Solution," *Trans. Am. Math. Soc.* **136** (1969), 219–229.
6. Owen, G., "Tensor Composition of Non-Negative Games," *Annals 52*.
7. Owen, G., "Discriminatory Solutions of *n*-Person Games," *Proc. Am. Math. Soc.* **17** (1966), 653–657.
8. Shapley, L. S., "Quota Solutions of *n*-Person Games," *Annals 28*.
9. Shapley, L. S., "A Solution Containing an Arbitrary Closed Component," *Annals 40*.
10. Shapley, L. S., "Solutions of Compound Simple Games," *Annals 52*.
11. Shubik, M., "Edgeworth Market Games," *Annals 40*.

Chapter 12

1. Banzhaf, J. F., III, "Weighted Voting Doesn't Work: A Mathematical Analysis," *Rutgers Law Review* **19** (1965), 317–343.
2. Banzhaf, J. F., III, "One Man, 3.312 Votes: A Mathematical Analysis of the Electoral College," *Villanova Law Review* **13** (1968), 304–332.
3. Coleman, J. S., "Control of Collectivities and the Power of a Collectivity to Act," *Social Choice* (B. Lieberman, ed.), London (Gordon and Breach), 1971, 269–300.
4. Dubey, P., "On the Uniqueness of the Shapley Value," *Int. J. Game Theory* **4** (1975), 131–139.
5. Dubey, P., and L. S. Shapley, "Mathematical Properties of the Banzhaf Power Index," *Math. Oper. Res.* **4** (1979), 99–131.
6. Hart, S., and M. Kurz, "Endogenous Formation of Coalitions," *Econometrica* **51** (1983), 1047–1064.
7. Owen, G., "Multilinear Extensions of Games," *Management Sci.* **18** (1972), P64–P79.
8. Owen, G., "Multilinear Extensions and the Banzhaf Value," *Naval Res. Logist. Quart.* 22 (1975), 741–750.
9. Owen, G., "Evaluation of a Presidential Election Game," *Am. Pol. Sci. Rev.* **69** (1975), 947–953.
10. Owen, G., "Characterization of the Banzhaf-Coleman Index," *SIAM J. Appl. Math.* **35** (1978), 315–327.
11. Owen, G., "A Note on the Banzhaf-Coleman Axioms," *Game Theory and Political Science* (P. C. Ordeshook, ed.), New York (New York University Press), 1978, 451–462.
12. Owen, G., "Values of Games with *a priori* Unions," *Mathematical Economics and Game Theory* (R. Henn and O. Moeschlin, eds.), Berlin (Springer Verlag), 76–88.
13. Owen, G., and E. Winter, "The Multilinear Extension and the Coalition Structure Value," *Games Econ. Behav.* **4** (1992), 582–587.
14. Shapley, L. S., "A Value for *n*-Person Games," *Annals 28*.
15. Winter, E., "A Value for Cooperative Games with Levels Structure of Cooperation," *Int. J. Game Theory* **18** (1989), 227–240.

Chapter 13

1. Aumann, R. J., and M. Maschler, "The Bargaining Set for Cooperative Games," *Annals 52*.
2. Billera, L. J., "Global Stability in *n*-Person Games," *Trans. Am. Math. Soc.* **144** (1972), 45–56.
3. Kalai, G., M. Maschler, and G. Owen, "Asymptotic Stability and Other Properties of Trajectories and Transfer Sequences Leading to the Bargaining Set," *Int. J. Game Theory* **4** (1975), 193–214.
4. Kohlberg, E., "On the Nucleolus of a Characteristic Function Game," *SIAM J. Appl. Math.* **20** (1971), 62–66.
5. Kohlberg, E., "The Nucleolus as a Solution of a Minimization Problem," *SIAM J. Appl. Math.* **21** (1972), 34–39.
6. Kopelowitz, A., "Computation of the Kernels of Simple Games and the Nucleolus of *n*-Person Games," Research Program in Game Theory and Mathematical Economics, Dept. of Mathematics, Hebrew University, Jerusalem, September 1967.
7. Littlechild, S. C., "A Simple Expression for the Nucleolus in a Special Case," *Int. J. Game Theory* **3** (1974), 21–30.
8. Littlechild, S. C., arid G. Owen, "A Simple Expression for the Shapley Value in a Special Case," *Manage. Sci.* **20** (1973), 370–372.
9. Maschler, M., "The Inequalities that Determine the Bargaining Set $M_1^{(i)}$," Research Program in Game Theory and Mathematical Economics, *Research Memorandum 17*, Hebrew University, Jerusalem, January 1966.
10. Maschler, M., and M. Davis, "The Kernel of a Cooperative Game," *Naval Res. Logist. Quart.* **12** (1965), 223–259.
11. Maschler, M., and B. Peleg, "A Characterization, Existence Proof, and Dimension Bounds for the Kernel of a Game," *Pacific J. Math.* **32** (1966), 289–328.
12. Milnor, J., "Reasonable Outcomes for *n*-Person Games," *RM-916*, The Rand Corporation, 1952.
13. Owen, G., "A Note on the Nucleolus," *Int. J. Game Theory* **3** (1974), 101–103.
14. Peleg, B., "On the Bargaining Set M_0 of *m*-Quota Games," *Annals 52*.
15. Peleg, B., "Existence Theorem for the Bargaining Set $M_1^{(i)}$," *Bull. Am. Math. Soc.* **69** (1963), 109–110.
16. Schmeidler, D., "The Nucleolus of a Characteristic Function Game," *SIAM J. Appl. Math.* **17** (1969), 1163–1170.
17. Stearns, R. E., "Convergent Transfer Schemes for *n*-Person Games," *Trans. Am. Math. Soc.* **134** (1968), 449–459.

Chapter 14

1. Aumann, R. J., "Markets with a Continuum of Traders," *Econometrica* **32** (1964), 443–476.
2. Aumann, R. J., and L. S. Shapley, *Values of Non-Atomic Games*, Princeton, New Jersey (Princeton University Press), 1974.

3. Billera, L. J., J. Heath, and D. C. Raanan, "Internal Telephone Billing Rates: A Novel Application of Non-Atomic Game Theory," *Oper. Res.* **27** (1978), 956–965.

4. Kannai, Y., "Values of Games with a Continuum of Players," *Israel J. Math.* **4** (1966), 54–58.

5. Milnor, J., and L. S. Shapley, "Values of Large Games II: Oceanic Games," *RM-2699*, The Rand Corporation, February 1961.

6. Neyman, A., and Y. Tauman, "The Partition Value," *Math. Oper. Res.* **4** (1979), 236–264.

Chapter 15

1. Asscher, N., "A Cardinal Bargaining Set for Games without Side Payments," *Int. J. Game Theory* **6** (1977), 87–114.

2. Aumann, R. J., "Acceptable Points in General Cooperative *n*-Person Games," *Annals 40*.

3. Friedman, J., *Oligopoly and the Theory of Games*, New York (North-Holland), 1977.

4. Harsanyi, J. C, "A Bargaining Model for the Cooperative *n*-Person Game," *Annals 40*.

5. Harsanyi, J. C., "A Simplified Bargaining Model for the *n*-Person Cooperative Games," *Int. Econ. Rev.* **4** (1963), 194–220.

6. Ichiishi, T, and S. Weber, "Some Theorems on the core of a Non-Sidepayment Game with a Measure Space of Players," *Int. J. Game Theory* **7** (1978), 95–112.

7. Isbell, J., "Absolute Games," *Annals 40*.

8. Miyasawa, K., "The *n*-Person Bargaining Game," *Annals 52*.

9. Nakamura, K., "The Core of a Simple Game with Ordinal Preferences," *Int. J. Game Theory* **4** (1975), 95–104.

10. Owen, G., "Values of Games without Side Payments," *Int. J. Game Theory* **1** (1972), 94–109.

11. Scarf, H., "The Core of An *n*-Person Game," *Econometrica* **35** (1967), 50–69.

12. Scarf, H., *The Computation of Economic Equilibria*, New Haven, Connecticut (Yale University Press), 1973.

13. Selten, R., "Valuation of *n*-Person Games," *Annals 52*.

14. Shapley, L. S., "A Value for *n*-Person Games without Side Payments," Proceedings of Conference at Princeton University, April 1965.

15. Stearns, R. E., "Three-Person Cooperative Games without Side Payments," *Annals 52*.

16. Stearns, R. E., "On the Axioms for a Cooperative Game without Side Payments," *Report 62-RL-3130E*, General Electric Laboratories, Schenectady, New York, September 1962.

Chapter 16

1. Black, D., *The Theory of Committees and Elections*, Cambridge, UK (Cambridge University Press), 1958.

2. Copeland, A. H., "A Reasonable Social Welfare Function," Unpublished manuscript, 1951.

3. Downs, A., *An Economic Theory of Democracy*, New York (Harper & Row), 1957.

4. Enelow, J. E., and M. J. Hinich, *The Spatial Theory of Voting*, Cambridge (Cambridge University Press), 1984.

5. Feld, S. L., and B. N. Grofman, "Necessary and Sufficient Conditions for a Majority Winner in *n*-Dimensional Spatial Voting Games," *Am. J. Pol. Sci.* **31** (1987), 709–728.

6. Ferejohn, J. A., R. D. McKelvey, and E. Packel, "Limiting Distributions for Continuous State Markov Models," *Social Choice Welfare* **1** (1984), 45–67.

7. Grofman, B. N., A Glazer, N. Noviello, and G. Owen, "Stability and Centrality of Legislative Choice in the Spatial Context," *Am. Pol. Sci. Rev.* **81** (1987), 539–553.

8. McKelvey, R. D., "Covering, Dominance, and Institution-Free Properties of Social Choice," *Am. J. Pol. Sci.* **30** (1986), 283–314.

9. McKelvey, R. D., P. Ordeshook, and M. Winer, "The Competitive Solution for *n*-Person Games without Transferable Utility," *Am. Pol. Sci. Rev.* **72** (1978), 599–615.

10. McKelvey, R. D., and R. E. Wendel, "Voting Equilibria in Multidimensional Choice Spaces," *Math. Oper. Res.* **1** (1976), 44–58.

11. Ordeshook, P., "Pareto-Optimality in Electoral Competition," *Am. Pol. Sci. Reu.* **65** (1971), 1141–1145.

12. Owen, G., "Political Games," *Naval Res. Logist. Quart.* **18** (1972), 345–354.

13. Owen, G., and Shapley, L. S., "Optimal Location of Candidates in Ideological Space," *Int. J. Game Theory* **18** (1989), 339–356.

14. Plott, C. R. "A Notion of Equilibrium and its Possibility under Majority Rule," *Am. Econ. Reu.* **57** (1967), 787–806.

15. Riker, W., and P. C. Ordeshook, *An Introduction to Positive Political Theory*, Englewood Cliffs, New Jersey (Prentice Hall), 1973.

16. Schofield, N., "The Theory of Dynamic Games," *Rev. Econ. Stud.* **45** (1978), 575–594.

17. Shapley, L. S., "A Comparison of Power Indices and a Non-symmetric Generalization," *Paper P-5872*, Rand Corporation, Santa Monica, California, 1977.

18. Wuffle, A., S. L. Feld, G. Owen, and B. N. Grofman, "Finagle's Law and the Finagle Point," *Am. J. Pol. Sci.* **33** (1989), 348–375.

Chapter 17

1. Bénoît, J. P., and V. Krishna, "Multiple-Object Auctions with Budget-Constrained Bidders," *Rev. Econ. Stud.* (2001), 155–179.

2. D'Aspremont, C., and L.-A. Gérard-Varet, "Incentives and Incomplete Information," *J. Public Econ.* (1979), 25–45.

3. Engelbrecht-Wiggans, R., P. Milgrom, and R. Weber, "Competitive Bidding and Proprietary Information," *J. Math. Econ.* (1983), 161–169.
4. Gale, D., and L. S. Shapley, "College Admissions and the Stability of Marriage," *American Mathematical Monthly* (1962), 9–15.
5. Gale, D., and M. Sotomayor, "Some Remarks on the Stable Matching Problem," *Discrete Applied Mathematics* (1985), 223–232.
6. Granot, D. "A Note on the Roommate Problem and a Related Revenue Allocation Problem," *Manage Sci.* (1984), 633–643.
7. Hurwicz, L., and S. Reiter, *Designing Economic Mechanisms*, Cambridge, UK (Cambridge University Press), 2006.
8. Knuth, D. E., *Mariages Stables*, Montreal: les presses de l'université de Montréal, 1976.
9. Krishna, V., *Auction Theory*, Burlington, Massachusetts (Academic Press), 2010.
10. Kuhn, H. W., "The Hungarian Method for the Assignment Problem", *Naval Research Logistics Quarterly*, 2, (1955) 83–97.
11. Maskin, E., "Nash Equilibrium and Welfare Optimality," *Rev. Econ. Stud.* (1999), 23–38.
12. Milgrom, P. R., and R. J. Weber, "A Theory of Auctions and Comptitive Bidding," *Econometrica* (1982), 1089–1122.
13. Milgrom, P., "Rational Expectations, Information Acquisition, and Competitive Bidding," *Econometrica* (1981), 921–943.
14. Myerson, R. B., "Optimal Auction Design," *Mathematics of Operations Research* (1981), 58–73.
15. Myerson, R., *Game Theory: Analysis of Conflict*, Cambridge, Massachusetts (Harvard University Press), 1991.
16. Roth, A. M., and M. Sotomayor, *Two-Sided Matching: A Study in Game-Theoretic Modeling and Analysis*, Econometric Society Monographs, No. 18, Cambridge University Press, 1990.
17. Saari, D. G., "A Method for Constructing Message Systems for Smooth Performance Functions," *J. Econ. Theory* (1984), 249–274.
18. Williams, S. R., *Communication in Mechanism Design: A Differential Approach*, Cambridge, UK (Cambridge University Press), 2008.

Appendix

1. Bonnesen, T., and W. Fenchel, "Theorie der Konvexen Körper," *Ergebnisse der Mathematik und Ihrer Grenzgebiete*, Vol. 3, No. 1, Berlin (Springer), 1934; New York (Chelsea), 1948.
2. Gale, D., "Convex Polyhedral Cones and Linear Inequalities," *Activity Analysis of Production and Allocation* (T. C. Koopmans, ed.), New York (Wiley), 1951.
3. Kakutani, S., "A Generalization of Brouwer's Fixed Point Theorem," *Duke J. Math.* **8** (1941), 457–459.
4. Weyl, A., "The Elementary Theory of Convex Polyhedra," *Annals 24.*

Index